T0318747

SERS for Point-of-care and Clinical Applications

SERS for Point-of-care and Clinical Applications

Edited by

Andrew Fales

Center for Devices and Radiological Health, U.S. Food and Drug Administration, Silver Spring, MD, United States

Elsevier
Radarweg 29, PO Box 211, 1000 AE Amsterdam, Netherlands
The Boulevard, Langford Lane, Kidlington, Oxford OX5 1GB, United Kingdom
50 Hampshire Street, 5th Floor, Cambridge, MA 02139, United States

ISBN: 978-0-12-820548-8

For information on all Elsevier publications visit our website at https://www.elsevier.com/books-and-journals

Publisher: Susan Dennis
Acquisitions Editor: Kathryn Eryilmaz
Editorial Project Manager: Maria Elaine D. Desamero
Production Project Manager: Kiruthika Govindaraju
Cover Designer: Greg Harris

Typeset by TNQ Technologies

Contents

Contributors

Sara Abalde-Cela, PhD
International Iberian Nanotechnology Laboratory, Avenida Mestre José Veiga s/n, Braga, Portugal

Akbar Ali
Department of Chemistry, Indian Institute of Technology, Bhilai, Chhattisgarh, India

Chrysafis Andreou, PhD
Department of Electrical and Computer Engineering, University of Cyprus, Nicosia, Cyprus

Claudia Beleites
Chemometrix GmbH, Wölfersheim, Germany

Alois Bonifacio
Raman Spectroscopy Lab, Department of Engineering and Architecture, University of Trieste, Trieste, Italy

Bridget Crawford
Department of Biomedical Engineering, Duke University, Durham, NC, United States; Fitzpatrick Institute for Photonics, Duke University, Durham, NC, United States

Alexander Czaja, M.S.
Department of Biomedical Engineering, University of Southern California, Los Angeles, CA, United States; USC Michelson Center for Convergent Bioscience, Los Angeles, CA, United States

Andrew Fales
Center for Devices and Radiological Health, U.S. Food and Drug Administration, Silver Spring, MD, United States

Stefano Fornasaro
Raman Spectroscopy Lab, Department of Engineering and Architecture, University of Trieste, Trieste, Italy

Yiota Gregoriou, PhD
Department of Biological Sciences, University of Cyprus, Nicosia, Cyprus

Hoan Thanh Ngo
School of Biomedical Engineering, International University, Vietnam National University, Ho Chi Minh City, Vietnam

Suchetan Pal, PhD
Department of Chemistry, Indian Institute of Technology, Bhilai, Chhattisgarh, India

Laura Rodríguez-Lorenzo, PhD
International Iberian Nanotechnology Laboratory, Avenida Mestre José Veiga s/n, Braga, Portugal

Valter Sergo
Raman Spectroscopy Lab, Department of Engineering and Architecture, University of Trieste, Trieste, Italy

Miguel Spuch-Calvar, PhD
CINBIO, Universidade de Vigo, Campus Universitario Lagoas, Vigo, Pontevedra, Spain

Pietro Strobbia
Department of Chemistry, University of Cincinnati, Cincinnati, OH, United States

Tuan Vo-Dinh
Fitzpatrick Institute for Photonics, Department of Biomedical Engineering, Department of Chemistry, Duke University, Durham, NC, United States; Department of Biomedical Engineering, Duke University, Durham, NC, United States; Fitzpatrick Institute for Photonics, Duke University, Durham, NC, United States; Department of Chemistry, Duke University, Durham, NC, United States

Hsin-Neng Wang
Department of Biomedical Engineering, Duke University, Durham, NC, United States; Fitzpatrick Institute for Photonics, Duke University, Durham, NC, United States

Cristina Zavaleta, PhD
Department of Biomedical Engineering, University of Southern California, Los Angeles, CA, United States; USC Michelson Center for Convergent Bioscience, Los Angeles, CA, United States

Editor biography

Andrew M. Fales received his BS in biochemistry and molecular biology from the University of Maryland, Baltimore County in 2010, and his PhD in biomedical engineering from Duke University in 2016. His dissertation focused on the development of theranostic nanoplatforms using SERS-based detection strategies. From 2016 to 2020, he conducted research at FDA's Office of Science and Engineering Laboratories, studying emerging biomedical optical technologies. In 2020, he joined the FDA's Office of Product Evaluation and Quality as a scientific lead reviewer. His research interests include plasmonics, pulsed laser—nanoparticle interactions, and Raman spectroscopy.

Disclaimer

The opinions presented by the editor and authors are their own and should not be construed to represent those of the Food and Drug Administration or the US Department of Health and Human Services. The mention of commercial products, their sources, or their use in connection with material reported herein is not to be construed as either an actual or implied endorsement of such products by the US Department of Health and Human Services.

CHAPTER

Data analysis in SERS diagnostics

1

Stefano Fornasaro[1], Claudia Beleites[2], Valter Sergo[1], Alois Bonifacio[1]

[1]*Raman Spectroscopy Lab, Department of Engineering and Architecture, University of Trieste, Trieste, Italy;* [2]*Chemometrix GmbH, Wölfersheim, Germany*

Introduction

This chapter focuses on the surface-enhanced Raman spectroscopy (SERS) data analysis workflow in biomedical/clinical studies. The most common characteristic for these studies is that they deal with the analysis of complex and information-rich datasets, composed of tens to hundreds of observations (i.e., spectra), each composed of hundreds or thousands of variables (i.e., wavenumbers). Biofluids (e.g., blood, plasma, saliva, tears, and urine) are typical samples used in biomedical SERS studies. All biofluids contain a wealth of biochemical information, which can be exploited by physicians to monitor many clinically relevant aspects, such as a patient's general health status, the immune system, the clinical response to a certain drug, or the delivery of nutrients (to name few examples). However, extracting the important information from these datasets is a puzzling process. Extracting a single number from each spectrum (e.g., the intensity of one band) so to have only one variable associated with each sample is, from many points of view, very convenient. Univariate data analysis methods such as difference (e.g., diseased minus control) spectral analysis, or changes in bands intensity (or area) values can be used to extract the useful information from the dataset. This approach, however, is often insufficient in SERS spectroscopy, and in particular in "label-free" SERS, since contributions from multiple analytes are present in each spectrum.

Thus, a multivariate analysis approach is often necessary. For that reason, it is of great importance to have a robust multistep data analysis workflow available, allowing for the correct use and interpretation of the spectral data.

Other spectroscopic techniques as Raman or IR present similar challenges, and can benefit to the same extent from the use of a multivariate data analysis. SERS, however, has some unique characteristics which make it more challenging, from a data analysis point of view. Differently from other techniques, a SERS spectrum always originates from a relatively small fraction of the molecules present in a sample, which have very specific interactions with a surface. In fact, it is now widely accepted that few molecules, i.e., those located at surface positions where the plasmonic amplification of the electromagnetic field is highest (often referred to as "*hot-spots*"), are accountable for most of the signal constituting a SERS spectrum,

SERS for Point-of-care and Clinical Applications. https://doi.org/10.1016/B978-0-12-820548-8.00002-3

whereas the majority of the other molecules have a negligible contribution. In other words, in SERS the contribution to the signal is not equally distributed at all. A first consequence of this fact is that the intensity of the signal observed is highly dependent on the spatial distribution of these "hot-spots" on the SERS substrate with respect to the size of the laser spot used for excitation. This dependence can lead to remarkable intensity fluctuations upon probing different areas of the same substrate, or upon investigating the same sample using two different substrates. In other words, this leads to an important source of variability which needs to be dealt with. Moreover, the fact that the signal also depends on the specific interaction between the analytes and the substrate can lead to other complications: the overall spectral shape can be concentration-dependent (i.e., different bands are observed for the same analyte, depending on the concentration), and slight differences in the physicochemical characteristics of the surface or its environment can bring drastic spectral changes (e.g., minor differences in pH or ionic strength can lead to quite different SERS spectra). This also means that small differences among different batches of the SERS substrates will further contribute to spectral variability. This analyte–metal interaction is also reflected in the occurrence of a characteristic intense and broad band in all SERS spectra between 100 and 300 cm^{-1}, which must be considered when considering the subtraction of a baseline. All these SERS-specific aspects further complicate the use of this technique for biomedical applications, and exalt the importance of data analysis.

The goal of this chapter is to introduce a general account of each step of this process, highlighting central ideas and specific challenges related to the analysis and interpretation of SERS spectral data. We believe that a general understanding of the key aspects of the problem can be of greater value than providing solutions for specific applications, since the success of a specific analysis is often limited to the application for which it was optimized. Thus, we try to give practical advice to the reader through a quick overview of the most relevant techniques for data visualization, modeling, classification, and calibration in SERS studies, with an emphasis on both their capabilities and weaknesses. The approach adopted is purposely avoiding in-depth mathematical treatments to make this chapter more readable by a broader audience. Equations are thus avoided in most instances, although the interested reader can find detailed mathematical explanations in the suggested literature.

General data processing workflow

The block diagram (Fig. 1.1) graphically represents the general data processing workflow for a typical SERS experiment, starting with the biological question, including design of experiments (DoEs), (meta)data collection, the selection and employment of appropriate techniques that will extract useful information from the samples, the verification/validation of the results, and ending with the biochemical, biological, or clinical interpretation. This interconnected path involves a

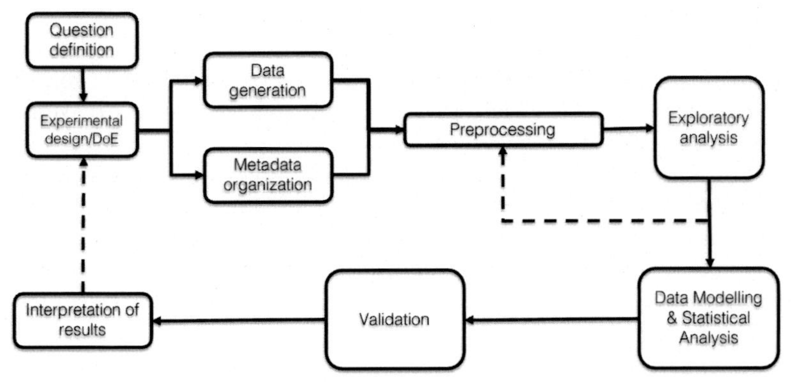

FIGURE 1.1

General data processing workflow for SERS studies.

Credit: Stefano Fornasaro.

constant feedback between experimental design, data processing, and statistical analysis: it's not unusual to return to data from an earlier stage (sometimes even to the raw data) to double-check key elements and resolve data ambiguities. In other words, it is extremely important to check the outcomes of each data analysis operation step by step, and if necessarily go back to a previous step, in a sort of self-consistent process.

Study definition and data collection

A typical SERS study begins with a biological/clinical question, which, for example, might involve two groups of individuals, often called case and control (e.g., patients vs. healthy). Even in this very simple scenario, *nested data structures* are often present (e.g., multiple samples per patient, many spectra per each SERS substrate). A characteristic of nested data structures is that the measurements within each nesting are more closely related to each other than the measurements between nesting. Here, the SERS intensities for each wavenumber within a substrate are more related or correlated to each other than the intensities between samples. And the intensities between samples are more correlated to each other than the intensities between patients. According to *Theory of Sampling* all these properties should be carefully considered before data are collected to ensure that the samples to be measured really represent the condition under investigation [1]. This is because a (biomedical) SERS experiment cannot totally be under control, since there are always unknown variations in replicates or individuals. Even so, this aspect is often overlooked until after the data are collected! On the contrary, a well-planned *experimental design* is the first element of a successful data analysis: an adequate number of replicates at the relevant influencing factors facilitate data analysis and, eventually, data interpretation. Therefore, the sample size should be always counted as the number of replicates/individuals rather than the number of spectra. If the dataset is composed of a large number of spectra in a nested data structure, the spectra belonging to the top-most hierarchical structure (the patient in the previous example) should be processed together. The importance of this point can never be overstated when it comes to get an honest assessment of a model's predictive ability on new cases (see "Verification of results" section).

DoE is a framework for systematic evaluation of complicated systems that are influenced by multiple factors [2]. The objective of DoE is to effectively plan and conduct experiments so that the experimental domain is systematically investigated, often with as few experiments as possible [3]. The key point is that some variation cannot be avoided, but experiments can be set up in such a way that it does not interfere with the results.

For example, many SERS studies make use of a blocking design to account for the effect of the use of different batches of Ag or Au colloids as substrates. In comparing the SERS response of a sample to several experimental variables or influencing *factors* (e.g., nanoparticles concentration, pH, incubation time), a common strategy is to perform all measurements using the same batch of SERS substrates (e.g., the same batch of metal colloid), studying the influence of one variable while keeping the others at a fixed value. Then, another variable is selected and modified to perform the next set of experiments, and so on and forth. In the end, everything is repeated for another colloidal batch to assess the variability between colloidal batches. This *one-variable-at-a-time* strategy has been shown to be inefficient, lacking the ability to address interactions between two or more variables and often needing many experiments to reach the optimum. A more comprehensive result can be achieved by studying several variables simultaneously and systematically by means of an appropriate experimental design. By using, for instance, a *factorial design* with *blocking*, where different batches of substrates contain an equal number of designed experiments, it is possible to detect the influence of all the variables and of their interactions within one batch and also the effect of different batches. A couple of designs useful for SERS are briefly described in Table 1.1. More information can be found in Refs. [4–7].

Data handling considerations (data structures, organization)

Typical research grade commercial Raman spectrometers will have a full scan range going from 50 to above 3500. With modern instruments, it is common to measure the entire individual spectral range with a single acquisition. The usual intensity scale used for spectral measurements is "photon counts" for SERS intensity, which may be further normalized or scaled. These data are usually arranged in the form of a matrix X, having as many rows as the number of spectra (each row of the X-matrix corresponds to a whole spectrum of a particular sample), and as many columns as the number of measured variables k (intensities at a particular wavenumber) over all the samples. Accordingly, the number of samples i may be a few ten to a few hundred; k may be a few hundreds to several thousands, depending on the density of data points. Fig. 1.2 shows a typical data matrix obtained from SERS measurement.

Table 1.1 Examples of designed experiments.

Design	Description
Full factorial design	Two or more factors, each with discrete possible values ("levels"). The experiments are performed on all possible combinations of these levels across all such factors. In this way, it is possible to study the effects of all the factors and of their interactions on the response variables. As an example, considering three independent two-level factors (treated/untreated, male/female, young/old) will lead to 23 = 8 experiments. In the presence of a large number of factors or factor levels, full factorial designs can become prohibitively expensive (in terms of time and costs). In these cases, it is possible to rely on partial/fractional designs where only a subset of the design space is analyzed.
Block design	A block is a group of samples (treated) measured in more or less homogeneous conditions (e.g., one colloidal batch). As an example, consider many biological replicates for each sample, and three colloidal batches to measure them all. The best option is to assign the samples in such a way that each batch corresponds to the analysis of one complete set of samples. Often, however, this cannot be arranged for several reasons. In such a case, an incomplete block design can be used.
Split-plot design	This type of design is used to handle nested structures, where it is more difficult to change the levels of a treatment during the experiment. The factors difficult to change are randomly used on each block, while the others are fully represented within each block.

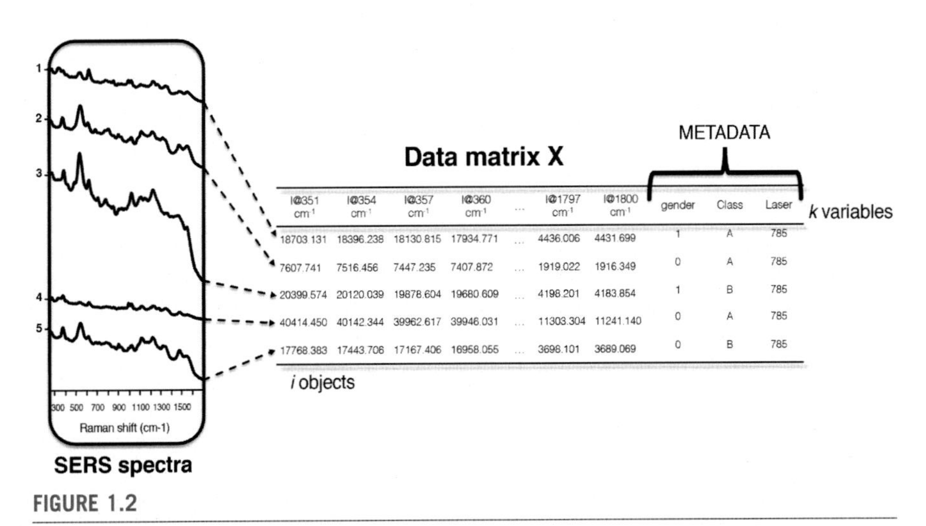

FIGURE 1.2

Data organization.

Credit: Stefano Fornasaro.

Metadata organization

Together with the spectral data, all the possible *metadata* should also be collected and organized. The term *metadata* refers to the set of information describing each sample and the whole experimental pipeline. They have to be considered during the design of the analytical workflow, in order to ensure proper randomization of the data collection. Moreover, they are extremely useful for a reliable interpretation of the results: organizing and storing a detailed record of the metadata is crucial to evaluate the presence (and the impact) of possible confounding factors, anomalies, or trends present in the dataset. Sample metadata often match the factors included in the DoE (e.g., type of treatment, reference diagnoses, gender, age, replicas), but they also include key characteristics of the samples. Some examples are

- date/time of sample collection
- date/time of sample analysis
- instrument on which the analysis has been performed
- type and power of laser
- laser power density on the sample
- number of accumulations
- operator who has performed the analysis
- type of SERS substrate

Recommendations

- Adopt good laboratory practice guidelines, possibly with the help of spreadsheets and laboratory notebooks, to record measurement parameters and results.
- organize the metadata in matrix form, in such a way that the import can be performed in a single step. If, by contrast, the data are spread in several files or sheets (e.g., one file for each sample or for each variable), then the import procedure would be much longer and more difficult.

A bit of statistics

To interpret the complex information hidden in the SERS spectra and connect it to specific biological phenomena, several data analyses and pattern-recognition algorithms are used. For any kind of analysis to be successful, it is essential to have an overall picture of the spectral dataset. Looking at all the SERS spectra at the same time (either stacked or superimposed) is not feasible for a large dataset, so one has to find other ways to represent its general characteristics. Descriptive data summarization techniques are commonly applied to visualize the typical properties of SERS spectra.

For many (pre)processing methods, it is essential to evaluate in advance the distribution of the data to ensure if the methods' assumptions will be met. *Descriptive statistics* for central tendency and dispersion are of great help for this purpose. Typical measures of central tendency include *mean* and *median*, while measures of data dispersion include quartiles, interquartile range (IQR), and variance. However, the mean is not always the best way of measuring the center of the data, especially for nonnormal distributions. Moreover, a major problem with the mean is its sensitivity to extreme values (e.g., *outliers*). Usually, the intensity distributions at each wavenumber are not-normal for a SERS dataset, and thus representing the dataset with the mean spectrum might be not entirely appropriate. For non-normal data distributions, a better measure of the center of data is the *median*, being the middle value of the variable when the data are aligned either in increasing or decreasing order. Thus, the *median spectrum* is a better option to represent the dataset in terms of "spectral profile," having the additional advantage of being less sensitive to *outliers*. Often, for SERS datasets the median and the mean spectrum are very similar, with only small differences, but if the dataset has many outliers, these two spectra can be quite different, so that using the median spectrum to represent the data is always a safer choice.

The spectra in a SERS dataset can be quite different from each other, that is the dataset present a certain *spectral variability*. The degree to which intensities at each variable tend to spread is called the *dispersion* of the data. A popular measure of the dispersion is the *standard deviation* (σ) defined as the square root of the *variance* (σ^2). Note that σ measures the spread about the mean and should only be only when the mean is chosen as the measure of center. Since we cannot assume our data to have a symmetric distribution, other options to describe the intensity dispersion (i.e., spectral variability) should be considered. The kth percentile of a dataset is the value x_i having the property that k percent of the data points lie at or below x_i. The *median* is, by definition, the *50th percentile*. The most commonly used percentiles other than the median are *quartiles*. The *first quartile* (Q1) is the *25th percentile*, whereas the *third quartile* (Q3) is the *75th percentile*. The quartiles, including the median, give some indication of the center, spread, and shape of a distribution. The distance between the first and third quartiles is a simple measure of spread that gives the range covered by the 50% of the data. It is called IQR and it is a common measure of dispersion used for nonnormal distributions. Thus, the *median* and IQR of the intensities for each wavenumber of the dataset give some indication of the center, spread, and shape of a distribution of intensities in the dataset. Mean ± standard deviation or median, 1st and 3rd quartiles cover only about 2/3 and 1/2 of the data, respectively. Plotting the 5th to 95th percentile of intensities gives a better visual impression of the actual range covered by the data. Box-and-whiskers plots are a very popular way of visualizing a distribution, using a box whose length is the IQR and whose width is arbitrary. A line inside the box shows the median, whereas whiskers are conventionally extended to the most extreme data point that is no more than 1.5 × IQR from the edge of the box (Tukey style) or all the way to minimum and maximum of the data values (Spear style) [8]. This graphical visualization can be used, for instance, to probe the distribution of the integrated areas of particular bands that are biochemically meaningful for each spectrum of a specific condition/batch/treatment. Measures such as median and IQR are more robust than mean and standard deviation, and thus they can be safely used to summarize the dataset in a figure (Fig. 1.3). Such a figure would immediately convey the overall spectral features of

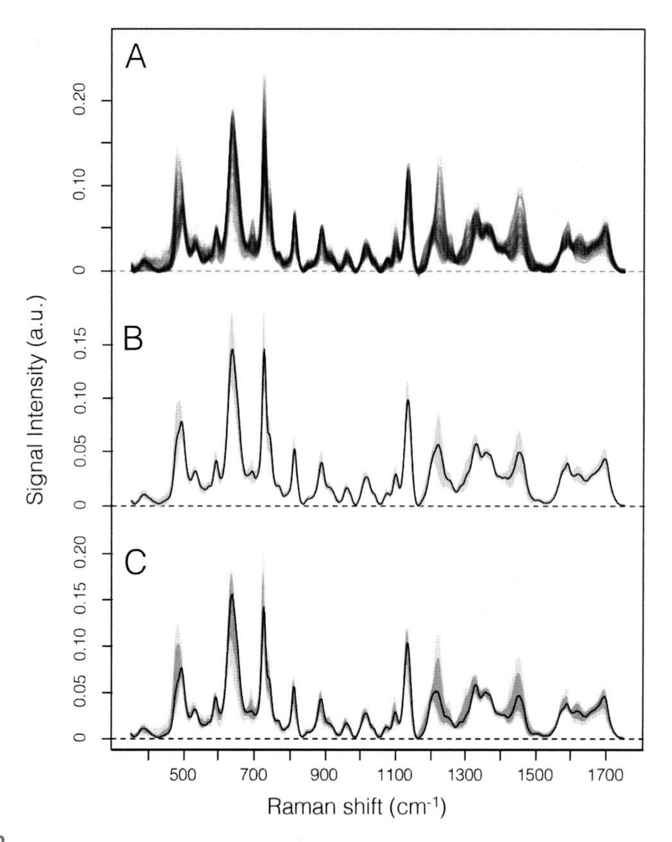

FIGURE 1.3

(A) Dataset composed of 200 SERS spectra of deproteinized serum; (B) mean spectrum ± 1 standard deviation; (C) 5th, 16th, 50th, 84th and 95th percentile.

Credit: Alois Bonifacio.

the dataset (as the median), at the same time giving an idea of how is the spectral variability for each Raman shift. Alternatively, a "functional boxplot" [9] approach could be used, with the immediate advantage of being able to find and visualize the spectra in the dataset which are significantly different from the others.

Survey of software available

One of the difficulties often encountered by researchers approaching SERS data analysis is to decide the software to be employed in importing, handling, and processing the data. These interconnected tasks require flexible and interactive tools.

Various software programs and packages are available, ranging from those for general-purpose use to those targeting specific data analysis tasks. Many spectrometer manufacturers offer software capable of data analysis, besides operating their instrument and collecting data. However, these software products are often limited with respect to data analysis, offering but few options. In spite of these limitations, and even if many commercial and noncommercial alternatives devoted to spectrum processing exist (see below), in the majority of the cases found in literature the first steps (or more often the whole sequence) of preprocessing are performed within the proprietary software provided by the instrument manufacturer. Then, the preprocessed data are imported in various statistical software (e.g., SPSS, Prism, Origin Pro, PLS Toolbox, and others available on the market) for further analysis. From an analysis of literature, a list of the more frequently used software in SERS studies was compiled and is given in Table 1.2. Most of these applications have a graphical user interface (GUI), resulting in a relatively shallow learning curve for the new user. These GUI-based software allow experimental scientists with little or no chemometric background to perform data processing and analysis of the measurements. However, these applications are frequently rather expensive. Moreover, the use of powerful multivariate analysis methods as "black-boxes" can lead to incorrect conclusions, and in general we discourage the reader to use them without a proper training. In addition, the use of a "point-and-click" approach in GUIs and the closed-source nature of these software usually limit the choice of algorithms, the user's ability to automate tasks and to develop new procedures within the software.

On the other hand, there is a growing community of spectroscopists that prefer to program their own methods using command line tools and short scripts written in different programming languages such as Matlab, R, or Python. The learning curve to manage these languages is steeper than those of most commercial software. The tradeoff for this time investment is a much tighter control and fine tuning on all the

Table 1.2 Results of search string "surface enhanced Raman" AND "diagnostics" AND (software name) from Google Scholar: most reported software together with the number of papers using the keyword.

Software	No. of papers on Google Scholar
Matlab	1750
SPSS	367
Prism	333
Excel	279
Origin Pro	262
Python	251
R	192
Unscrambler	107
Minitab	46
SIMCA	42
JMP	28
Stata	26

different steps of data handling, processing, and validation. These steps can be developed independently (also with different programming languages), or constitute the "nodes" of a unique pipeline, leading to several advantages. Since it is not easy to compare results from different software, having a uniform pipeline addressing all the tasks is a definite advantage. Moreover, a well-written workflow script is already an ordered description of the data analysis process, contributing to make the results transparent and reproducible, as requested by the FAIR principles of scientific data management [10]. A list of recommended software/packages is reported in Table 1.3.

The data format adopted is also relevant in the light of the FAIR principles. Raw data are usually stored in vendor proprietary formats that can usually be read only by proprietary software. An instrument manufacturer may have more than 10 different data formats (including older format versions). A closed data format obstructs data comparison and sharing, which is instead made possible adopting "open" data formats with publicly available specifications. For this reason, all the vendor software allows data export in open file format (usually as ASCII files). It is important to mention that not all the analytical metadata are usually included in the exported open source version of the data. It is therefore suggested to always store a copy of the original "closed" files, which can be inspected only with the vendor software, to avoid important information loss.

Data integrity

The first step after spectra collection is always the visual assessment of the quality and integrity of the recorded raw data. This is usually accomplished by examination of individual spectra, checking for anomalies that occurred during the measurement process (e.g., spikes/cosmic rays; presence of saturated channels, very poor signal-to-noise ratio), and for spectral contamination from unknown/unwanted chemicals.

A first way to check data, as already mentioned, is to plot the median together with the 16th and 84th percentile for each Raman shift (Fig. 1.3). Note that for normally distributed data, 16th, 50th, and 84th percentile are equal to mean $\pm$ one standard deviation (see "A bit of statistics" section).

Since errors can also occur in the metadata values (e.g., concentrations, class membership, etc.), it is important to carefully check also the correct association between spectra and the table containing this set of "numbers" before starting the preprocessing step. This data/metadata checking step is critical to maximize the possibility of performing a robust and successful analysis: many algorithms used for the analysis of SERS spectra are sensitive to the presence of "corrupted" spectra affected by the issues listed above, or to a mismatch between data and metadata. Whether the aim is to develop a multivariate calibration model to quantify a specific analyte in a complex matrix, or trying to correctly classify a patient from the analysis of a biofluid, in many cases just few corrupted spectra or mismatched metadata can degrade the performance of the model. For instance, a wrongly assigned concentration label to a SERS spectrum can seriously affect the performance of a regression model.

Table 1.3 Recommended software/packages.

Software	License	Package/ Toolbox	Handling	Preprocessing	Analysis	Purpose/description
R[a]	O					General purpose statistics platform using packages to provide specific functionality (https://www.r-project.org)
	O	hyperSpec	+	(+)	+[b]	Infrastructure for handling of spectra; works together with preprocessing and modeling packages (https://github.com/r-hyperspec/hyperSpec)
	O	ChemoSpec	+	+	+	High-level convenience functions wrappers for widely used data analysis such as PCA or HCA (https://github.com/bryanhanson/ChemoSpec)
	O	prospectr	-	+	-	Preprocessing, and sample selection calibration sampling (https://github.com/antoinestevens/prospectr)
	O	mdatools	-	+	+	Projection-based methods for preprocessing, exploring, and analysis (https://github.com/svkucheryavski/mdatools)
	O	baseline	-	+	-	Various baseline correction algorithms (https://github.com/khliland/baseline)
	O	EMSC	-	+	-	Package for EMSC (https://github.com/khliland/EMSC)
	O	pls, MASS, chemometrics, unmixR, ...	-	-	+	R has a wide variety of packages that provide statistical models

Continued

Table 1.3 Recommended software/packages.—*cont'd*

Software	License	Package/ Toolbox	Handling	Preprocessing	Analysis	Purpose/description
Python[a]	O					General-purpose programming language using packages (modules) to provide specific functionality (https://www.python.org)
	O	Quasar/Orange	+	+	+	Collection of data analysis toolboxes for spectroscopy (https://quasar.codes) expanding the orange data mining and machine learning software suite (https://orange.biolab.si)
	O	Scikit	-	(+)	+	Framework for statistical data analysis and machine learning (https://github.com/scikit-learn/scikit-learn)
	O	Keras	-	-	+	Deep learning library (https://github.com/keras-team/keras)
MATLAB	P					Multiparadigm programming language and numeric computing environment (https://www.mathworks.com)
	P	PLS_Toolbox	+	+	+	Suite of multivariate and machine learning tools (https://eigenvector.com/software/pls-toolbox/)
	O	MCR_ALS	-	-	+	GUI for Multivariate Curve Resolution-Alternating Least Squares (MCR-ALS) algorithm (http://www.mcrals.info)
	O	Biodata toolbox	+	+	+	Framework for handling of spectra; works together with preprocessing and modeling packages (https://www.mathworks.com/matlabcentral/fileexchange/22068-biodata-toolbox)
Unscrambler	P		+	+	+	Stand-alone[c] program for chemometrics (https://www.aspentech.com/en/products/msc/aspen-unscrambler)
Origin for spectroscopy	P		+	+	+	Stand-alone program for chemistry-related data (https://www.originlab.com/index.aspx?go=Solutions/Applications/Spectroscopy)

+ recommended; (+) use with caution; (-) not recommended/not available/impossible; O, open source; P, proprietary.
[a] *R and python allow mutual interaction, e.g., a python module providing file import for a specific file format can be used from R and the resulting dataset then analyzed in R (or both R and python).*
[b] *Does not provide data analysis, but seamless use of modeling packages like PLS, MASS, etc.*
[c] *Current version offers integrated python scripting support.*

Outliers

In some circumstances, visual inspection of all the spectra in the dataset can highlight clear outliers, including substantial spectral contamination. Ideally, such samples should be measured again, but if that is not feasible then they should be removed from the dataset and that exclusion reported. Other spectra could be identified as outliers using more complex outlier detection methods based on several popular distance measures, such Mahalanobis or Euclidean distances [11,12].

Despite the differences between all the existing methods, they have one feature in common: most of the detection algorithms assume a (approximately) multivariate normal distribution of the data. Unfortunately, SERS data frequently fail to meet this assumption.

Other methods, based on so-called robust estimates, like Minimum Covariance Determinant [13], or one-class classification approaches like one-class support vector machines (SVMs) have also been proposed [14]. These methods are based on the assumption that the dataset represents a sample from a single "good" population, contaminated by outliers from different populations. This assumption is not always correct, especially in clinical studies, where the samples do actually come from multiple populations. The *functional box-plots* proposed by Sun and Genton [15] could provide a useful strategy for reliable and robust multiple outlier detection (even when data are not normally distributed), by implementing the idea of the "depth of a curve" (Fig. 1.4).

Data preprocessing

Data preprocessing, also known as data pretreatment, is to be broadly intended as the set of procedures performed on raw data to make them more adequate for the analysis planned. Most often, this set of procedures include methods to compensate for the effects originated from measurement noise (random variation) and from systematic errors (interferences from optical and physical effects) that are uncorrelated to the biochemical fingerprint in the spectra. These methods are usually applied before the actual analysis to reduce the influence of the random variation and remove instead the systematic variation. However, it is not always easy to exactly define what these two variations are or how to quantify them. It must be clear that although preprocessing can be very helpful in "cleaning up" a SERS signal, it does not, by itself, improve its quality. On the other hand, the risk of removing pertinent variation to spectral data is always present when applying preprocessing: an excessive noise removal or background correction can suppress important spectral information.

Preprocessing of SERS data could be pursued along different routes, and a wide range of methods exist in the literature. It is important to remember that there is no one and only correct way to choose which preprocessing technique to use, or a sequence of different methods that always works with any dataset. The choice of which preprocessing technique to use highly depends on the goal of the analysis, the quality of the spectra, the availability of methods and software, as well as on

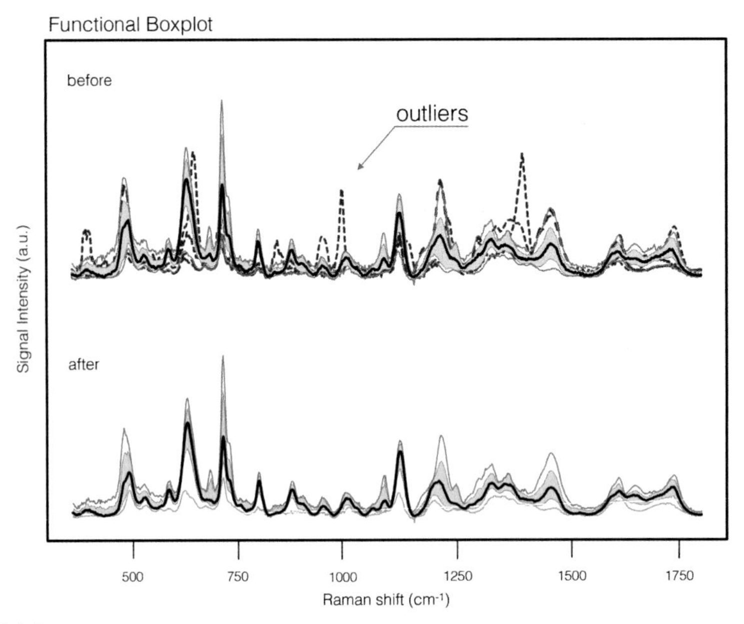

FIGURE 1.4

Outlier detection using functional boxplot. (A) Before; (B) after. The *black line* in the center is the median in the standard boxplot. The light gray region corresponds to the box in a standard boxplot. Here it is the envelope of the central 50% of the *curves*. The limits of the central 50% envelope are moved out by a distance equal to 1.5 times the width of this envelope at each wavenumber. Any spectrum not completely contained in this extended band is considered as an outlier and plotted separately as a *red dashed line*. Then the outer *dark gray lines*, corresponding to the ends of the whiskers in the standard boxplot, are the envelope of the remaining spectra.

Credit: Stefano Fornasaro.

a prior knowledge of the physical and chemical phenomena related to the dataset. Is the background due to fluorescence or to the formation of amorphous carbon by photo-thermal degradation? Are some interfering bands due to instrumental artifacts or to the normal Raman signal of the solvent? Looking at the dataset, asking such questions and, possibly, knowing the answers can make the difference between a good or a bad choice of preprocessing methods. On the other hand, many different approaches can work equally well for the same dataset, as long as they are used correctly. At the same time, a high number of preprocessing combinations will lead to a worse performance compared with doing no preprocessing. Thus, it is necessary to evaluate the suitability of the preprocessing, which is far from trivial. An "optimal" preprocessing sequence for a specific dataset can be determined by trial and error or by DoE [16]. The "optimal" preprocessing finds the "best" compromise between flexibility and reliability. In this sense, the amount of preprocessing can be considered a parameter to be tuned, and treated consequently (see "Hyper-parameter optimization" section).

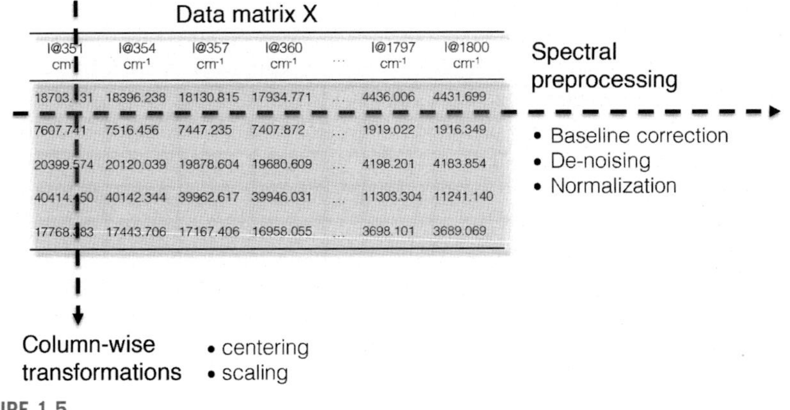

FIGURE 1.5

A visual representation of preprocessing on a data matrix.

Credit: Stefano Fornasaro.

This section will focus on practical suggestions and hints on how to correctly perform preprocessing of SERS data. For a more detailed description of preprocessing methods, the reader is referred to Refs. [17,18]. The most frequently used preprocessing techniques for SERS spectra are the same as those employed in classical Raman spectroscopy, and can be grouped essentially in row-wise and column-wise methods (Fig. 1.5). With row-wise (also known as spectral/signal preprocessing) methods each spectrum is transformed individually, one sample at a time over all variables (wavenumbers). Column-wise methods (e.g., centering and scaling), on the other hand, operate on one variable (wavenumber) at a time over all spectra. Therefore, if a sample is added or deleted from a dataset, column-wise preprocessing, must be repeated, while row-wise calculations will not be affected.

Spectral preprocessing (row-wise methods)

Row-wise methods are a very heterogeneous group of mathematical procedures applied to remove from the SERS signals many different unwanted interferences that can create challenges in the subsequent qualitative or quantitative analysis. In general, a SERS spectrum can essentially be considered as the sum of the true signal multiplied by an intensity scaling factor of interest plus a background (usually caused by residual Rayleigh scattering at low wavenumbers and/or by the fluorescence of organic molecules intrinsic to the analyzed sample), both convolved with an instrumental blurring function, possibly shifted along the spectral axis and with noise added to the result. Row-wise methods can thus be divided in techniques that (i) use no prior information (regarding the noise, baseline, signal), or (ii) do use some prior information (e.g., spectrograph calibration). In both cases, there are procedures applied to filter just one kind of unwanted feature separately (e.g., either background contribution or shot noise), and procedures applied to correct all the interferences in one step.

"Baseline-correction" or "background-correction" techniques eliminate the effects of fluorescence or *low frequency* components (i.e., those components whose signal intensity shows few, broad variations in the spectral region considered) present in the SERS spectra. These methods generally involve complicated and time-consuming processes to select appropriate key parameters. This selection process, as well as the relative outcome, is strongly dependent on the researcher's experience. An excellent overview of these methods is given by Schulze et al. [19]. In addition, *high frequency* components of the SERS signal (i.e., those components whose signal intensity shows frequent, sharp variations in the spectral region considered), such as random instrumental noise (e.g., "shot noise"), can be partly removed by using smoothing interpolation techniques. This smoothing step can be performed separately (before or after) or at the same time of baseline estimation. If applied before the baseline estimation, the implicit assumption is that the baseline estimation is not depending on the noise level. It should also be stressed that noise in SERS datasets can be efficiently handled by most multivariate analysis methods, so that a noise reduction is often not strictly necessary. On the contrary, if not carried out correctly, noise reduction can subtract relevant information from the dataset (see below). A better approach is to combine smoothing with down-sampling: by decreasing the overall number of data points, trading some spectral resolution for higher signal to noise ratio, the multivariate algorithms are presented with spectra having less variables *and* less noise. The amount of "spectral information per data point" increases, easing the task of the algorithms.

Manual or automatic polynomial fitting algorithms are the most widely used to estimate linear and curved baselines. This strategy is simple and convenient in terms of spectral interpretation, but it has a poor performance for spectra with a low signal to noise/background ratio. Moreover, the baseline for a spectrum with many overlapped peaks and a nonlinear background is hardly fitted well by these methods. Polynomial baselines might also cause some artifacts (tails or false peaks) at the borders of the spectral region, requiring a "cropping" of the region to leave these artifacts out of further analyses. Different variants of the polynomial baselines have been proposed to overcome these issues. The most famous are the *modified multi-polynomial fitting* method, proposed by Lieber and Mahadevan-Jansen [20], the *Iterative Polynomial Fitting*, proposed by Gan et al. [21], and the *Vancouver Raman Algorithm*, proposed by Zhao et al. [22], which has the advantage of incorporating a statistical approach to reduce the effects of noise. Liu et al. introduced the *Goldindec* algorithm to solve the low correction accuracy problem in presence of elevated number of peaks [23]. Methods based on the *Whittaker smoother* and *penalized least squares* are also frequently applied for baseline correction. Eilers et al. first obtained an effective baseline estimator which combined an iterative least squares smoothing with asymmetric weights for positive and negative intensity values [24]. Their *Asymmetric Least Squares* (AsLS) is one of the most frequently used methods in SERS. To improve the accuracy of the AsLS baseline correction method, Zhang et al. [25] proposed the *Adaptive Iteratively Reweighted Penalized Least Squares* (airPLS) methods. He et al. [26] proposed the *Improved Asymmetric Least Squares* (IALS) method, an extension of AsLS that imposes extra smoothness constraints on the baseline.

With all these options available, and in absence of defined criteria for method selection, the choice of a method might seem a challenging task. Moreover, for many methods it is necessary to carefully optimize their parameters on a case by case basis, which can be very time consuming. As a rule of thumb for SERS datasets, however, a simple approach using a polynomial baseline fitting or an AsLS method will work reasonably well in many cases (Fig. 1.6).

A commonly used smoothing method for reducing noise in SERS spectra is the Savitzky–Golay (SG) filter. Although this method is useful in noise reduction, it may also suffer from the degradation of the underlying SERS spectral features. For that reason, special care must be taken in the selection of the spectral sub window, since large sub windows will eliminate informative peaks, while small windows might generate more noise. The spline-fitting method uses predesigned smooth curves to fit a noisy dataset. Although this method has long been used for baseline correction of Raman spectra, its performance is heavily dependent on the choice of the positions of knots, the type of spline functions, and it is greatly affected

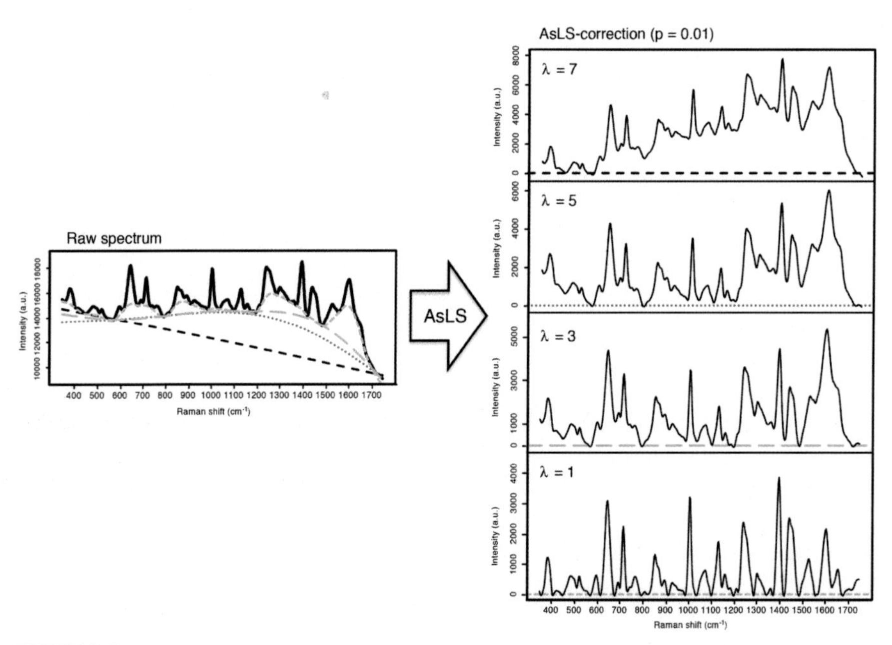

FIGURE 1.6

The effect of different λ values on the results of asymmetrical least squares (AsLS) baseline correction. The *P* value was kept at the value of .01 for all plots. Low λ values produce baselines that follow the data too closely, resulting in overfitting. As a result, spectral band round 1600 cm^{-1} was artificially diminished. High λ values generate a linear baseline, resulting in underfitting. As a result, the broad baseline features are not removed.

Credit: Stefano Fornasaro.

by noise. Another category of smoothing methods such as wavelet and Fourier transform (FT) filtering are conducted in the frequency domain [27]. Overfiltering or underfiltering are the main drawbacks to FT and wavelet filtering when it is not easy to separate the frequency components of SERS spectrum from noise. In addition, the associated optimization is complex to implement.

Convenient multistep all-in-one methods are the 4S Peak Filling baseline estimation by iterative mean suppression, proposed by Liland [28], and the SABARSI (Statistical Approach of Background Removal and Spectrum Identification for SERS Data), proposed by Wang et al. [29].

Recommendations

- We recommend visual inspection of spectra after each stage.
- All decisions should be taken by a priori spectroscopic knowledge of the system. Although such a strategy may not be the optimal one, it will lead to a better model performance.

Normalization

After baseline correction and denoising, the SERS spectra can sometimes be analyzed directly. In most situations, however, a normalization of the entire spectrum is necessary. SERS spectra collected from the same sample can present huge intensity fluctuations (multiplicative effects), because of the highly localized nature of the hot spots responsible for the signal enhancement. If more substrate batches are used, batch-to-batch variability can also be a source of intensity variations for the same sample. Furthermore, other intensity fluctuations can be the result of different sample preparations and varying collection parameters (e.g., spectra recorded at different times and under different instrument conditions, such as alignment and laser power levels). Normalization is the process which takes care of removing spectral differences due to intensity fluctuations from SERS data.

Basically, normalization is performed by dividing each variable of a spectrum by a constant, and normalization procedures differ from each other by the choice of this constant. A common approach to normalization is to use the height of a single reference band (measured as the range between the baseline and the maximum intensity) as the normalization constant. This band can be an internal or external reference. Note that this reference should be a SERS signal originating from the same substrate as the analytes responsible for the other bands in the spectrum. Normal Raman bands of solvents or substrate components cannot be used for normalization purposes, unless the intensity fluctuations are mostly due to problems related to instrument instability (e.g., fluctuations in the laser power, unstable optical alignment).

SERS bands can be used for normalization, but the choice of different SERS bands will lead to rather different results. Analysts have to decide which specific SERS band is reasonable as reference for their application. Ideally, an internal standard should be used, i.e., a chemical species yielding a well-defined, characteristic

SERS signal in the same chemical context of the analytes. Such internal standards can be very effective to compensate for SERS intensity fluctuations and also present several challenges which prevent the adoption of specific molecules as universal standards.

When no internal standard reference band is available, a SERS band due to substrate components (e.g., coatings) or to solvent is often used [30]. An interesting alternative was proposed by the Wei et al. [31]. They exploit surface plasmon–enhanced elastic (Rayleigh) scattering signals as internal standards for SERS signal normalization, by using the low-wavenumber *pseudo band* arising from the interaction of the amplified spontaneous emission of the laser and the edge filter as the normalization constant.

Instead of using the intensity or area of a single reference band, the so-called *vector normalization* can be applied. In this type of normalization different vector-norms can be used as normalization constant. It is a potentially useful approach when the relevant differences among samples are related to the shape of the signal and not to the global signal intensity (e.g., for many *classification* studies). The most common to use is total sum of absolute values of all variables (*1-Norm*), or to use the square root of the sum of squared elements (Euclidian). The former is commonly referred as total intensity normalization. Alternatively, the vector-norm called *2-Norm* can be applied, where variables with larger values (higher intensities) will be more heavily weighted in the scaling [18]. Regardless of the chosen norm (or area), signals used to determine the normalization factor need to have a well-defined baseline (uncertainty in the baseline will be amplified in the normalization factor) and should be of sufficient and stable intensity.

Another approach to normalization is *standard normal variate* (SNV). This method, originally proposed by Barnes et al. [32] for near-infrared diffuse reflectance spectra, has proved to be very suitable for SERS data [33]. SNV is a weighted normalization method, in which different weights are assigned to each spectral data point, with respect to the normalization factor. First, individual mean value is subtracted from each sample spectrum. Then, the resulting value is divided by the standard deviation of all pooled variables for the given sample. This normalization will give the spectrum a unit standard deviation of 1. Mathematically, SNV correction is identical to autoscaling of the rows instead of the columns of the matrix. It must be noted that SNV is applied to every spectrum individually, and not on an entire set of spectra.

Recommendations

- Normalization should be performed before any column-wise preprocessing steps and after background removal

Model-based methods

Alternative to baseline correction and normalization approaches, a *model-based preprocessing* can be used. The *multiplicative signal correction* (MSC), one of the most widely used preprocessing techniques, was also originally developed to separate physical and chemical variances of NIR spectra [34], but can be used for SERS data as well. MSC is performed by calculating the slope and intercept of the regression between each individual sample spectrum and a reference spectrum, which is typically the mean spectrum of the dataset. However, determining the best reference spectrum is important to have a good correction. The mean spectrum of the calibration set is often used, but it is not always representative of the new data. The corrected spectra are obtained by subtracting intercept from each variable in all the sample spectra followed by a division with the slope. The technique has been further developed into *extended multiplicative signal correction* (EMSC) [35]. In EMSC additional terms are included to correct for wavelength-dependent spectral variations from sample to sample. The EMSC method also allows for including a priori information regarding the analytical system under investigation (e.g., spectra of pure constituents), as well as extra terms describing additional physical or chemical interfering features (background). A typical example is the inclusion of an extra term describing the average baseline of the calibration set as estimated by a polynomial fitting. An additional advantage of this method is that it performs both baseline correction and normalization in one step and the replicate variations within the data can be considered. The EMSC methodology and extensions have been described in the EMSC Tutorial of Afseth and Kohler [36].

Column transformations

In general, column-wise transformations alter each sample according to the characteristics of the entire set of spectra. Thus, adding or subtracting a single sample from the dataset affects the transformation of other samples as well.

Mean centering is one of the most commonly used column-based methods. The mean spectrum for a set of spectra is subtracted from all the spectra in the set. From a spectroscopic point of view, this technique is frequently used to remove the average chemical composition of the sample from the dataset, to better exploit the remaining spectral information. However, for very complex samples, such as biofluids, the chemical meaning of the "average spectrum" may be questionable. Depending on the particular dataset, a more suitable option would be to subtract the fifth percentile spectrum, instead of the mean. This will hopefully have almost the same benefit on the data analysis, since many data analysis techniques, such as *principal component analysis* (PCA) or *partial least squares* (PLS), work much better on centered data. In most software solutions, PCA and PLS functions automatically center data by default. For SERS data, *centering* is usually the only column-wise pretreatment performed, although sometimes *autoscaling* (also known as *column standardization*, or *variance scaling*) is also employed.

Autoscaling implies *mean centering* and dividing the intensity values of each wavenumber by its corresponding column standard deviation, and only makes sense for centered data. In this way, all intensities at each wavenumber have a mean equal to zero and a standard deviation equal to one. This type of scaling is often used in multivariate analysis to compensate the influence of variables with typically

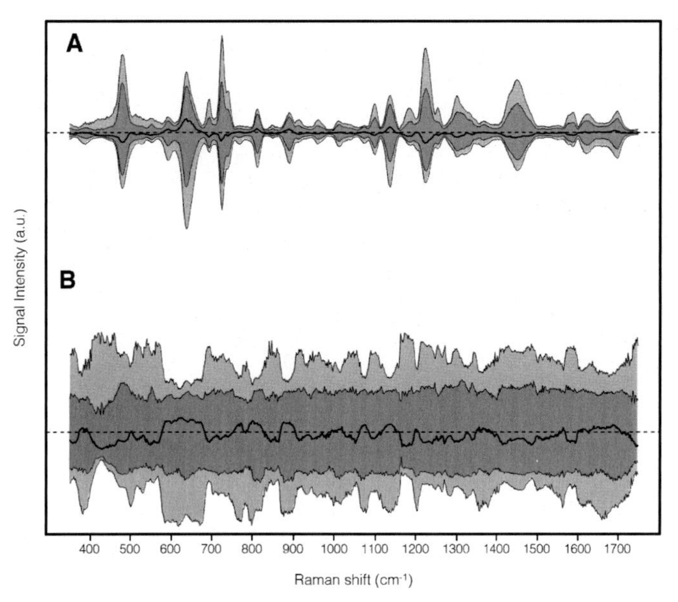

FIGURE 1.7

An example of (A) centered and (B) autoscaled SERS spectra of serum (data shown as 5th, 16th, 50th (median), 84th and 95th percentile).

Credit: Stefano Fornasaro.

different physical scales [37]. In a spectroscopic context, autoscaling can be applied to obtain an idea about the relative importance of the variables (SERS intensities already do have the same scale), but it is not recommended. In fact, one has to trade the expected numeric benefit with the fact that autoscaling will inflate the intensity (and thus the importance with respect to the further analysis) of noise for wavenumbers with low signal (baseline regions). Moreover, the transformation of the intensities obtained by autoscaling might completely change the "look" of a spectrum, making the visual interpretation more difficult (Fig. 1.7).

Models

Once a SERS spectral dataset has been checked and preprocessed, many different methods from chemometrics and machine learning may be used to relate the spectral signal to the analytical property to be determined, such as class membership, concentrations of analytes, distribution of substances, disease markers, or sample types. If no training data with reference values are available, *unsupervised methods*, such as PCA, are the only techniques which can be applied. If, on the other hand, training data with reference values are available, *supervised methods*, like regression or classification models, can be used (Fig. 1.8). In general, *linear models* are the most

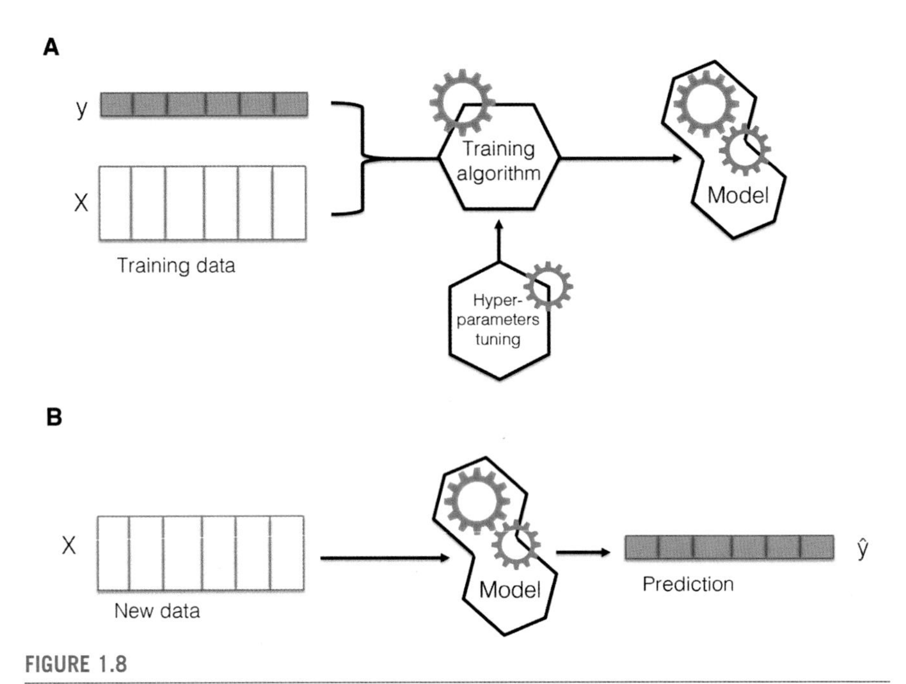

FIGURE 1.8

(A) supervised training: a training algorithm takes training data (dependent X and labels Y) as well as possibly hyperparameters and calculates a model (dark). (B) Prediction: The model takes new data without labels and predicts (calculates) labels $\widehat{Y}$.

Credit: Stefano Fornasaro and Claudia Beleites.

frequently applied due to their simplicity and robustness, while the use of machine learning algorithms can be effective in the presence of complicated nonlinear interactions in the data. However, despite convincing prediction results, the lack of an explicit model can make machine learning solutions difficult to directly relate to existing biochemical knowledge. Most common or useful models to be used with SERS datasets are briefly discussed below. Going in-depth on every available algorithm is beyond the scope of this chapter, and more in-depth resources are available elsewhere [38]. A comparison of different classification and regression models is reported later in Table 1.4.

Exploratory data analysis (unsupervised learning)

When a dataset consists of few spectra, we can simply plot them together, stacked or not, and compare them to have a direct idea of differences and similarities. However, already with more than 10 spectra this inspection becomes challenging, especially if we don't know yet where to look (e.g., we don't have a specific marker band on which to focus our attention). For larger datasets, in fact, we have no idea about the differences occurring in our data. We are used to consider differences between

Table 1.4 A comparison of different classification and regression models.

Model/model type	Regression	Classification		
		Discriminant	Threshold	One-class
PCA	p	p	p	(p)
PCR	+	-	c	-
LDA	-	+	-	-
QDA	-	+	-	-
SIMCA	-	-	-	+
PLA	+	p	+	-
PLSR	+	-	c	-
PLS-DA	-	(+)	+	-
PLS-LDA	-	+	-	-
PLS-LR	+*	+	+	-
LR	+*	+	+	-
SVM	+	+	+	+
SVM classifier	-	+	+	-
SVM regression	+	-	(+)	-
One-class SVM	-	-	-	+
k-NN	+	+	+	-
Distance based k-NN	+	+	+	+
Decision tree	+	+	(+)	-
Random forest	+	+	(+)	-
ANN	+	+	+	+

+ recommended; (+) use with caution; - not recommended/not available/impossible; p *for pre-processing,* c *with calibration (regression) labels; *regresses probabilities.*

just few simple objects, and even trained spectroscopists cannot grasp differences in large spectral dataset at a glance. This is when an exploratory data analysis can help us. The aim of exploratory data analysis is to investigate what the data can tell us with no statistical inference or underlying probabilistic model. Because of the very well-resolved bands found in the spectra, SERS provides a very powerful tool for distinguishing between samples with only minimal differences. However, the variations between spectra can be very small and difficult to identify, by just looking at the data. Approaches used for finding patterns and clusters within a dataset are called *unsupervised* techniques because no prior information needs to be presented to the model prior to analysis. In other words, this type of analysis even be done without knowing anything about the dataset: the dataset tells a few or some important things about itself. The most commonly used unsupervised multivariate technique is PCA, a mathematical method, first described by Pearson in 1901 [39] and by Hotelling in 1933 [40]. More in-depth resources about less commonly used unsupervised techniques are available elsewhere [41]. Simply put, PCA allows you to look at your dataset "from a different perspective," like when you try to guess the shape of an object by looking at the shadows it projects along different directions. In particular, the privileged "perspective" of the PCA is the one that better evidences spectral differences in the dataset. In PCA, the data matrix, X, can be decomposed into a score matrix (T) and a loading matrix (P), which capture the main variation in data, leaving unmodeled the unsystematic variation in the residuals, E.

$$X_c = TP^T + E$$

where X_c is mean centered. Systematic variations in samples and variables are found in the score matrix (T) and loading matrix (P), respectively. By PCA, the main variation in a multidimensional dataset is found by creating new linear combinations of the original variables (i.e., the new "perspectives"). These linear combinations are called principal components (PCs) and they are a smaller set of uncorrelated variables, chosen to maximize the variance of the data. The first PC explains the maximum variance and the subsequent PCs capture the remaining variance in the decreasing order. One can also describe it as a rotation that takes a viewpoint showing the measurement points spread out as far as possible. This enables us to describe the dataset with considerably fewer variables than those originally present. From this perspective, PCA is a way to drastically simplify the dataset by focusing on those spectral features which are responsible for most of the differences within the dataset. This capability is the reason why PCA is often employed to "reduce the dimensionality" of data when it is used before a classification or regression modeling step (see below). As such, PCA is an unsupervised method, that is it considers only the spectral data (and its variance) and not the labels (metadata) that may be available. Rather than going into the mathematics (see Ref. [42] for that), we show how to use PCA and interpret the results, for the analysis of SERS data. In case of spectral data such as SERS spectra, where the variables are highly collinear, the data matrix is likely to be decomposed into a considerably lower number of PCs than the number of variables. In a common scenario, the first few PCs (instead of the hundreds of data points in a spectrum) retain a very high percentage of the information content, often more than 99%. The result of a PCA model can be graphically illustrated as score plots and loading plots. Information from the score matrix is usually presented in two-dimensional scatter-plots of one PC versus another. The scores are the coordinates of the samples in the new coordinate system where the axes are defined by the PCs. From a score plot, sample similarities (trends and groupings), as well as outliers can be identified. Samples that are close to each other on a given score plot are similar with respect to the original measurements. When metadata are available, one can use different colors and/or symbols to indicate groups of metadata and score plots can be used to check if the variance in the data reflects a true biochemical difference (Fig. 1.9A).

The loadings matrix shows the importance of the original spectral bands for each PC. This information can be visualized in the loadings plots, by plotting each loading vector versus the wavenumber axis (Fig. 1.9B). The loading plot is used to interpret the relationship between the scores and the original variables, evidencing which variables are responsible for the separation between sample groups or for a particular sample to be an outlier. If the focus is to find clusters in the dataset,

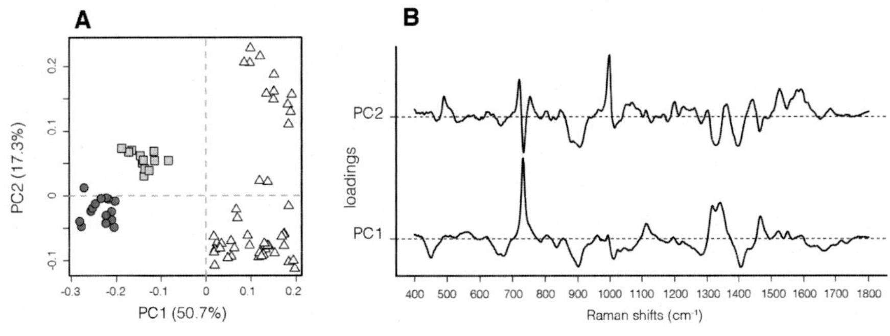

FIGURE 1.9

Examples of (A) score plot and (B) loading plot. The dataset used for the PCA is constituted of SERS spectra of different cells lysates [89].

Credit: Stefano Fornasaro.

loading plots can be useful for identifying which variables characterize the different groups. They are a powerful tool when they are applied when compared to structure-correlation locations of spectra. Loadings quite near 0 have little diagnostic influence, whereas loadings that are intense (either negative or positive) suggest wavenumber that are quite important for that PC. The interpretation of the loadings in terms of SERS bands should be performed with a grain of salt, and even this can be tried only for unscaled data, as SERS spectral patterns in the loadings of scaled datasets are different from real spectra or real difference spectra.

From a spectroscopic point of view, loadings can also be used to help determine the inherent dimensionality of the dataset (choose how many PCs retain in the model). This is accomplished by examining loadings plots for each PC and determining for which of them the loadings begin to exhibit noise.

Recommendations

- Always show PC numbers, and percentage of variance explained for that PC when plotting the scores and the loadings. Loadings cannot be interpreted without scores and vice versa.
- Avoid ellipses in score plots as automatically printed by some software. These convey a shape and size of the scores point cloud that can be quite misleading.
- Sometimes the PCs actually do measure real variables, but sometimes they just reflect patterns of covariance which have many different causes. Don't expect to see "pure" SERS spectra in the loadings, although you should expect to see features from them.
- PCA is not a classification method. Any method that is aimed at classifying needs a step of training and a step of validation (see "Verification of results" section). In PCA, those stages do not exist.

Regression

The aim of a quantitative SERS method is to provide accurate and reliable determination of the concentration of a target analyte (usually a drug, metabolite, or biomarker) in a medium such as complex biological matrix (e.g., biofluids, cell lysates, extracts).

There are different approaches to quantitative analysis of SERS spectra, but all the models presented here are based on learning a prediction rule from an observed dataset, which is also called calibration set, and then apply that rule to predict the concentration of the target analyte in unknown samples. These models, however, have different learning and prediction strategies. In the easiest scenario, the intensity at a single band or the ratio of bands can be related to the concentration of a specific compound in the system under investigation [43]. The data are usually fitted by a straight line (calibration curve) based on least-squares principle, a procedure known as *univariate calibration* that is easily performed with most data analysis software. The obtained calibration curve is then used to calculate standard errors and confidence intervals for the parameters, as well as for the concentrations of unknown samples. This type of calibration is often used for SERS data obtained by an indirect detection, i.e., when SERS-tags or in Raman reporters in general are used to reveal the presence of an analyte [44]. Since the signal due to the reporter is usually very intense and dominates the SERS spectrum, the most intense band works well for the purpose of a univariate calibration. This approach is also useful in label-free SERS methods, provided that the method is very specific, i.e., the analyte, and only the analyte, gets selectively adsorbed on the nanostructure surface. This scenario, however, is rather uncommon when targeting analytes in complex matrices, mostly because of interferences from the matrix having its own variability due to interindividual differences. More frequently, multivariate calibration algorithms are employed to learn the quantitative relationship between the constituent concentration and the whole SERS spectrum. They are distinguished into *direct calibration* methods and *inverse calibration* methods. In the direct models, the signal is considered to be directly proportional to the concentration, as dictated, for example, by Beer–Lambert–Bouguer law. In the inverse models, on the other hand, the concentration is considered to be directly proportional to the signal. In univariate calibration, no substantial difference appears to exist in calibrating a regression line in a direct or in an inverse way, with the inverse univariate model appearing more efficient for the analyte prediction in cases of small datasets with high noise level [45]. In multivariate calibration, on the other hand, the difference between direct and inverse models is crucial, and inverse multivariate models are preferred [46]. Multiple linear regression (MLR) is one of the most widely used *direct* calibration methods. However, it requires that the number of samples should be higher than the number of variables (a requirement practically impossible to meet for SERS data), and that the variables themselves should be as uncorrelated as possible. Both limitations do not allow the application of MLR for SERS. Suitable *inverse* methods employed to build regression models in cases where MLR is not applicable are principal component regression (PCR) [47,48], and PLS regression [49].

Principal component regression

After a PCA dimension reduction step (i.e., the decrease of the number of variables describing the dataset to few PCs), it is possible to use the scores of the data matrix X, T, in a regression equation to predict the concentration, considered as a dependent variable Y.

$$Y = TB + E$$

Here, B is the matrix of regression coefficients, which can be estimated with least-squares regression easily: the columns of T are orthogonal, and the number of columns is usually quite a lot smaller than the number of rows. This kind of model is called PCR. It must be noted that in many real cases (e.g., when there are many uninformative sources of variability in the data), the directions of maximum explained variance as extracted by the PCA may not be relevant for the prediction of Y. To overcome this drawback, an alternative approach is represented by the PLS regression.

Partial least squares regression

PLS is probably the most widely used and well-described method for multivariate regression [50–52]. It uses the same trick as PCA: a combination of the original variables into a smaller number of new ones, in this context called *latent variables* (instead of *PCs*). Again, the regression model has the form

$$Y = TB + E$$

This formulation is identical to that for PCR; however, the calculated components and the model coefficients are not the same. Rather than focusing on capturing the variance in X, in PLS the variance in X and Y and their mutual correlation are simultaneously maximized (i.e., each latent variable is obtained by maximizing the covariance between Y and all possible linear functions of X), and many different algorithms have been proposed for its implementation [53].

In most cases the loadings will be similar to the loadings that would have been obtained by using PCA. PCA and PLS loadings differ only if the spectral information related to Y is not at the most variable wavenumbers in the spectrum. PLS compromises between modeling variance in X and correlation between Y and X, so that large loadings values for a wavenumber may correspond to high variance, high correlation, or both.

When predictions are required over a large range of concentrations, nonlinearity is common for SERS datasets, and linear models fail to keep their accuracy across the range. In such situations, *nonlinear* regression models, such as SVMs or *artificial neural networks* (ANNs), can be more appropriate. SVMs are a very popular method based on the *structural risk minimization* principle within the statistical learning theory. ANNs are also very popular, although, the majority of the ANN methods are very complex and present some inherent drawbacks associated with the training process. The theory of both methods is beyond the scope of this chapter, and is described in Refs. [54,55].

It should be also mentioned that while SVMs or ANNs can be used intrinsically for both classification and regression, also PLS and PCR, can be used as classification tools to model class affinities. One of the key aspects that must be fulfilled to build an accurate and reliable calibration, no matter which method is used, is the development of proper *validation*, as will be explained in Section "Verification of results". Many *figures of merit* (FoMs) can be used to validate a regression model, such as the *root mean square error* (RMSE), or *bias* among others (see "Regression" section). For a comprehensive review on different approaches to calculate the FoMs in regression see Ref. [56].

Classification

Classification models assign *cases* (i.e., spectra) to prespecified *groups* or *classes*. In biomedical SERS applications, each time you want to exploit spectral information to "label" a spectrum as belonging to a specific group of subjects sharing a common diagnosis, you need a classification model.

In *discriminative* classification (Fig. 1.10A), each case must belong to exactly one of the mutually exclusive classes. Sometimes this is slightly relaxed to allow "uncertain" as outcome for cases close to the class boundaries. In medical contexts, this corresponds to differential diagnosis, i.e., finding out which of the differential diagnoses applies. Examples of discriminative classifiers include *bilinear* models like *linear discriminant analysis* (LDA) and *quadratic discriminant analysis* (QDA), *decision tree*–based methods, most SVM, *k-nearest neighbors* (kNN), and many more.

A subgroup of discriminative classification methods may be described as "regression methods in disguise" or *threshold-type* classification (Fig. 1.10B).

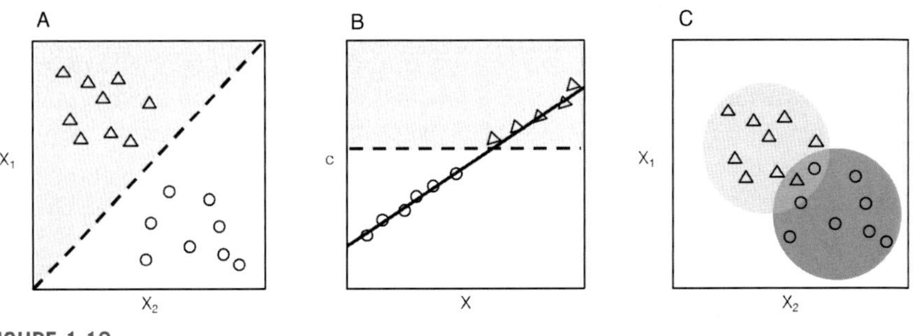

FIGURE 1.10

(A) Discriminative classifier separates feature space into disjoint regions for the classes, (B) threshold-type classification: the underlying nature of the data is a regression whose outcome is divided into classes according to a threshold. (C) one-class classification models each class independently of the others, possibly leading to overlap.

Credit: Stefano Fornasaro.

This type of classification is closely related to traditional qualitative analysis or checking whether a case/sample is above or below a certain concentration limit for an analyte or property. The underlying nature of the property (e.g., analyte concentration) and the signal are metric in nature, thus making a regression model appropriate in the first place to which then a threshold is applied. This may be done by setting up a regression model and postprocessing its prediction with a threshold, but also discriminative classification algorithms that disregard samples far from the class boundary can be used. Examples are PLS-DA (Partial Least Squares regression Discriminant Analysis), *logistic regression* (LR), or SVM (including SVM classifiers).

In contrast to discriminative approaches, *one-class classification* (Fig. 1.10C) evaluates whether a spectrum belongs to a specific class or not, considering each class independently. Consequently, a spectrum may belong to more than one class or to no class at all (situations which are not possible for discriminative classification). In medical diagnosis, one should consider that a patient may have a tumor *and* an unrelated infectious disease. Or a patient sample may deviate from the normal/control but it is still different from any known class.

One-class classification is particularly useful if a well-known and well-defined class is to be distinguished from "everything else": it is often used to detect outliers, or, in a medical context, for a checkup against normal values. We refer to these well-defined classes as *positive classes*. *Negative classes*, comprising everything but the positive class, need not be modeled in one-class classification, and their being *ill-defined* does not disturb the recognition of the positive classes. As each class is independent from other classes, one can also check for additional well-specified conditions, e.g., particular diseases, without influencing or disturbing the check against the normal values in any way. The outcome may also be that the spectrum is ambiguously classified, e.g., not as "normal" and maybe a tumor and/or an infectious disease. In general, *one-class classifiers* work by considering a "distance" between a spectrum and all the spectra of a specific class, where the distance between spectra can be calculated using various metrics [57]. As a distance between two points in space can be calculated using a function of the three x, y, z variables, the distance between spectra are calculated considering as if each spectrum is a point in an n-dimensional space, where n is the number of data points. If a spectrum is sufficiently "close" to the class in question, it is predicted to belonging to that class. Typical examples of *one class* classification methods are *Soft Independent Modeling of Class Analogies* (SIMCA), kNN, and SVM.

Instead of directly predicting the membership to a class for an "unknown" sample processed by the model, many classification algorithms predict more-or-less continuous scores or class membership probabilities. Class labels are then assigned from scores by setting a threshold, i.e., spectra are assigned to a class if the probabilities to be in that class is above a certain value (e.g., 60%). For such threshold classifiers, the threshold value needs to be optimized for the specific application (see "Hyperparameter optimization" section).

Linear discriminant analysis and quadratic discriminant analysis

LDA and QDA work by defining *class boundaries* in the feature space which for SERS dataset is usually constituted by a number of PCs of the spectral dataset (PCA-LDA and PCA-QDA, see below). The *class boundaries* are estimated using descriptive statistical parameters (see "A bit of statistics" section) that define their position and their dispersion: namely, mean/centroids as location parameters, variances, and covariances as dispersion parameters. The underlying heuristic of discriminant analysis is to assign each case to the class whose mean it is closest to. Distance is measured according to the *covariance* observed within the classes (*Mahalanobis distance*). LDA uses one *pooled covariance* matrix for all classes, resulting in linear *class boundaries* in feature space. If *individual* covariance matrices are used for the classes, the resulting *class boundaries* are quadratic functions (e.g., spherical, elliptical, hyperbolic) in the feature space and the model is called *quadratic* (QDA) instead of *linear.*

LDA is *optimal* if all classes in fact share the same covariance matrix and follow Gaussian distributions. In practice, the heuristic performs well also if these optimal conditions are not met as long as the distribution doesn't have too heavy tails and the classes form compact clusters. Note that LDA and QDA are not suited for classification tasks of the threshold type: cases far away from the class boundary heavily influence the model.

LDA and QDA are the classification equivalent of MLR, thus, suffering from the same collinearity effects, and requiring more samples than input features (wavenumbers). As for MLR, this requirement is seldom met for SERS datasets. However, it can be overcome by reducing the number of variables. This reduction can be done by selecting the wavenumbers of specific SERS bands or a number of *PCs* or *latent variables* as extracted by a PCA or PLS, respectively (see below).

Logistic regression

Logistic Regression (LR) is a *binary* model (i.e., for discriminative situations in which only two classes are considered) that actually models the probability that a case belongs to one of the two classes. Probabilities are between 0 and 1, and each sample either belongs to class 1 or to class 2 leading to binomial rather than Gaussian residuals. Behind the scenes, LR performs a *least squares regression* on the *log-odds* $L = ln\frac{p}{1-p}$ or *logit* of the probability p and is thus a so-called *generalized linear model*. Generalized linear models use transformations of the independent y to get a variable that is better suited for least squares regression. The primary output of LR are the log-odds or the probability of a case to belong to the class in question, which can be then postprocessed into class labels (e.g., by setting a probability threshold). In contrast to LDA, cases far from the class boundary do not carry much weight in LR. LR is thus suitable for *threshold*-type classification.

Principal component analysis and partial least squares as preprocessing: PCA-LDA, PLS-LDA, PLS-LR, etc.

As previously mentioned, for SERS datasets classification models like LDA or LR are mostly computed by using PCA or PLS variables (i.e., *principal components* or *latent variables*), thus reducing the number of variables from several hundreds to just a few. This results in "hyphenated" models such as PCA-LDA (LDA in PCA score space), PLS-LR (logistic regression in PLS score space), etc. Due to the close relationship between PLS and LDA [58], performing an LDA in PLS score is a natural choice to obtain a regularized variant of LDA (i.e., PLS-LDA). PLS scores may not be the optimal projection for dimensionality reduction before LR, and algorithms exist to adjust the PLS projection, see, e.g., Ref. [59], but in practice PLS-LR is very effective.

PCA-LDA and PLS-LDA are often used on label-free (in the chemical sense) SERS datasets from biofluids [60–62]. Such methods have the advantage over other methods that the loadings plot of the PCA or PLS components can be still interpreted in terms of spectral and thus biochemical information. At each step of the process (including the verification, see below), these plots can be inspected to check if results are meaningful or not, for instance, by verifying that no components carrying noise or unrelevant spectral information (e.g., artifacts) are used in the discriminant analysis. In other words, the researcher has a tighter control over the model, especially compared to nonlinear models (see below). Moreover, the spectral interpretability of the loadings plot might allow a biochemical interpretation of results, adding a scientific value to a classification tool. For instance, if the second PC of a SERS dataset turns out to be crucial to get a good classification, the corresponding loadings plot will tell us which SERS bands are the most "diagnostic." In spite of these advantages, there are pitfalls as well for these methods. For instance, the dimensionality reduction step causes dependence between all involved cases, so that performance must be verified on data that are independent of the PCA or PLS step (see "Verification of results" section). The same precaution should be adopted during the optimization of model parameters as well (see "Hyperparameter optimization" section). Also, the spectroscopic interpretation needs to consider some crucial differences between coefficients of a predictive model and difference spectra (which have similar appearance): while difference spectra will always show all bands of the underlying substances, that is not necessarily the case for model coefficients where some bands may appear while physically related band do not (e.g., if they are subject to more noise). Plus, the coefficients may perform something similar to a baseline subtraction, so it will not always be clear whether a particular coefficient probes a band or a baseline.

Soft independent modeling of class analogies

SIMCA is the *one-class* analogue of LDA and QDA, meaning that it enables to reject samples as not belonging to any class, rather than put the sample into the closest class. SIMCA usually works by using PCA models (one for each modeled class) to form a library where each individual model defines a classification rule and the SIMCA architecture defines how to apply the library models to each sample. Whether a case belongs to a particular (positive) class is judged by two distances, the *in-model distance* and the *out-of-model distance*. *In-model distance* measures types of variation from the class mean that normally occurs within this class. *Out-of-model distance* measures variation that is not described by the model for this class. SIMCA thus detects both normal variation that happens to an unusual extent and unusual variation that was possibly never seen in any of the training data. Due to the highly graphical and informative nature of SIMCA, there are a multitude of diagnostic tools available for training and applying SIMCA models to new samples, and a variety of heuristics to decide the threshold for what is considered belonging to a class [63].

Nonlinear models

k *nearest neighbors*

kNN is one of the simplest *nonparametric* machine learning methods. A nonparametric technique does not make any assumptions on the underlying data distribution. In other words, the model structure is determined from the data. The kNN algorithm only assumes that similar things are near to each other. kNN works by taking the *k* closest training data points to a data point selected for prediction. For classification, membership is predicted from the proportions of the various classes among these *k* neighbors, either as probability or as class label with a threshold or majority vote (assigning the most prevalent class), kNN regression computes a regression model of the *k* neighboring training data points on the fly. kNN can thus be implemented practically without training but prediction may be slow and may have high computational demand. Implementations that preorganize the training data for faster lookup of the neighbors shift computation from prediction toward training. kNN are highly nonlinear and local models and can thus adapt to arbitrarily complex situations. This flexibility is bought at the price of requiring large sample sizes.

The usual implementation considers the *k* nearest neighbors regardless of their distance and is thus discriminative. However, variants that evaluate either only the up-to-*k* neighbors that are at most a prespecified distance away from the case in question or all neighboring training data points within a given distance of the case in question are obvious. With these, one-class models can be built.

Support vector machines

SVMs are a family of methods based on statistical learning, frequently used in bioinformatics and other fields that use pattern recognition. The most commonly used SVM for classification are *binary* discriminative models. They find a *hyperplane* that maximizes the margin, i.e., the space without data, between two classes. This margin is defined by training data points close to the class boundary, the *support vectors*. If the classes overlap, a *soft margin* is used.

SVMs are frequently used with *kernel functions* that transform the classification problem nonlinearly in order to make it linearly separable. Note that such kernel transformations exist for other models as well (Kernel PCA, Kernel PLS, Kernel LR) but are less standard than for SVM. Again, this increased complexity comes at the price that more training cases are required to obtain a stable model.

As for other inherently binary classifiers, multiclass problems can be tackled by formulating multiple binary decisions, e.g., of the one-vs.-all type or by pairwise class separations (one-vs.-one). A situation where SVM performs well is when some classes are partly overlapping, i.e., where classical methods, will result in ambiguities and thus may not be effective as a classification rule.

SVMs are also available for *one-class* classification, e.g., the Support Vector Domain Description where the smallest hypersphere including a prespecified fraction of each positive class is found, and for regression where instead of all training data points a few support vectors are retained to describe the regression function. For measurements that are subject to high instrumental noise (like shot noise in SERS), it is desirable to have rather large numbers of support vectors compared to the model complexity: the decision or regression function will then be subject to less noise.

We recommend the tutorial [64] for further reading.

Artificial neural networks

ANNs are computational models inspired by the functionality of the human brain. They can be described as several layers of interconnected nodes (artificial equivalents of neurons). Nodes are the basic units of an ANN and they are characterized by a so-called *activation function* that transforms the input of the node to its output. The nodes are organized by layers and all neurons in the same layer share the same activation function. Typically, the information is processed and passed in a single direction, from an input layer, though one or more hidden layers, toward an output layer. The more the hidden layers, the deeper is the neural network [65]. During training phase, an NN learns the relationship between SERS intensities at specific wavenumbers, which serve as inputs to the network, and class membership, which are assigned as outputs of the network. When a training set, composed of spectra and the class labels for each sample, is presented to the network, the NN weights (corresponding to the learning parameters) are adjusted to minimize differences between the predicted and the true output. After the NN weights are adjusted, the model can be used to predict the class membership of spectra not part of the original training set. These computational models are very well suited to deal with highly nonlinear problems and do not rely on data distribution, making them suitable for analyzing large, complex, and noisy datasets which are typically difficult to analyze by conventional linear data analysis methods [66]. Nevertheless, ANNs present two main issues that prevented their routine use in SERS data analysis until recently. When just a few samples are available to train all weights, the ANN can easily overfit the training data, leading to poor performance on new data. Moreover, it is not clear how ANNs approach the solution, making often impossible to interpret the classifiers, and as such, they are sometimes called "black boxes." Recently, advances in deep learning enabled researchers to address both of these issues, by implementing Convolutional Neural Networks (CNNs). CNN needs fewer parameters than traditional NN and are more robust to overfitting when coupled with embedded regularization techniques. Moreover, CNNs can also be used to identify important regions of spectra from the classifier after it has been trained. For that reasons CNNs is considered an important emerging technique for both classification and regression problems in SERS [67].

Verification of results

It is not sufficient to build a model of the SERS data: in addition, we need to verify its performance and validate it. Unfortunately, terminology in this context is often used in an ambiguous and even misleading manner. We will stick as closely as possible to the Eurachem definitions [68]: *verification* means demonstrating that the model fulfills prespecified performance requirements, and *validation* of an analytical method shows that it is fit for purpose. Validation thus comprises of both developing appropriate experiments to measure whether the method is fit for its purpose and then verifying method performance according to the specified criteria. We will focus here on a few criteria and experimental setups widely used for verification. For more detailed discussion of the subject, see the literature, e.g., Refs. [69–71].

These activities can be further distinguished into *internal* that is within the developing lab only and *external* where an external auditor guarantees that the lab cannot know the correct prediction for the test cases.

In general, we need two ingredients for the verification of a predictive model (Fig. 1.11): (i) an experiment (scheme) that provides us with suitable cases (samples), the so-called *test cases* which are submitted to the model; and (ii) one or more figures of merit that are calculated from comparing predictions to reference values (ground truth) of the tested cases.

Bias and variance in verification

Verification results are subject to systematic and random error (bias and variance) like any other measurement.

Optimistic bias, i.e., results that systematically look better than the model really is are caused by the test cases not being statistically independent from the training cases. Typical causes are

- An inappropriate verification scheme, namely *autoprediction* (see below).
- An inappropriate splitting procedure for resampling and hold-out schemes. This happens frequently with *nested data*: data that have a grouping (clusters) such as

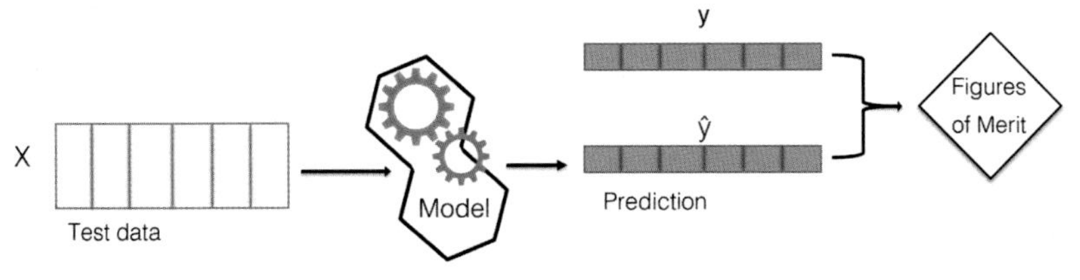

FIGURE 1.11

Verification: a model takes test data without labels X and predicts labels $\widehat{Y}$. Predictions are $\widehat{Y}$ compared with the true labels Y and the result is summarized as particular figure of merit.

Credit: Stefano Fornasaro and Claudia Beleites.

repeated measurements or multiple samples for some patients. All cases of a cluster need to be either in training or in testing, but never in both. The solution here is to distinguish training patients from test patients rather than speaking of training and test spectra. Statistically speaking, the rule is to decide training versus testing at the uppermost random factors (confounders) of the data hierarchy. *Hasse diagrams* can help to visualize this [72].

- Column-wise transformations and, in general, preprocessing and data reduction methods that involve multiple cases. For instance, centering, PCA, or feature selection, all cause dependence between cases [73]. Such procedures are best seen as part of the model and thus as being trained on the training data and applied to test data in the same way they are an applied for production use prediction.
- Using intermediate verification results used for model optimization or selection. This makes these results and the selected/optimized model dependent and a further verification of the selected/optimized model will need a fresh independent test set.
- Temporal dependence: Consecutive measurements may be more similar to each other than measurements taken a long time apart. This may happen due to drift in measurement conditions or instrument performance. Residual substance of previous sample can in some cases also cause such dependence. If such temporal dependence happens when samples are measured in a systematic manner (e.g., one class after the other, increasing concentration series), the effect of such drift cannot be distinguished from real variation due to the sample composition. As in production use, class or concentration are unknown and cases will arrive in arbitrary order, also test cases must be measured in randomized order (and also training will benefit from randomized measurements).

With complex data or experimental designs with many crossed influencing factors, independence between training and test sets may be achievable only by excluding part of the data from *both* training and testing.

Resampling schemes typically have a slight **pessimistic bias**. They approximate model performance measurements by training surrogate models on somewhat smaller data subsets and verify performance for those surrogate models. Due to the smaller training set, the surrogate models are on average a bit worse than the complete model trained on the whole available dataset.

Variance of verification results depends on three factors: (i) the true performance of the model; (ii) the figure of merit; and (iii) the number of statistically independent cases that are tested.

Of these factors, the last is the most important for planning a study: the true performance is naturally unknown when the verification experiments are planned and thus can at best be approximately considered. The figure of merit is mostly chosen according to which properties of the model are to be verified. The number of samples should then be planned according to the required precision (see, e.g., Ref. [74] for classification; regression requires preliminary experiments as basis for verification sample size planning). In contrast to model training where small sample sizes can be taken into account by restricting model complexity, no such adaptation is possible in verification: for verification, the *absolute* number of test cases is relevant.

Verification schemes

Experimental schemes to obtain test cases range from reusing training cases (so-called autoprediction or training error) over various resampling schemes all the way to designing separate validation studies.

Validation studies and assessing ruggedness

Validation studies obtain cases for the sole purpose of assessing various aspects of behavior of the model and/or its predictions in various situations. They start after model training is completely finished. Ideally, a validation study contains a very close simulation of real use of the model and it may also contain deliberate simulation of certain error sources. As independent test cases are used for the single purpose of assessing predictive performance, the estimates are unbiased.

Validation studies can measure certain aspects that no other verification scheme can tackle, for example drift in the modeled system [75]. A validation study can be designed to collect new cases over a specified period of time after modeling is finished. If drift occurs, the predictive performance will deteriorate over time and the validation study will thus answer questions like

- How often needs the analytical method to be checked? How often should control cases be run to demonstrate ongoing good performance?
- How long can the method be used as it is, and when is instrument/model maintenance needed?

Validation studies can also be used to check the ruggedness of the predictions against particular known sources of error, such as

- variation in substrates
- variation/drift in excitation laser wavelength and/or intensity
- probable contaminants
- artifacts on the signal (e.g., spikes due to cosmic rays)
- patient population, e.g., problematic comorbidities

Unfortunately, validation studies are very expensive because they need large sample sizes (see below) and come at considerable experimental effort. In practice, a full validation study is not efficient in the early method development stages where all verification and validation activities are still internal to the developing lab.

However, various aspects of ruggedness can and should be assessed already during model development. This may also be done by simulating distortions to the spectra due to known sources of error [76].

Recommendations

- Full method validation studies are needed before real use in clinics.
- Due to the experimental effort and cost, combine this with external validation.
- Use dedicated small internal studies already during method development to assess ruggedness.

Hold out/independent test sets

These approaches are the smaller brothers of validation studies: from the available dataset, a part is set aside (*split off*) for testing of the model once training is finished. Any training activities take place only on the training subset.

This early separation into training and test subsets makes it easy to ensure independence of training and test data by organizational means such as the lab that takes the samples keeping the ones reserved for testing until the data analyst signals "training finished." This possibility is particularly valuable for complex experimental designs that take various influencing factors and confounders into account. If the underlying structure of influencing factors means that independent subsets can only be obtained by excluding part of the data from both training and testing, it may be more efficient to decide a fixed training/test set assignment as part of the experimental design development.

In practice, however, the hold out set is frequently obtained by randomly selecting a fraction of the available cases and the same splitting principles as for resampling procedures (see below) are used. Such a randomly split off hold out is subject to the same risks of training cases leaking into the test data thus causing dependence as resampling. Resampling offers larger numbers of tested cases in the same situation, at the cost of some computational effort and a slight pessimistic bias. Hold-out verification is unbiased, but the model is worse about the same amount that resampling is pessimistically biased. For the typical sample sizes in SERS studied from biological samples, the larger sample size outweighs this pessimistic bias, and hold out by randomly splitting the data is typically inferior to resampling with the same random splitting even if it is unbiased.

Autoprediction (training error)

Autoprediction error is easy to calculate and no new cases are required: the training cases are reused as test cases. By doing so, autoprediction cannot detect overfitting and thus misses one of the most important sources of error in practice. To detect overfitting, we need cases that are statistically independent from the training dataset. Using autoprediction is therefore acceptable only in two situations:

1. Very large sample size ($n \gg k$): If the number of truly independent cases is very large compared to the complexity of the model, the model cannot overfit. Statistically speaking, after fitting the model, lots of degrees of freedom are left and can be used to assess the fit. In practice, this is the case with a univariate linear calibration (regression, so $k = 1$) with more than $n = 10$ calibration samples.

2. As additional indicator together with verification schemes that do detect overfitting: The comparison between training error and cross-validation results can help to find out whether and where overfitting occurs.

If an autoprediction error is abused as approximation to predictive ability for unknown cases (the so-called *generalization error*), it will have an optimistic bias. That is, autoprediction error systematically underestimates generalization error and overestimates predictive ability. Particularly with small sample sizes n and highly multivariate data (such as models of full spectra), this optimistic bias can become so large that training error becomes totally useless.

Resampling: cross validation and out-of-bootstrap

In general, any sample that can be used for testing can also be used for model training. This creates an inherent conflict between using as many samples as possible for training to obtain the best possible model and reserving samples for testing to obtain more precise test results. As long as model development is still ongoing and assessment is internal to the developing labs, *resampling strategies* offer an attractive way out of this conflict. Resampling produces subsets of the available dataset ("whole dataset") and train a so-called *surrogate model* on each such subset (Fig. 1.12). The exact procedure for subsampling varies between different resampling methods: the *bootstrap* draws with replacement, while *cross validation* draws without replacement.

The subsets contain all but a few of the original cases. Those cases are independent of the surrogate model if the splitting procedure correctly accounts for clusters in the data, e.g., by splitting patient-wise instead of spectra-wise if multiple spectra of the same patient are in the study. These left-out cases are used to assess the surrogate model's performance in an unbiased fashion. As the surrogate models are trained on almost the same training cases as the complete model trained on the whole dataset, their performance should also be approximately the same. Thus, the test results for the surrogate models approximate the predictive performance of the complete model. As the surrogate models are trained on slightly fewer training cases than the complete model, we expect them to have somewhat worse predictive ability. This is the cause of a pessimistic bias, which in practice turns out to be small [77–79].

If the training set gets too small compared to the model complexity, the models become unstable. Resampling validation allows to measure model (in)stability, see "Model stability and overfitting" section.

Leave-one-out cross validation (LOO) is the special case of k-fold cross validation where k equals the number of cases n. LOO has two undesirable properties compared to cross validation that leaves out more cases for each fold: firstly, calculating these $k = n$ surrogate models is *exhaustive*, i.e., no further combinations are possible. In particular, there is no way to get one case left out by more than one surrogate model and thus no possibility to measure model stability (but the surrogate models are of course still subject to any such instability). Secondly, LOO cannot be stratified (see "Study definition and data collection" section). Particularly some

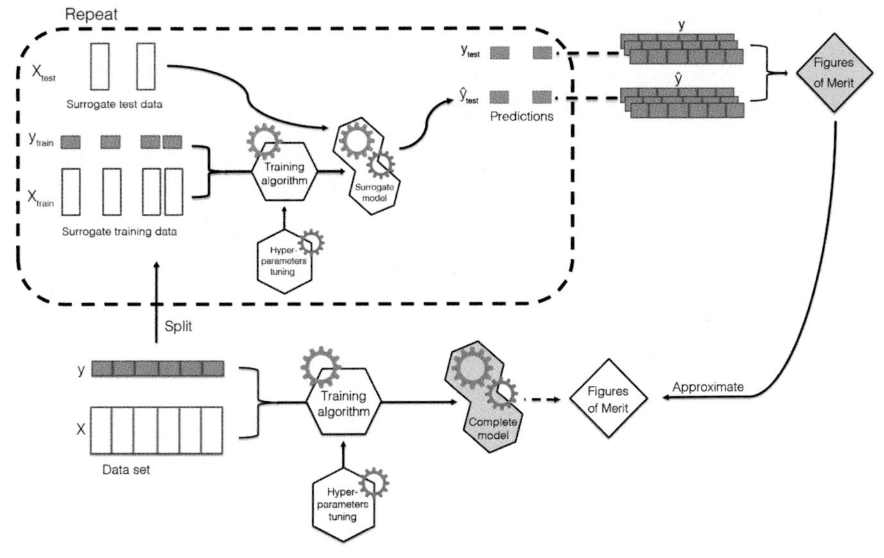

FIGURE 1.12

Resampling verification: (bottom) the available labeled dataset (as well as some hyperparameters) is used to train the *complete model*. We want to measure the *complete model*'s performance. We therefore split the dataset into *surrogate training set* and *surrogate test set*. A *surrogate model* is trained using the same training algorithm and the same hyperparameters as for the *complete model*. This *surrogate model* is next tested with the *surrogate test data*. The procedure from splitting to predicting surrogate test cases is repeated: b times for out-of-bootstrap and $i \cdot k$ times for repeated k-fold cross validation (*dashed box*). The predictions of all *surrogate models* can be pooled because the surrogate models are almost equal since they were obtained with the same training algorithm and the same hyperparameters from almost equal surrogate training data. The comparison is summarized as the figure of merit (full diamond). Not only are the *surrogate models* almost equal but are also almost equal to the *complete model* since again the respective training data were almost equal and training algorithm and hyperparameters were the same. We can thus use the full diamond figure of merit as approximation for the complete model's empty diamond figure of merit.

Credit: Stefano Fornasaro and Claudia Beleites.

classification algorithms are quite sensitive to relative class frequencies. In LOO, the test case will always belong to a class that is underrepresented in the surrogate training set compared to the whole dataset. This combination can cause a relatively large pessimistic bias instead of the expected very slight pessimistic bias (e.g., Ref. [78]). We therefore recommend not to employ LOO unless the clustering of the data leaves the analyst with very few possibilities to split the data at all.

For *double* or *nested cross validation* please see Section "Hyperparameter optimization".

Recommendations

- Designed validation study for full external validation
- For complicated systems of influencing factors, designed experiment with prespecified training/test assignment also for internal verification
- Repeated (also known as "iterated") k-fold cross validation or out-of-bootstrap for internal verification if splitting can be done randomly
- Always split at the highest level of a nested data hierarchy, e.g., patient-wise rather than spectra-wise
- Indicate the verification scheme with the figure of merit: $RMSE_{CV}$ refers to RMSE calculated from a cross-validation experiment, $RMSE_P$ is RMSE for actual predictions, $RMSE_C$ the training (calibration) error, etc.

Figures of merit

Regression

A figure of merit is a quantity used to characterize the performance of a device, system, or method, relative to its alternatives [80]. For regression, mean squared error (MSE) and its root (RMSE) are the most widely used figures of merit (Table 1.5). MSE is a natural choice since it is exactly the term what least squares regression minimizes. RMSE is on the same scale and unit as the outcomes y and is thus easier to interpret in terms of its practical relevance. Both MSE and RMSE penalize both systematic and random error in the prediction and give thus a combined measure of accuracy (trueness and precision), i.e., they measure total error.

Relative RMSE, that is RMSE divided by the true concentration is often useful. One should always specify what is meant by "true" concentration: referencing to the largest concentration yields a best possible relative error similar to accuracy classes of various instruments. The relative error at varying concentrations is closely related to the *limit of quantitation* (LOQ), which is often specified as relative error of 10%.

Table 1.5 Figures of merit (FoMs) for regression models.

Figure of merit	Formula
MSE	$\text{MSE} = \frac{1}{n}\sum_{i=1}^{n}(\hat{y}_i - y_i)^2$
RMSE	$\text{RMSE} = \sqrt{\text{MSE}}$
rel.RMSE	$r\text{RMSE} = \frac{1}{y}\text{RMSE}$
R^2	$R^2 = \frac{\sum_{i=1}^{n}(\hat{y}_i - \bar{y})^2}{\sum_{i=1}^{n}(\hat{y}_i - \bar{y})^2}$

A similar specification puts the *limit of detection* (LOD) at 33% relative error [81]. In order to measure LOD or LOQ, the RMSE needs to be determined in that concentration range. Extrapolating from other concentration ranges will typically allow only very rough guesstimates of these limits.

The coefficient of determination R^2 (fraction of explained variance) is another frequently (mis)used figure of merit to characterize regression models. Despite being popular, it does not in itself convey good information about the quality of the model and should therefore be avoided [82]. For some striking examples, please see the seminal paper [83] which demonstrates three clearly problematic situations having the same R^2 as a perfectly fine albeit somewhat noisy prediction. Some key difficulties with R^2 are that neither does it detect overfitting nor can it be relied to detect underfitting. Yet, it can be affected by systematic error, so neither can it be used alone as measure for precision. Diagnostic plots give a much better picture of the situation.

Diagnostic plots

The *calibration plot* shows predictions $\widehat{y}$ over reference y. Both axes have the same scale and range, leading to a square plot (set aspect ratio to 1 if necessary). In addition, ideally predictions match reference, thus the bisectrix $\widehat{y} = y$ is an appropriate guide. This graph (Fig. 1.13) gives a very good overview over the model's predictive ability.

If the predictions are very good and/or the calibration range is very large, it may be difficult to spot important differences or patterns in the calibration plot. Plotting $\widehat{y} - y$ (i.e., testing residuals) over y can then help.

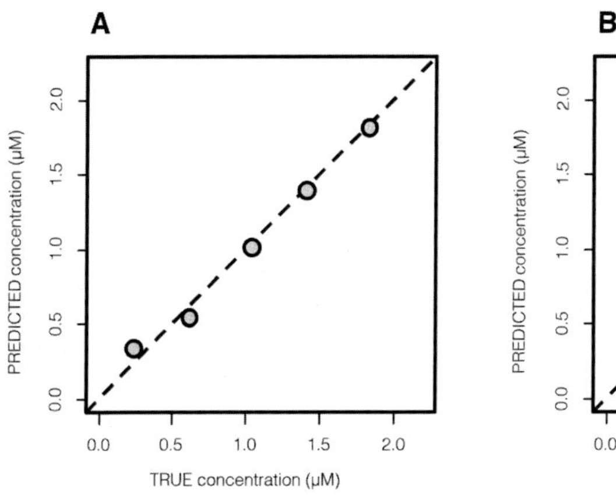

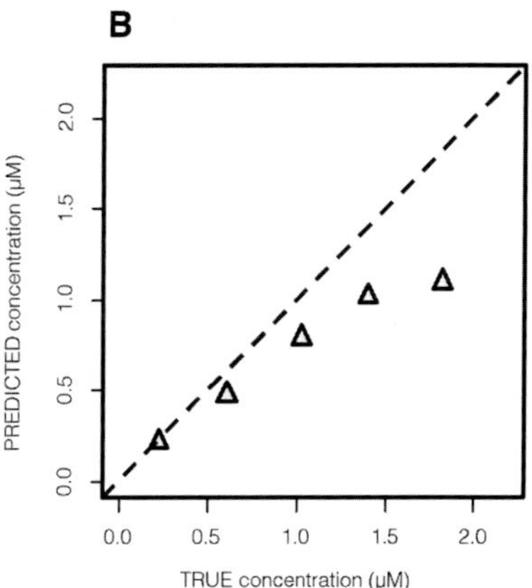

FIGURE 1.13

Diagnostic plots of predicted values against the reference values for the evaluation of the adequacy of the calibration model. (A) Model with perfect predictions situated on the diagonal of the plot. (B) Predictions with large departures from the line denoting a large bias.

Credit: Stefano Fornasaro.

Classification

Sensitivity, specificity, predictive values, and similar proportions

Classifier test results based on predicted class labels are often tabulated as so-called *confusion matrix* (Fig. 1.14A) which is then further summarized in a number of figures of merit (Fig. 1.14B) that each answer particular questions with respect to the performance of a particular class or across all classes:

- Sensitivity ($Sens_A$): of all cases that truly belong to class A, what fraction was correctly recognized belonging to class A by the classifier? Note that in calibration/regression sensitivity has a different meaning: it refers to the slope of the calibration function.
- Specificity ($Spec_A$): of all cases that truly do *not* belong to class A, what fraction was correctly recognized as not belonging to class A by the classifier? Again, in general analytical chemistry, specificity has a different meaning of a method not being subject to cross sensitivity.
- Positive predictive value (PPV_A): of all cases that the classifier predicted to belong to class A, which fraction truly belongs to class A?
- Negative predictive value (NPV_A): of all cases that the classifier predicted to *not* belong to class A, which fraction truly does not belong to class A?
- Accuracy: fraction of correctly classified cases across all classes.
- Error rate: fraction of misclassified cases across all classes.

From a verification design point of view, sensitivity is the easiest to measure: all that is needed is a sufficient number of cases that do truly belong to the class in question. The other figures of merit combine results for (potentially) multiple classes. Whenever that is the case, care needs to be taken that the relative frequencies of

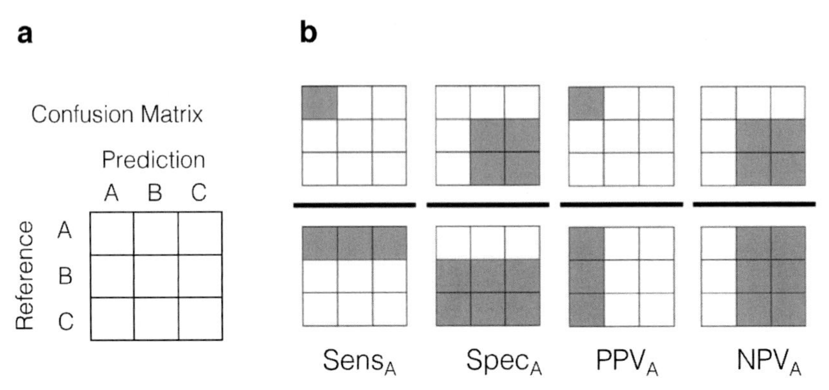

FIGURE 1.14

(A) Confusion matrix counting each case in the row of its reference label (true class) and the column of the class it was predicted to belong to. (B) typical figures of merit are calculated as proportions that sum particular regions of the confusion matrix in enumerator and denominator.

Credit: Claudia Beleites.

the classes correspond to the relative frequencies of the classes in the application, i.e., the relevant incidence or prevalence of class A. The available samples often do not reflect this prevalence (or incidence), and in addition to the observed proportion of cases also the figure of merit for relevant prevalence should be calculated and reported as they can be drastically different [84]. The confusion table is reweighted for the application prevalence by dividing the values in each row (same true class) by the (true) prevalence of this class in the test data and multiplying them by the prevalence (incidence) of this class in the application. Application-relevant figures of merit can then be calculated as usual from this reweighted confusion matrix. It is perfectly valid and useful to give figures of merit for specific subgroups of cases/patients such as sensitivity of spectra with cosmic rays, specificity for particular failure modes for a fault-detection one-class classifier, figures of merit for patients with particular comorbidities, etc.

For one-class classification, the determination of specificity and negative predictive value requires special effort as it is often very hard to obtain a sufficiently large representative set of cases not belonging to a particular class A. Here it is recommended to use cases that are expected to be difficult to recognize/similar to the class but outside (e.g., when detecting food fraud, specificity for *Parmigiano* should be tested with *Grana Padano* rather than *Gouda*). Still, similar experimental verification design is highly relevant for discriminative classification as well, e.g., specificity for recognition of breast cancer in older women needs to be tested with a population of women without breast cancer but similar age distributions, and not with "healthy volunteers from the lab" having a very different age and sex distribution.

A general rule of thumb is that one must be careful not to produce artificially easy test datasets. Consider a classifier to help classify borderline (unclear) tumor cases. If consensus diagnosis of a number of pathologists is an inclusion criterium, the borderline cases which are the target of the method will tend to be excluded from the study—leading to an artificially easy classification problem (for including borderline cases see Ref. [85]). If instead the application scenario is to relieve the pathologists from the burden of a large number of clear cases, that would be a more appropriate dataset (one would still want to verify that borderline cases are correctly referred to the pathologist as such, though).

In general, the predictive values are more important for the application than sensitivity and specificity, but they are highly dependent on prevalences/incidences and these may change orders of magnitude depending on the particular target population of the method: after the test predicted "disease," both patient and doctor need to know the probability that *this particular prediction* is correct.

From a statistical point of view, these proportions have unfortunately rather undesirable characteristics: they are subject to large variance that mostly depends on the number of independent cases in the denominator of the proportion. As a rule of thumb, measuring a proportion with a precision of about 10%-points requires about 100 cases (patients, not spectra) in the denominator. The observed case counts or at the very least the denominator counts for the test data should therefore always be reported in addition to the resulting Figures of Merit.

A ROC (*receiver operating characteristic*) curve is a graph showing the performance for predicting a particular class as its discrimination threshold is varied. General classifier performance is often summarized as *Area Under the receiver operating* Curve. While this is a sensible figure of merit when comparing classification algorithms, it is less suitable for reporting the performance of the obtained classifier for a given application: the application will almost always require a particular tradeoff between sensitivity and specificity and performance far from this working point is not relevant.

Model stability and overfitting

Models that are not sufficiently complex to capture the underlying relevant structure in the data will have large bias (systematic error). Models that are too complex for the amount of training data at hand will be subject to variance, i.e., slight changes in the training data such as adding a new case lead to large changes in the model or its predictions. This is referred to as *model instability*. Adjusting model complexity is thus a trade-off between bias and variance.

Out-of-bootstrap- and repeated [86] (also known as "iterated" [87]) cross validation offer a convenient way to quantify this instability: many surrogate models are computed, so that individual cases are test cases for multiple such surrogate models. Considering a particular case, the training data of all surrogate models that were used to test this case differs in a few other cases that were left out together with the case in question. As the case in question is always the same for this subset of the surrogate models, any difference in the prediction for this case must be caused by instability in the training procedure together with exchanging *a few* other cases. For a detailed description, see, e.g., Refs. [87,88].

Recommendations

- For regression, RMSE allows best judgment of the performance with reference to the particular application
- For classification, proportion-type figures of merit are easy to interpret.
- Multiple figures of merit are needed to characterize performance
- Positive and negative predictive values are most important
- If the relevant prevalence or incidences in the target population are unknown, several realistic guesstimates may be used to calculate predictive values for specific application scenarios.
- Proportions need large numbers of test cases
- Take care to design and obtain meaningful verification datasets
- Always check stability

Hyperparameter optimization

Many models include so-called *hyperparameters*. Hyperparameters steer the general behavior of the model in a way that is highly dependent on the data at hand and therefore left to the user by the programmer of the training algorithm. Examples are hyperparameters that steer model complexity such as the number of PCs or latent variables to consider for a PCA/PLS-LDA or the classification threshold for LR but also the choice of various preprocessing techniques or the selection or exclusion of particular spectral ranges.

Sometimes the hyperparameters can be fixed by *external knowledge*, i.e., knowledge about the application that does not need to be extracted from the dataset. Examples would be deciding the appropriate preprocessing strategy and spectral ranges based on spectroscopic and biochemical knowledge about the data, specifying the number of PCs from detailed chemical knowledge including the number of chemical species in the system under study or fixing the classification threshold to a particular predicted probability that results from the application. Sometimes spectroscopic knowledge allows to restrict the choice, e.g., if the coefficients or loadings of a linear model of full spectra look noisy this is a sure sign of overfitting, i.e., the model is too complex. Whenever such guidance for the choice of hyperparameters is available, this should be used.

A different approach is to try and choose good hyperparameters based on predictive performance, a so-called *data-driven model optimization*. Data-driven model optimization is still subject of ongoing research and suffers from confusing terminology (such as *validation* sets that must *not* be used for model validation or verification). A few points can be made though: the convenience of data-driven model optimization is not for free, it is paid for by requiring larger sample sizes. In SERS studies where the available sample sizes are typically small ($\ll 100$ patients), this is an important drawback. The optimization typically uses estimates of predictive performance similar to the ones discussed in the "Verification of results" section. That performance estimate needs to be sufficiently precise to distinguish good from bad models. The widely used proportion-type figures of merit for classification are unfortunately subject to high variance uncertainty. They are further unsuitable for optimization because they do not react continuously to (small) changes in the model: consider a test case close to the class boundary and a slight change in a hyperparameter that slightly shifts the class boundary. A proportion like accuracy or error rate will not detect any change until the class boundary reaches the test case. Then a sudden jump in the figure of merit results. With imprecise performance estimates, selecting the model with apparently lowest error cannot be recommended as the decision is not only rather uncertain but the increasing variance with increasing model complexity (instability) results in a bias toward too complex models. A heuristic that counteracts this bias is the one-standard-error-rule: selecting the least complex model that is within one standard error of the apparently best model [73]. Last but not least, it is important to realize that any performance estimate that enters data-driven model optimization has become part of the model training and therefore verification needs additional test data that is independent of the whole training procedure including the optimization.

Concluding remarks

Because of its characteristics, SERS has a huge potential for POC applications. SERS datasets obtained from biomedical samples are rich in information. But this wealth of information comes at a cost: SERS spectra are complex objects, and the information in there is not always easy to get. The complexity arises from the fact that there are different, overlapping sources of variability involved: variability among different batches of SERS substrates, naturally occurring interindividual variability in the concentration of different biomolecules as well as variability associated with a specific condition (e.g., varying concentrations of drugs or disease markers). Extracting the right information from this complexity, be it the concentration of a drug or the presence of a specific medical condition, is a challenging task. Preprocessing procedures and multivariate analysis methods are extremely powerful tools to help us in this task. These tools, however, are as powerful as dangerous, if not correctly used, and can easily lead to wrong conclusions. As human beings, we are subject to common psychological behaviors, including a selective attitude toward favorable results. An incorrect use of the data analysis tools easily leads to incorrect, overoptimistic results, which we are inclined to accept as true. In other fields of science, an improper use of scientific tools leads to uninteresting results (e.g., wrong use of an instrument gives no signal, a wrong reagent in a chemical synthesis does not yield the product desired, and so on), so that data analysis is quite unique with respect to this aspect. At a time where the scientific community is putting the reliability and reproducibility of results as a priority, a proper knowledge of advantages and limitations of the available data analysis methods is can easily make the difference between good and bad SERS studies. Thus, researchers aiming at the development of a POC SERS applications should familiarize themselves with the techniques, methods, and concepts mentioned in this chapter, in order to select the most appropriate to their case. A conscious, responsible, and transparent use of data analysis is a necessary attitude for any researcher aiming at bringing SERS to the clinics.

References

[1] K.H. Esbensen, C. Wagner, Theory of sampling (TOS) versus measurement uncertainty (MU) – a call for integration, Trends Anal. Chem. 57 (2014) 93–106, https://doi.org/10.1016/j.trac.2014.02.007.

[2] R. Leardi, Experimental design in chemistry: a tutorial, Anal. Chim. Acta 652 (2009) 161–172, https://doi.org/10.1016/j.aca.2009.06.015.

[3] H. Fisk, C. Westley, N.J. Turner, R. Goodacre, Achieving optimal SERS through enhanced experimental design, J. Raman Spectrosc. 47 (2015) 59–66, https://doi.org/10.1002/jrs.4855.

[4] N. Altman, M. Krzywinski, Split plot design, Nat. Meth. 12 (2015) 165–166, https://doi.org/10.1038/nmeth.3293.

[5] M. Krzywinski, N. Altman, Analysis of variance and blocking, Nat. Meth. 11 (2014) 699–700, https://doi.org/10.1038/nmeth.3005.

[6] M. Krzywinski, N. Altman, P. Blainey, Nested designs, Nat. Meth. 11 (2014) 977–978, https://doi.org/10.1038/nmeth.3137.

[7] B. Smucker, M. Krzywinski, N. Altman, Two-level factorial experiments, Nat. Meth. 16 (2019) 211–212, https://doi.org/10.1038/s41592-019-0335-9.

[8] M. Krzywinski, N. Altman, Visualizing samples with box plots, Nat. Meth. 11 (2014) 119–120, https://doi.org/10.1038/nmeth.2813.

[9] T. Fearn, Functional boxplots, NIR News 22 (2011) 19–20, https://doi.org/10.1255/nirn.1260.

[10] M.D. Wilkinson, M. Dumontier, I.J. Aalbersberg, G. Appleton, M. Axton, A. Baak, et al., The FAIR guiding principles for scientific data management and stewardship, Sci. Data 3 (2016), https://doi.org/10.1038/sdata.2016.18.

[11] R. Todeschini, D. Ballabio, V. Consonni, F. Sahigara, P. Filzmoser, Locally centred mahalanobis distance: a new distance measure with salient features towards outlier detection, Anal. Chim. Acta 787 (2013) 1–9, https://doi.org/10.1016/j.aca.2013.04.034.

[12] O.Y. Rodionova, A.L. Pomerantsev, Detection of outliers in projection-based modeling, Anal. Chem. 92 (2019) 2656–2664, https://doi.org/10.1021/acs.analchem.9b04611.

[13] P.J. Rousseeuw, K.V. Driessen, A fast algorithm for the minimum covariance determinant estimator, Technometrics 41 (1999) 212–223, https://doi.org/10.1080/00401706.1999.10485670.

[14] B. Schölkopf, J.C. Platt, J. Shawe-Taylor, A.J. Smola, R.C. Williamson, Estimating the support of a high-dimensional distribution, Neur. Comput. 13 (2001) 1443–1471, https://doi.org/10.1162/089976601750264965.

[15] Y. Sun, M.G. Genton, Adjusted functional boxplots for spatio-temporal data visualization and outlier detection, Environmetrics 23 (2011) 54–64, https://doi.org/10.1002/env.1136.

[16] J. Gerretzen, E. Szymańska, J.J. Jansen, J. Bart, H.-J. van Manen, E.R. van den Heuvel, et al., Simple and effective way for data preprocessing selection based on design of experiments, Anal. Chem. 87 (2015) 12096–12103, https://doi.org/10.1021/acs.analchem.5b02832.

[17] P. Lasch, Spectral pre-processing for biomedical vibrational spectroscopy and microspectroscopic imaging, Chemom. Intell. Lab. Syst. 117 (2012) 100 114, https://doi.org/10.1016/j.chemolab.2012.03.011.

[18] R. Gautam, S. Vanga, F. Ariese, S. Umapathy, Review of multidimensional data processing approaches for Raman and infrared spectroscopy, EPJ Tech. Instrument. 2 (2015) 8, https://doi.org/10.1140/epjti/s40485-015-0018-6.

[19] G. Schulze, A. Jirasek, M.M.L. Yu, A. Lim, R.F.B. Turner, M.W. Blades, Investigation of selected baseline removal techniques as candidates for automated implementation, Appl. Spectrosc. 59 (2005) 545–574, https://doi.org/10.1366/0003702053945985.

[20] C.A. Lieber, A. Mahadevan-Jansen, Automated method for subtraction of fluorescence from biological Raman spectra, Appl. Spectrosc. 57 (2003) 1363–1367, https://doi.org/10.1366/000370203322554518.

[21] F. Gan, G. Ruan, J. Mo, Baseline correction by improved iterative polynomial fitting with automatic threshold, Chemom. Intell. Lab. Syst. 82 (2006) 59–65, https://doi.org/10.1016/j.chemolab.2005.08.009.

[22] J. Zhao, H. Lui, D.I. McLean, H. Zeng, Automated autofluorescence background subtraction algorithm for biomedical Raman spectroscopy, Appl. Spectrosc. 61 (2007) 1225–1232, https://doi.org/10.1366/000370207782597003.

[23] J. Liu, J. Sun, X. Huang, G. Li, B. Liu, Goldindec: a novel algorithm for Raman spectrum baseline correction, Appl. Spectrosc. 69 (2015) 834–842, https://doi.org/10.1366/14-07798.

[24] P.H.C. Eilers, Parametric time warping, Anal. Chem. 76 (2004) 404–411, https://doi.org/10.1021/ac034800e.

[25] Z.-M. Zhang, S. Chen, Y.-Z. Liang, Baseline correction using adaptive iteratively reweighted penalized least squares, The Analyst 135 (2010) 1138, https://doi.org/10.1039/b922045c.

[26] S. He, W. Zhang, L. Liu, Y. Huang, J. He, W. Xie, et al., Baseline correction for Raman spectra using an improved asymmetric least squares method, Anal. Meth. 6 (2014) 4402–4407, https://doi.org/10.1039/c4ay00068d.

[27] H. Chen, W. Xu, N. Broderick, J. Han, An adaptive denoising method for Raman spectroscopy based on lifting wavelet transform, J. Raman Spectrosc. 49 (2018) 1529–1539, https://doi.org/10.1002/jrs.5399.

[28] K.H. Liland, 4S Peak Filling – baseline estimation by iterative mean suppression, MethodsX 2 (2015) 135–140, https://doi.org/10.1016/j.mex.2015.02.009.

[29] C. Wang, L. Xiao, C. Dai, A.H. Nguyen, L.E. Littlepage, Z.D. Schultz, et al., A statistical approach of background removal and spectrum identification for SERS data, Sci. Rep. 10 (2020) 1460, https://doi.org/10.1038/s41598-020-58061-z.

[30] S. Patze, U. Huebner, K. Weber, D. Cialla-May, J. Popp, TopUp SERS substrates with integrated internal standard, Materials 11 (2018) 325, https://doi.org/10.3390/ma11020325.

[31] H. Wei, A. McCarthy, J. Song, W. Zhou, P.J. Vikesland, Quantitative SERS by hot spot normalization surface enhanced Rayleigh band intensity as an alternative evaluation parameter for SERS substrate performance, Farad. Discus. 205 (2017) 491–504, https://doi.org/10.1039/c7fd00125h.

[32] R.J. Barnes, M.S. Dhanoa, S.J. Lister, Standard normal variate transformation and detrending of near-infrared diffuse reflectance spectra, Appl. Spectrosc. 43 (1989) 772–777, https://doi.org/10.1366/0003702894202201.

[33] C.-C. Lin, C.-Y. Lin, C.-J. Kao, C.-H. Hung, High efficiency SERS detection of clinical microorganism by AgNPs-decorated filter membrane and pattern recognition techniques, Sens. Actuat. B Chem. 241 (2017) 513–521, https://doi.org/10.1016/j.snb.2016.09.183.

[34] P. Geladi, D. MacDougall, H. Martens, Linearization and scatter-correction for near-infrared reflectance spectra of meat, Appl. Spectrosc. 39 (1985) 491–500, https://doi.org/10.1366/0003702854248656.

[35] H. Martens, E. Stark, Extended multiplicative signal correction and spectral interference subtraction: new preprocessing methods for near infrared spectroscopy, J. Pharm. Biomed. Anal. 9 (1991) 625–635, https://doi.org/10.1016/0731-7085(91)80188-f.

[36] N.K. Afseth, V.H. Segtnan, J.P. Wold, Raman spectra of biological samples: a study of preprocessing methods, Appl. Spectrosc. 60 (2006) 1358–1367, https://doi.org/10.1366/000370206779321454.

[37] R. Bro, A.K. Smilde, Centering and scaling in component analysis, J. Chemom. 17 (2003) 16–33, https://doi.org/10.1002/cem.773.

[38] R.G. Brereton, Pattern recognition in chemometrics, Chemom. Intell. Lab. Syst. 149 (2015) 90–96, https://doi.org/10.1016/j.chemolab.2015.06.012.

[39] K. Pearson, LIII. On lines and planes of closest fit to systems of points in space, Lond. Edinb. Dub. Philos. Mag. J. Sci. 2 (1901) 559–572, https://doi.org/10.1080/14786440109462720.

[40] H. Hotelling, Relations between two sets of variates, Biometrika 28 (1936) 321, https://doi.org/10.2307/2333955.

[41] B.K. Lavine, N. Mirjankar, Clustering and classification of analytical data update based on the original article by barry k. Lavine, encyclopedia of analytical chemistry, © 2000, john wiley & sons, ltd, in: Encyclopedia of Analytical Chemistry, American Cancer Society, 2012, https://doi.org/10.1002/9780470027318.a5204.pub2.

[42] R. Bro, A.K. Smilde, Principal component analysis, Anal. Meth. 6 (2014) 2812–2831, https://doi.org/10.1039/c3ay41907j.

[43] S.E.J. Bell, N.M.S. Sirimuthu, Quantitative surface-enhanced Raman spectroscopy, Chem. Soc. Rev. 37 (2008) 1012, https://doi.org/10.1039/b705965p.

[44] Y. Wang, J. Sun, Y. Hou, C. Zhang, D. Li, H. Li, et al., A SERS-based lateral flow assay biosensor for quantitative and ultrasensitive detection of interleukin-6 in unprocessed whole blood, Biosens. Bioelectron. 141 (2019) 111432, https://doi.org/10.1016/j.bios.2019.111432.

[45] J. Tellinghuisen, Statistical error calibration in UV-visible spectrophotometry, Appl. Spectrosc. 54 (2000) 431–437, https://doi.org/10.1366/0003702001949537.

[46] A.C. Olivieri, The classical least-squares model, in: Introduction to Multivariate Calibration, Springer International Publishing, 2018, pp. 19–38, https://doi.org/10.1007/978-3-319-97097-4_2.

[47] H. Hotelling, The relations of the newer multivariate statistical methods to factor analysis, Br. J. Statis. Psychol. 10 (1957) 69–79, https://doi.org/10.1111/j.2044-8317.1957.tb00179.x.

[48] T. Næs, H. Martens, Multivariate calibration. II. Chemometric methods, Trends Anal. Chem. 3 (1984) 266–271, https://doi.org/10.1016/0165-9936(84)80044-8.

[49] S. Wold, M. Sjöström, L. Eriksson, PLS-regression: a basic tool of chemometrics, Chemom. Intell. Lab. Syst. 58 (2001) 109–130, https://doi.org/10.1016/s0169-7439(01)00155-1.

[50] S. Fornasaro, A. Bonifacio, E. Marangon, M. Buzzo, G. Toffoli, T. Rindzevicius, et al., Label-free quantification of anticancer drug imatinib in human plasma with surface enhanced Raman spectroscopy, Anal. Chem. 90 (2018) 12670–12677, https://doi.org/10.1021/acs.analchem.8b02901.

[51] B. Deng, X. Luo, M. Zhang, L. Ye, Y. Chen, Quantitative detection of acyclovir by surface enhanced Raman spectroscopy using a portable Raman spectrometer coupled with multivariate data analysis, Coll. Surf. B Biointerf. 173 (2019) 286–294, https://doi.org/10.1016/j.colsurfb.2018.09.058.

[52] A. Subaihi, L. Almanqur, H. Muhamadali, N. AlMasoud, D.I. Ellis, D.K. Trivedi, et al., Rapid, accurate, and quantitative detection of propranolol in multiple human biofluids via surface-enhanced Raman scattering, Anal. Chem. 88 (2016) 10884–10892, https://doi.org/10.1021/acs.analchem.6b02041.

[53] M. Andersson, A comparison of nine PLS1 algorithms, J. Chemom. 23 (2009) 518–529, https://doi.org/10.1002/cem.1248.

[54] C. Cortes, V. Vapnik, Support-vector networks, Mach. Learn. 20 (1995) 273–297, https://doi.org/10.1007/bf00994018.

[55] J. Zupan, J. Gasteiger, Neural Networks in Chemistry and Drug Design, second ed., Wiley-VCH, 1999.

[56] A.C. Olivieri, Practical guidelines for reporting results in single- and multi-component analytical calibration: a tutorial, Anal. Chim. Acta 868 (2015) 10–22, https://doi.org/10.1016/j.aca.2015.01.017.

[57] P. Oliveri, Class-modelling in food analytical chemistry: development, sampling, optimisation and validation issues a tutorial, Anal. Chim. Acta 982 (2017) 9–19, https://doi.org/10.1016/j.aca.2017.05.013.

[58] R.G. Brereton, G.R. Lloyd, Partial least squares discriminant analysis: taking the magic away, J. Chemom. 28 (2014) 213–225, https://doi.org/10.1002/cem.2609.

[59] B. Ding, R. Gentleman, Classification using generalized partial least squares, J. Comput. Grap. Stat. 14 (2005) 280–298, https://doi.org/10.1198/106186005X47697.

[60] S. Cervo, E. Mansutti, G.D. Mistro, R. Spizzo, A. Colombatti, A. Steffan, et al., SERS analysis of serum for detection of early and locally advanced breast cancer, Anal. Bioanal. Chem. 407 (2015) 7503–7509, https://doi.org/10.1007/s00216-015-8923-8.

[61] C.D. Bocsa, V. Moisoiu, A. Stefancu, L.F. Leopold, N. Leopold, D. Fodor, Knee osteoarthritis grading by resonant Raman and surface-enhanced Raman scattering (SERS) analysis of synovial fluid, Nanomed. Nanotechnol. Biol. Med. 20 (2019) 102012, https://doi.org/10.1016/j.nano.2019.04.015.

[62] A. Stefancu, M. Badarinza, V. Moisoiu, S.D. Iancu, O. Serban, N. Leopold, et al., SERS-based liquid biopsy of saliva and serum from patients with sjögren's syndrome, Anal. Bioanal. Chem. 411 (2019) 5877–5883, https://doi.org/10.1007/s00216-019-01969-x.

[63] A.L. Pomerantsev, O.Y. Rodionova, Concept and role of extreme objects in PCA/SIMCA, J. Chemom. 28 (2013) 429–438, https://doi.org/10.1002/cem.2506.

[64] R.G. Brereton, G.R. Lloyd, Support vector machines for classification and regression, Analyst 135 (2010) 230–267, https://doi.org/10.1039/b918972f.

[65] J. Acquarelli, T. van Laarhoven, J. Gerretzen, T.N. Tran, L.M.C. Buydens, E. Marchiori, Convolutional neural networks for vibrational spectroscopic data analysis, Anal. Chim. Acta 954 (2017) 22–31, https://doi.org/10.1016/j.aca.2016.12.010.

[66] J. Yang, J. Xu, X. Zhang, C. Wu, T. Lin, Y. Ying, Deep learning for vibrational spectral analysis: recent progress and a practical guide, Anal. Chim. Acta 1081 (2019) 6–17, https://doi.org/10.1016/j.aca.2019.06.012.

[67] F. Lussier, V. Thibault, B. Charron, G.Q. Wallace, J.-F. Masson, Deep learning and artificial intelligence methods for Raman and surface-enhanced Raman scattering, Trends Anal. Chem. 124 (2020) 115796, https://doi.org/10.1016/j.trac.2019.115796.

[68] Eurachem guide: terminology in analytical measurement – introduction to, in: V. Barwick, E. Prichard (Eds.), VIM 3 (2011).

[69] C. Adams, K. Cammann, H.A. Deckers, Z. Dobkowski, D. Holcombe, P.D. LaFleur, et al., Quality Assurance for Research and Development and Non-routine Analysis, 1988.

[70] B. Magnusson, U. Örnemark (Eds.), Eurachem Guide: The Fitness for Purpose of Analytical Methods – a Laboratory Guide to Method Validation and Related Topics, second ed., 2014.

[71] S. Kromidas (Ed.), Handbuch validierung in der analytik, second ed., Wiley-VCH, Weinheim, 2011. http://www.wiley-vch.de/publish/en/AreaOfInterestCH00/bySubjectCH10/availableTitles/3-527-32938-2/?sID=21141s2gibi76obnp4hh9r82f7.

[72] G.D. Oehlert, A First Course in Design and Analysis of Experiments, Electronic Version, W. Freeman, 2010. http://users.stat.umn.edu/gary/book/fcdae.pdf.

[73] T. Hastie, R. Tibshirani, J. Friedman, The Elements of Statistical Learning; Data Mining, Inference and Prediction, second ed., Springer Verlag, New York, 2009.

[74] C. Beleites, U. Neugebauer, T. Bocklitz, C. Krafft, J. Popp, Sample size planning for classification models, Anal. Chim. Acta 760 (2013) 25–33, https://doi.org/10.1016/j.aca.2012.11.007.

[75] K.H. Esbensen, P. Geladi, Principles of proper validation: use and abuse of re-sampling for validation, J. Chemom. 24 (2010) 168–187, https://doi.org/10.1002/cem.1310.

[76] M. Sattlecker, N. Stone, J. Smith, C. Bessant, Assessment of robustness and transferability of classification models built for cancer diagnostics using Raman spectroscopy, J. Raman Spectrosc. (2010) 897–903, https://doi.org/10.1002/jrs.2798.

[77] R. Kohavi, A study of cross-validation and bootstrap for accuracy estimation and model selection, in: C.S. Mellish (Ed.), Artificial Intelligence Proceedings 14th International Joint Conference, 20 – 25. August 1995, Montréal, Québec, Canada, Morgan Kaufmann, USA, 1995, pp. 1137–1145.

[78] C. Beleites, R. Baumgartner, C. Bowman, R. Somorjai, G. Steiner, R. Salzer, et al., Variance reduction in estimating classification error using sparse datasets, Chemom. Intell. Lab. Syst. 79 (2005) 91–100.

[79] J.-H. Kim, Estimating classification error rate: repeated cross-validation, repeated hold-out and bootstrap, Computational. Stat. Data Anal. 53 (2009) 3735–3745, https://doi.org/10.1016/j.csda.2009.04.009.

[80] A.C. Olivieri, Analytical figures of merit: from univariate to multiway calibration, Chem. Rev. 114 (2014) 5358–5378, https://doi.org/10.1021/cr400455s.

[81] M. Belter, A. Sajnóg, D. Barałkiewicz, Over a century of detection and quantification capabilities in analytical chemistry historical overview and trends, Talanta 129 (2014) 606–616, https://doi.org/10.1016/j.talanta.2014.05.018.

[82] F. Raposo, Evaluation of analytical calibration based on least-squares linear regression for instrumental techniques: a tutorial review, Trends Anal. Chem. 77 (2016) 167–185, https://doi.org/10.1016/j.trac.2015.12.006.

[83] F.J. Anscombe, Graphs in statistical analysis, Am. Statis. 27 (1973) 17–21. http://www.jstor.org/stable/2682899.

[84] L. Buchen, Cancer: missing the mark, Nature 471 (2011) 428–432, https://doi.org/10.1038/471428a.

[85] C. Beleites, R. Salzer, V. Sergo, Validation of soft classification models using partial class memberships: an extended concept of sensitivity & co. Applied to grading of astrocytoma tissues, Chemom. Intell. Lab. Syst. 122 (2013) 12–22, https://doi.org/10.1016/j.chemolab.2012.12.003.

[86] P. Filzmoser, B. Liebmann, K. Varmuza, Repeated double cross validation, J. Chemom. 23 (2009) 160–171, https://doi.org/10.1002/cem.1225.

[87] C. Beleites, R. Salzer, Assessing and improving the stability of chemometric models in small sample size situations, Anal. Bioanal. Chem. 390 (2008) 1261–1271, https://doi.org/10.1007/s00216-007-1818-6.

[88] L. Breiman, Out-of-bag Estimation, 1996.

[89] E. Genova, M. Pelin, G. Decorti, G. Stocco, V. Sergo, A. Ventura, A. Bonifacio, SERS of cells: what can we learn from cell lysates? Anal. Chim. Acta 1005 (2018) 93–100, https://doi.org/10.1016/j.aca.2017.12.002.

CHAPTER

Label-free SERS techniques in biomedical applications

2

Laura Rodríguez-Lorenzo, PhD [1], Miguel Spuch-Calvar, PhD [2], Sara Abalde-Cela, PhD [1]

[1]*International Iberian Nanotechnology Laboratory, Avenida Mestre José Veiga s/n, Braga, Portugal;* [2]*CINBIO, Universidade de Vigo, Campus Universitario Lagoas, Vigo, Pontevedra, Spain*

Introduction

Label-free approaches have been long pursued by researchers in different analytical techniques as they are, a priori, the most straightforward way of detecting analytes of interest. The fact of using a label-free approach implies that the transduced signal is a direct consequence of the presence of the biological event, organism, or material. SERS first observation [1] is indeed an example of a label-free detection strategy. In this case, the nanoparticulated silver present in the solution enhanced the Raman spectra of the pyridine that Fleischman and coworkers were trying to detect. The next two decades after the observation of this unexpected high Raman signal of the pyridine were devoted by researchers in the field to explain the SERS phenomenon and the mechanisms behind it. It is only at the beginning of the 1990s that SERS applications started to boost. This was intrinsically linked to advances in synthesis of nanoparticles (NPs) having different compositions, shapes, and sizes. First proof-of-concept works for label-free detection using SERS started by pursuing the so-called "average" SERS. Variable concentrations of the molecule of interest were added to the as-synthesized nanoparticle colloidal suspensions, and the SERS signature of the molecules were analyzed (Fig. 2.1). Often, a full assignment of the molecular vibrations of the analyte under study was pursued. As SERS substrates evolved, solid SERS substrates appeared to be excellent candidates for the detection of analytes [2]. However, the first decades after the discovery of SERS were mainly dedicated to the establishment of SERS as a powerful analytical technique, but with noticeable limitations on the demonstration of its potential to be translated into a real application [2]. One of those limitations is the difficulty of analyzing complex matrixes using label-free SERS. In particular, biological matrixes are a special challenge due to the presence of all types of cells, proteins, genetic material, extracellular vesicles, tissues, and salts. A panoply of strategies for the detection of bioanalytes have been developed by researchers in the field, which may overcome this limitation such as (i) sample preparation and purification pre-SERS analysis; (ii) molecular sieves; (iii) host—guest substrates, or (iv) affinity capture [3]. The use of

SERS for Point-of-care and Clinical Applications. https://doi.org/10.1016/B978-0-12-820548-8.00007-2

FIGURE 2.1

Schematic representation of the concept of label-free SERS analysis of different biological species (purple and green) in the presence of plasmonic nanoparticles (yellow).

Credit: http://miguelspuch.wix.com/mspuchdesign.

codified NPs as SERS reporters (non—label-free approaches) arose from the need of being specific when using SERS as an analytical technique. The very complicated biological matrixes, together with the low stability and reproducibility of SERS substrates, made label-free approaches less popular among the SERS community to apply this technique to the biomedicine field [4].

As a consequence of these bottlenecks and the direction of the field into SERS-codified strategies and indirect detection [5], no SERS substrate or diagnostic test has been approved to be used in a clinical setting [6]. The synthesis of codified SERS NPs though is challenging, not reproducible and with low yields, and despite promising, still far from gathering the requirement from the clinical market. Considering the later, label-free analysis stands a better chance considering the current advances in SERS substrate synthesis and fabrication [7]. In order to unlock the access of SERS to this clinical market, it is also crucial the ability of data interpretation and spectra deconvolution when dealing with those complex matrixes (blood, saliva, tears, or tissues, among others) [8]. Statistics, chemometrics, and machine learning have been a game changer for label-free SERS analysis in biomedicine. As a result of the flourishing of data analytics and digital science since the early 2000s, the interest into label-free approaches has been gaining *momentum* due to advanced statistical algorithms that are capable of extracting information from complex data in a very efficient manner [9].

Chemometrics has been a widely used term when it comes to the statistical analysis of Raman spectral acquisitions. Different mathematical models have been applied to discern the sensitivity, sensibility and accuracy of Raman and SERS measurements. The calculation of these three parameters is crucial when developing an analytical method that may eventually be used to diagnose diseases. The most popular mathematical functions for the determination of these parameters when using SERS have mainly been partial least squares (PLS) and linear discriminant analysis (LDA) including their different variants. More recently, with the flourishing of machine learning (ML) and artificial intelligence (AI), more and more advanced algorithms are being used to train data sets and apply these models for the understanding of biological mechanisms driving disease evolution as well as diagnostic and even prediction tools [10]. These latest advances may unlock the use of SERS substrates as a point of care (POC) testing method (Fig. 2.2). The ability of testing at the bedside using POC tests brings clinicians, patients, and healthcare systems benefits such as reducing the time of analysis, higher connectivity, or no need of large infrastructures or specialized personnel. Some of the most known POC systems are the glucose sensor, the pregnancy test, or pathogen screening. These methods rely on reading and output systems that are easy to implement and interpret with no need, or very minimal, of instrumentation. However, as known through the COVID-19 pandemic, in order to comply with these requisites POC testing often sacrifices on sensitivity or multiplexing [11].

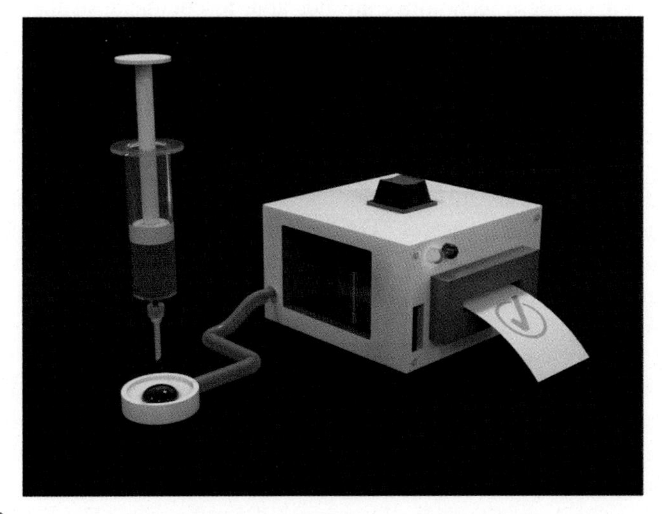

FIGURE 2.2

Example of a POC system: a sample of a body fluid (blood) is analyzed in a microfluidic device offering continuous and real-time assessment of the presence/absence of a disease in a miniaturized and automated manner.

Credit: http://miguelspuch.wix.com/mspuchdesign.

Oncological diseases

The challenge to understand, diagnose, treat, and monitor cancer is still, and unfortunately, very much a hot topic. The reason behind the high mortality of cancer (9.6 million deaths in 2018) is that it is a heterogeneous and dynamic disease, and that an average of 60% of patients diagnosed with a primary tumor will relapse, having other tumors spread in their body [12,13]. However, 90% of cancer related deaths are due to metastasis in secondary organs different to that of the primary tumor [14]. Due to the complicated nature of cancer, a panoply of different inherent factors have been defined as the hallmarks of metastasis: motility and invasion, ability to modulate the secondary site or local microenvironments, plasticity, and ability to colonize secondary tissues [14]. The metastatic cascade has been very well described [15], in which the circulating tumor cells appear as one of the main actors on tumor propagation. Intravasation into circulation allows these invasive cells to reach distant organs, where some of them are able to adapt to a new microenvironment, survive, and eventually create a macrometastatic lesion [16]. This ability of colonizing secondary tissues is related to the malignancy of a single specific cell and the starting point of metastasis [17]. Nowadays, by using the omics (genomics, transcriptomics, and/or metabolomics) it is possible to obtain information about the disease at many different levels, from tissues to single-cells and from protein expression to specific gene mutations [18]. Many technologies created in the lab made their path to the clinic, starting with the traditional ELISA, protein expression analysis of PSA for prostate cancer, and even gene sequencing [19]. Despite the advances are amazing, some of these techniques are not able to conjugate in the same sensor sensitivity, specificity, or multiplexing. Remarkably, SERS is able to combine those properties in the same sensor, and as such several promising approaches are demonstrating its huge potential as sensors for POC and clinical testing [20,3].

Chemical modifications such as methylation, oxidation, or hydroxylation of DNA may be applied in disease diagnosis [21,22]. Label-free SERS has already demonstrated the ultrasensitive and fast detection of hyper- or hypomethylation, DNA oxidation of hydroxylation [23]. Several SERS substrates in combination with chemometrics-based data analysis have been used for the identification of those chemical modifications with remarkable sensitivities in a rapid manner. Besides DNA/RNA chemical modifications, also DNA point mutations, single-base mutations, and sequence modification have also been analyzed using a plethora of SERS substrates based on spherical Ag NPs [24–26], Au electrodes [27], Au and Ag nanorod arrays [28–30], hybrid graphene-Au NPs [31], nanowave substrates [32], Au nanosunflowers [33], or gold nanoplate films [34], among others.

For example, Halas and coworkers have reported the detection of DNA and disease-driven DNA modifications by using SERS substrates. The SERS substrate used for the detection of adenine vibrational signature was Au nanoshells [23,35]. These Au nanoshells have a core of a silica surrounded by a gold shell and combined to obtain spherical dimers [36,37]. In these works, they used adenine-free probes to detect the presence of adenine in DNA by monitoring DNA hybridization.

Recently, Wu et al. presented a novel approach for the use of SERS beacons to report hybridization of DNA when the KRAS mutation occurring in colorectal cancer is present [29]. Despite this work being a proof-of-concept of the potential of miniaturized systems to detect cancer mutations, its efficacy relies on the simplicity of the preparation of the plasmonic material, based on the assembly of silver-coated gold nanorods on a glass substrate, followed by the incorporation of microfluidic channels for the injection of the samples to be analyzed (Fig. 2.3). In this specific case, the authors demonstrated that this highly sensitive and uniform SERS substrate could achieve an enhancement factor of 1.85×10^6. The strategy behind was the use of molecular beacon probes complementary to the wild-type *KRAS* genes, allowing the indirect detection of the mutated *KRAS* genes among the lysed material of cancer cells, all of it in just 40 min. In a second work the authors demonstrated the multiplex capacity of this approach by profiling DNA mutation patterns from colorectal cancer patients [30]. In this case they implemented a supervised learning algorithm for the classification of the cancer profiles obtained through SERS fingerprinting. This is a perfect example of a potential POC testing as the SERS substrate is small and simple to use [38]. However, to achieve this level of sensitivity still requires the use of a standard Raman spectrometer, and proof-of-concept with a handheld or a portable Raman would be necessary in order to translate this approach as a pure POC testing system.

Despite all the range of different proof-of-concept reports available in the literature, it is still not a common practice to present works in which real samples are

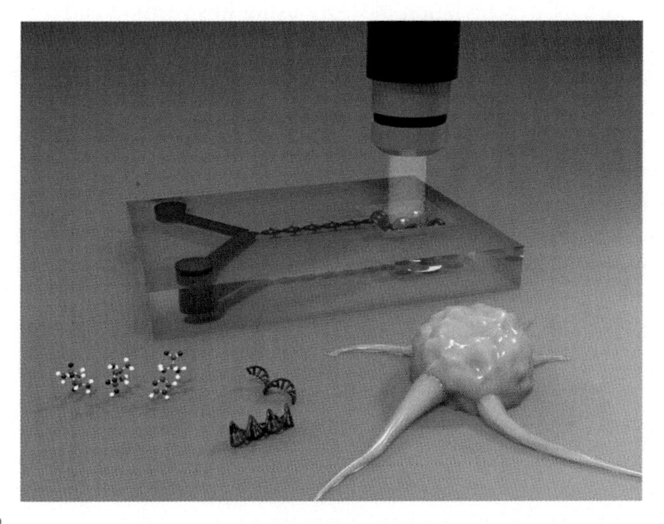

FIGURE 2.3

Schematic representation of the analysis of (from left to right) metabolites, genetic material, or cells from cancer patients. The analysis takes place in a miniaturized microfluidic device enabling standardization and automation of the analysis.

Credit: http://miguelspuch.wix.com/mspuchdesign.

used. However, when real samples are analyzed using label-free SERS spectroscopy in biomedicine, it is almost always the case that chemometric and multivariate analyses are being used to unravel important changes in the spectrum of interest. Remarkably, Chen et al. present one of the few preclinical studies using SERS for the identification of ribonucleic acid analysis (RNA) using real samples from colorectal cancer patients [39]. The key of this work was the use of chemometrics to perform the assignments of the RNA contributions to the spectra of the patient population ($n = 55$) with respect to the healthy population ($n = 45$) selected for the study. For this, they used 3D Ag nanofilms for the direct enhancement of the signal of the RNA present in the sample after preparation and extraction steps from serum. Using these substrates and statistical analyses the authors were able to achieve a sensitivity and specificity of 89.1% and 95.6%, respectively.

Several works report the label-free detection of methylation based on SERS using gold nanoshells, Ag NPs, or aluminum nanoantenna arrays [23,40–42]. For example, Li and coworkers have shown how by using Ag NPs it is possible to identify methylation for a panel of selected genes [43]. They analyzed the plasma of 48 non–small cell lung cancer (NSCLC) patients including 51 healthy control individuals. With a prediction accuracy of 88% they could identify methylation in the p16, MGMT, and RASSF1 genes. The analyses were performed in plasma samples and the SERS spectra were deconvoluted using multiple linear regression (MLR) to classify the methylation levels of different patient populations. In a different configuration of the SERS substrate, using a nanohole gold array, Luo et al. took advantage of the hot spots generated in the nanogaps of their substrate to perform the ultrasensitive analysis of DNA methylation [44]. By using this method authors were able to detect methylation changes with a sensitivity of 1%. Besides detection, this method offers dynamic information on the methylation changes by combining principal component analysis (PCA) and 2D correlation spectroscopy analysis.

Very recently, single-molecule detection limits using SERS for DNA detection were reported by De Angelis and coworkers [45]. In order to overcome the typical low residence time of the analytes within hot-spots, which in turn endows single-molecule resolution in SERS, they have applied an electroplasmonic trapping strategy. In this way, single nucleobases could be discriminated in a single oligonucleotide.

For instance, Chen and coworkers analyzed the proteins present in the serum of 103 colorectal cancer patients and 103 healthy volunteers [46]. In this case, to enable the label-free detection the authors used acetic acid to trigger the aggregation of the NPs and spectra of serum proteins were acquired. The authors were able to relate changes in the spectrum related to the structure of the proteins for cancer patients. In this case, standard silver nanoparticles (AgNPs) were used as SERS substrate and mixed with the purified proteins from the serum of the cohort of individuals selected for the study. Afterward, a multivariate statistical analysis based on PCA, LDA, and the PLS was applied to demonstrate the diagnostic capability of this approach. Authors reported diagnostic accuracies of 100% (sensitivity: 100%; specificity: 100%) when the analysis was based on albumin SERS spectroscopy and

99.5% (sensitivity: 100%; specificity: 99%) based on globulin SERS spectroscopy. Furthermore, they introduced a partial least PLS approach to predict unidentified subjects which yielded a diagnostic accuracy of 93.5% and 93.5% for albumin and globulin, respectively.

This same research group already presented before other works in which they interrogated different body fluids for the detection of different types of cancer such as nasopharyngeal [47], gastric [48], and hepatocellular carcinomas [49]. These approaches relied on a presample treatment which involved the separation of the serum or plasma from the blood. After the separation, as in the work described above, Ag NPs are added to the samples coming from cancer patients and analyzed by different chemometric methods.

Wang et al. [50] have also identified purified serum proteins from human serum samples of colorectal cancer patients (103 cancer patients and 103 healthy volunteers). They used as a guiding markers the dysregulation of serum albumin and globulin know to be found in a cancerous status. Once more, Ag NPs were the chosen SERS substrates to enhance the signal of this proteins, and after a label-free analysis of the samples, different chemometric analyzes were performed. In this case, they used instead of salts, acetic acid as aggregating agent to increase the SERS efficiency and signals of the studied proteins. To analyze the capability distinguishing the colorectal cancer group from the normal group, their approach was based on a combination of a multivariate statistical method using PCA and LDA. The authors report impressive prediction power both for albumin and globulin for unidentified subjects with a diagnostic accuracy of 93.5%.

De Angelis et al. [51] have challenged the most current technologies applied in the field of metabolomics. Despite techniques like mass spectrometry, chromatography or nuclear magnetic resonance have been at the forefront of the analysis of metabolites, they are very expensive, highly specialized, and with very low potential for miniaturization toward POC. In a proof-of-concept work these authors have analyzed the extracellular media of a cancer cell culture over time. They were able to identify and provide a kinetic study of the variable concentrations of L-tyrosine, L-tryptophan, glycine, L-phenylalanine, L-histidine, and fetal bovine serum proteins. They expanded the approach to analyze reactions from a macrophage cell line after stimulation with lipopolysaccharides. After a PCA analysis they were able to determine functional change of cells toward the activated proinflammatory state induced by the lipopolysaccharide. In this case, the SERS substrate used for the analysis is a solid Ag NP substrate, fabricated through an electroless process, involving just immersion of silicon wafers into precursor solutions. Silver nanoislands were then formed, and applied to the metabolomics analysis of the cell extracts at different time points.

In situ and in vivo detection of Raman signatures have been broadly reported in the literature, mostly for the identification of aberrant mutations in tissues and cancerous areas (Fig. 2.4) [52,53]. Inspired by the famous mass spectroscopy pen, Raman probes have also been designed to use at the operation theater [54,55]. However, these reports are mostly based in conventional Raman and not SERS. Though,

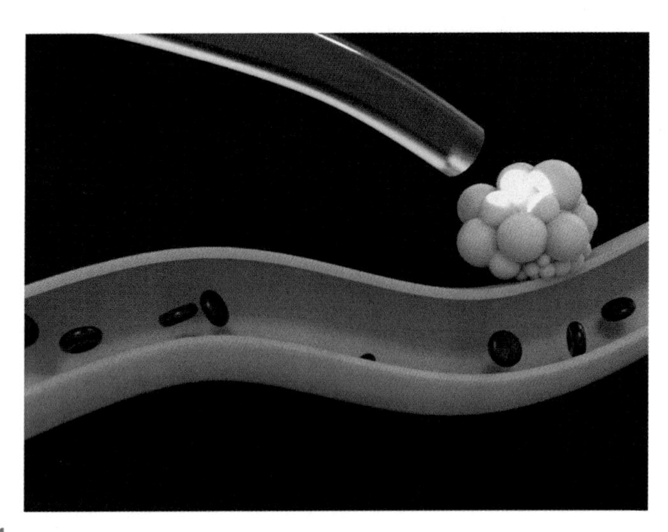

FIGURE 2.4

Conceptual representation of an in vivo analysis of biological material by using optical probes.

Credit: http://miguelspuch.wix.com/mspuchdesign.

a few examples of the combination of these strategies with NPs can still be found in the literature. For example, Huang et al. [56] used silver NPs for the identification of diabetic and normal pancreatic tissue. The normal known tissue signals such as that of present proteins, lipids, DNA, and RNA bases are the usual signals that can be directly identified in these type of studies. However, the differentiating features are the tissue structures and their modification at a disease status. In this specific case the authors distinguished between normal and diabetic tissue, applying the same concept used in their previous paper to identify thyroid tissue [57]. For the SERS analysis of the tissue, the silver NPs were drop casted on top of the preprepared tissue samples, and then analyzed. In order to identify the difference in the tissues, the authors compared the intensities of the SERS signal obtained. Despite the number of samples used in this study is fairly small (<10 samples), it was one of the first and promising studies related to the use of NPs for tissue structural identification. In a similar approach, also using Ag NPs but introducing PCA-LDA multivariate analysis, Feng et al. showed an outstanding sensitivity of 90.9% and specificity of 97.8% for the classification of esophageal cancer tissue slices [58]. In this case, the number of samples was much higher (88 paired tissue samples), and as such this study is much more robust toward the demonstration of the potential of SERS to complement or substitute immunostaining protocols performed at the pathology laboratory in hospitals. Different reports benefit from the use of hybrid materials, combining plasmonic NPs with alternative materials to provide more stability and reproducibility to the SERS substrates. For example, Kajimura and coworkers combined gold NPs on top of bohemite to perform SERS large area imaging of brain ischemia [59]. The focus of this work was the fabrication of a highly controlled SERS substrate and its optical properties to provide a robust SERS substrate with high SERS enhancement factor with high two-dimensional homogeneity. Using this substrate they demonstrated the presence of metabolic alterations in the brain when ischemia is present. Impressive SERS imaging maps of the mouse brain were shown in this study, in which they can clearly identify the ischemic region of the brain by the 518 ± 5 and $736 \pm 15\ cm^{-1}$ bands. By combining both signals they were able to identify the core region of the ischemic injury and a metabolically delineated boundary being the interface between the ischemic core and the nonischemic region.

Also, by using a hybrid material approach, Koyakutty and coworkers combined Ag NPs with TiO_2 nanostructures for the detection of oral cancer [60]. This work shows a systematic study of the combination of different TiO_2 hierarchical structures with Ag NPs and their SERS enhancement. After the optimization, the authors analyzed eight oral cancer patients (squamous cell carcinoma of tongue) comprising a total of 24 normal and 32 tumor tissue sections and the recorded spectra were analyzed by principal component analysis and discriminant analysis. After applying the multivariate methods, they reported a specificity and sensitivity of 95.83% and 100%, respectively. In an attempt to overcome the limitation of conventional SERS to measure at significant depths in tissue samples, Graham and coworkers demonstrated the application of SESORS, surface-enhanced spatially offset Raman spectroscopy, in tissue analogues [60,61]. Remarkably, they were able to obtain SESORS signals at tissue thicknesses of >6.75 mm. The capacity of measuring at these depth ranges overcomes the acquisition of spectra dominated by contributions from the surface layers of the tissues. Since then, several other works described the application of SESORS for the analysis not only of metabolic and phenotypic changes in tissues but also to measure temperatures at subsurface locations of tissues or through the skull of monkeys [62,63]. Despite the promising nature of this technology, it has still to be rolled out in larger amounts of samples from real patients, and combined with multivariate and chemometrics analysis.

Paper-based SERS sensors arise as promising candidates, as they are cheap, easy to produce, and affordable. Also, the market uptake might be easier as they resemble to other widely accepted POC sensors, such as the glucose one. Teixeira et al. presented an approach that gathers a priori all the requirements for a POC testing sensor [64]. A SERS-paper sensor was developed using gold nanostars assembled following a very controllable process. As one of the main concerns when preparing SERS substrates is the reproducibility of fabrication, a microfluidic chamber was used in order to achieve a higher control during a self-assembly process. As a result, a close packed sensor on a paper was produced. For the proof-of-concept the authors tested the lysed contents of two types of cells. Different cells components were detected and a cancer versus a noncancer line were discriminated based on the dysregulation of different cell components. Bamrungsap [65] and coworkers reported another paper-based SERS sensor with 98% accuracy for the identification of cancer cells. A downside of this approach is that it was combined with magnetic separation, whereas for POC and clinical applications, the least steps possible, the higher chances of technology translation and uptake.

Matteini and coworkers have also been active in the field of paper-SERS sensors. They have used silver nanowires to create silver spotted substrates, and applied them to the analysis of proteins relying on machine learning classification [8]. The authors developed an effective machine learning classification of proteins species with similar spectral profiles. They highlighted the importance of standardization of data analysis in order to bring SERS a step closer to the clinic. Furthermore, authors pointed the limitation of traditional PCS analysis as being unable to relate the obtained classified datasets to relevant chemostructural insights to use in biological and clinical evaluation.

Olivo et al. [66] reported a very interesting preclinical proof of concept for the detection of the protein A1AT, a potential biomarker for bladder cancer, in clinical urine samples. For that they used an assay which after protein binding detects the presence of the protein through structural changes in a Raman reporter previously bonded to the SERS substrate, similar to the approach of Wu et al. described above [29].

Neurological diseases

The World Health Organization (WHO) compiles statistics on the leading causes of death worldwide and, as of 2015, the top 10 causes of death included neurological diseases, diabetes, cardiovascular diseases, cancers, and viral diseases [67,68]. Neurological diseases will serve as the focus of this chapter section. Mental disorders, such as Alzheimer, Parkinson, schizophrenia, depression, or dementia, have become Europe's largest health challenge of the 21st century with 38% of Europeans suffering this kind of health issues at some point in their lives. At present, diagnoses of metal diseases are made based on a full psychiatric evaluation, medical history, and physical exam. However, currently there are not valid biochemical tests to support physicians' decisions and therapeutic strategies. Brain imaging analyses such as a magnetic resonance imaging (MRI) or computerized tomography (CT) scan to visualize brain abnormalities are possible, but not applicable for routine examination. Therefore, the introduction of valid biochemical POC systems to support physicians' decisions in the diagnosis of neuronal diseases would have an important social impact component due the large number of people that suffer these type of diseases.

In relation to Alzheimer, the latest statistics made by the international federation Alzheimer's Disease International (ADI) show that nearly 44 million people worldwide have Alzheimer's or a related dementia and only 1-in-4 people with Alzheimer's disease have been diagnosed. The global cost of Alzheimer's and dementia is estimated to be $605 billion, which is equivalent to 1% of the entire world's gross domestic product. More than 50,000 new cases of Parkinson's disease are reported in the United States each year, but there may be even more, since

Parkinson's is often misdiagnosed [69,70]. On the other hand, depression has become the second leading global cause of years lived with disability [71] and the WHO states that every year about 1 out of 15 people suffer from major depression [72]. Depressive disorders often start at a young age; they reduce people's functioning and are often recurring and even chronic. A recent study of the World Mental Health Survey found that, on average, about 1 in 20 people reported having an episode of depression in the previous year. Lifetime prevalence varies widely, from 3% in Japan to 17% in United States. For these reasons, depression is the leading cause of disability worldwide in terms of total years lost due to disability.

Schizophrenia is also a highly prevalent disorder affecting approximately 1% of the world's population. Patients with schizophrenia often require antipsychotic medication throughout their lifetime. According to the WHO, the prevalence rate for schizophrenia is approximately 7 per 1000 of the population over the age of 18 or, in other words, at any one time as many as 51 million people worldwide suffer from schizophrenia. According to the US National Institute for Mental Health the cost per schizophrenia patient rounds $28 500 per year due to their chronical nature and relapses. For the healthcare system and for the patients, relapse can have devastating repercussions. As the situation worsens, financial burden on both healthcare system and families is enlarging. To minimize this burden on the patients and their families, it is vital to get the proper treatment in the early stages, before the first relapses, and properly follow up the prognosis of the treatment to act quickly in the case the drug is inadequate.

It is clear that with support, medication, and therapy, many people with these diseases are able to increase their life's quality for longer time periods. However, the outlook is best when mental diseases are early diagnosed and treated in a right away. If you spot the signs and symptoms of these diseases and seek help without delay, the patients can take advantage of the many treatments available and improve the chances of recovery. Therefore, sensitive and specific detection and monitoring of a biomarker is a prerequisite for early clinical diagnosis, allowing a molecular correlate which supports physicians' diagnosis is highly demanded. There is scientific evidence that with several biomarkers it is possible to detect these neurological diseases (Fig. 2.5). It has been reported that ultrasensitive detection and spatially resolved mapping of neurotransmitters, such as dopamine (DA), noradrenaline (NET), and serotonin (SER), are critical to facilitate the understanding of brain functions and investigate the processing of information in neural networks. In fact, a decrease in peripheral blood platelet serotonin transporter and dopamine receptor binding is one of the few well-characterized biomarkers of diverse neurological diseases [73]. It has been reported that multifocal microscopy determination of serotonin transporter (SERT) in lymphocytes predicts the therapeutic efficacy of selective serotonin reuptake inhibitors, and they are currently studying whether these finding can be extended to other neurotransmitter transporters like DAT and NET. By measuring the clustering of these transporters in the membrane of the T cells, the therapeutic response to different antidepressant drugs can be predicted.

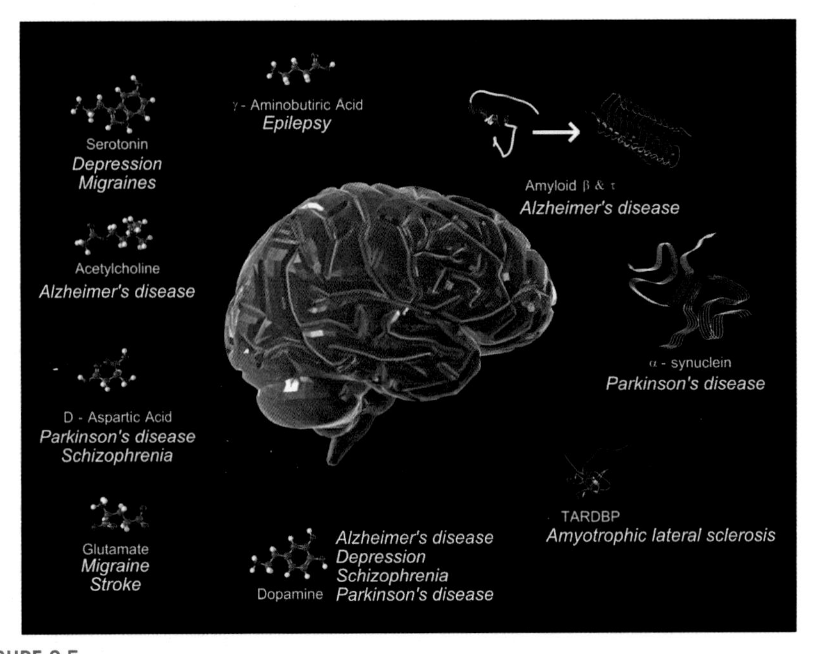

FIGURE 2.5

Chemical structure of the representative biomarkers identified for neurological diseases.

Credit: http://miguelspuch.wix.com/mspuchdesign.

In this line, SERS demonstrated to be one of the effective alternatives for multiplex monitoring of these neurotransmitters and other biomarkers, since SERS has the ability to detect concentrations down to the nanomolar range both in the brain and in biological fluids (i.e., blood, plasma, serum, cerebrospinal fluid, and urine) while maintaining a good resolution [67]. Thanks to these feature, SERS may open a real possibility for developing a simple and inexpensive POC assays for brain diseases supporting their diagnosis and monitoring and will therefore improve the clinical practice and foster the sustainability of the mental sector of the healthcare system. However, SERS-based sensing technologies are at the very preliminary stages in the area of neurological diseases yet. Therefore, most of the references related to SERS applications in this area reported proof-of-concept studies on the monitoring and imaging of neurotransmitters in nonclinical settings.

The feasibility of SERS for the label-free direct detection of neurotransmitters is dependent on the chemical structure of these molecules, since these biomarkers must be close enough to the metallic nanostructures to enhance their Raman signal [74–76]. Therefore, the optimization of SERS strategies for this application must be performed as a function of the noble metal selected, the SERS substrate architecture, the excitation laser line, and neurotransmitters' nature. Sundin et al. [77] and Manciu et al. [78] reported the ultradetection of adenosine, SER, and DA down to 0.1 pM. They also demonstrated changes on the SERS band intensities that may be correlated with the neurotransmitter' orientation on silver surface. They showed that the Raman cross sections of neurotransmitters and their grafting density on Ag surface have an important role in preferential SERS enhancement. This study not only provided direct evidence of label-free SERS detection of these neurotransmitters but it further advanced the knowledge of their interactions at the interface with metal NPs. The strength of the analyte—substrate interaction plays an important role on the evaluation of the SERS enhancements. Several reports have demonstrated a correlation of surface chemistry of neurotransmitters with the measured SERS signal. SER monolayers can be formed on silver electrode surface via O from hydroxyl group and N in the ring and these adsorbed SER molecules can undergo a reorientation on the surface with the applied potential [79]. Relatively good agreement between the calculated and experimentally determined SERS spectra of DA has indicated the presence of different DA molecular forms, such as uncharged $DA^{\pm}$, anionic DA^{-}, and dopaminequinone, which allowed a better understanding for the DA detection at physiological levels [80]. The adsorption behavior of melatonin on Au surface was deduced from the SERS selective rules and analysis of the calculated molecular electrostatic potential (MEP). Comparison of Raman and SERS spectra of melatonin suggested an almost flat orientation of the indole-anisole aromatic ring of melatonin on the surface and possible partial contribution of the chemical-transfer mechanism [81]. Bailey et al. [82] demonstrated that the molecular structure and surface affinity influence the sensitivity based on COMSOL model and experimental results, such that neurotransmitters with the strongest affinity for the metal nanostructured surface will have the highest signal-to-noise in SERS experiments in flowing solutions. Moody et al. [83] investigated the effect of metal NPs nature and wavelength of the excitation laser line in the detection of seven neurotransmitters: melatonin, SER, glutamate, DA, γ-amino butyric acid (GABA), norepinephrine, and epinephrine. They reported the maximum SERS enhancement factor ($\approx 10^5 - 10^6$) for Au NPs at an excitation wavelength of 785 nm and at 633 nm for Ag NPs. In addition, the results demonstrated that the best detection of catecholamines (i.e., DA, norepinephrine, and epinephrine = and amino acids neurotransmitters (i.e., glutamate, GABA)) was achieved using Au NPs and Ag NPs, respectively.

Several reports have shown the detection of neurotransmitters using metal noble nanoparticles-based active SERS substrates with excellent limit of detections (LODs). The most studied neurotransmitter is DA due to the association between the loss of dopamine and several neurological diseases such as epilepsy, Alzheimer's disease, Parkinson's disease, and schizophrenia [84]. The detection of glutamate and γ-amino butyric acid was performed using silver nanoparticles (Ag NPs) on a crystal fiber reported LODs down to 0.1 μM for glutamate and 0.1 mM for γ-amino butyric acid in aqueous solution [85]. Another report demonstrated the detection of choline, the precursor for acetylcholine, and catecholamine neurotransmitters including acetylcholine, DA, and epinephrine using Ag NPs electrodeposited on tin-doped indium oxide. This substrate exhibited an enhancement factor of 10^7 for p-aminothiopenol and extremely good reproducibility, which enabled LODs down to 2 μM for choline, 4 μM for acetylcholine, 10 μM for DA, and 0.7 μM for epinephrine [86].

El-Said and Choi [87] applied a high sensitivity single-step and label-free SERS method to determine the *in* situ effect of cisplatin, bisphenol-A, and cyclophosphamide on the extracellular DA level released from rat pheochromocytoma PC12 cells by monitoring the intensity of the Raman peak of DA at 1270 cm^{-1} (amide III and catecholamines, e.g., DA). A significant increase in the amount of DA released was detected after PC12 cells treatment with three drugs. In addition, they analyzed the changes in the biochemical composition of the PC12 cell lysates to determine the intracellular DA level after toxic drug exposure. P. Wang et al. [88] detected DA and SER in the presence of simulated body fluids in just 1 s and with a LOD of 0.1 nM using a graphene-gold nanopyramid heterostructure platform. The presence of the single layer graphene superimposed on the Au nanostructure not only can locate SERS hot spots (enhancement factor of $\approx 10^{10}$) but can also modify the surface chemistry to realize selective enhancement Raman yield. The SERS quantification of DA in plasma and urine using iron–nitrilotriacetic acid-functionalized AgNPs as substrate and 4-mercaptobenzoic acid as internal standard was also reported. The calibration model based on multiplicative effects model for SERS (MEM_{SERS}) achieved quite accurate and precise concentration predictions for DA with recovery rates varying within the range of 91.9%–112%. In addition, LOD of MEM_{SERS} in combination with SERS technique for DA in plasma and urine was estimated to be 0.04 μM. This report demonstrated that MEM_{SERS} had effectively corrected the multiplicative effects on the SERS signals of DA caused by the uncontrollable variations in physical properties of SERS substrate [89]. Interesting, Manciu et al. [78] reported the formation of a hydrogen-bonded complex between DA and SER, which further complicated potential analyte discrimination and quantification at concentrations characteristic of physiological levels (e.g., nanomolar) [89]. All these findings represent a step forward in enabling in-depth studies of neurological processes including those closely related to brain activity mapping (BAM).

There are other type of biomarkers for the monitoring of neurological diseases (Fig. 2.5). On one hand, neuro proteases are essential for the biosynthesis of peptide neurotransmitters for neurological functions. On the other hand, neurotoxic peptides produced by brain proteases converting protein precursors accumulate in brains of Alzheimer's, Huntington's, Parkinson's, and other neurodegenerative diseases [90].

The mammalian bombesin-like peptides are structurally related neuropeptides with a wide spectrum of biological activities such as smooth muscle contraction, stimulation of secretion, modulation of neural activity, and growth regulation. Bombesin (BN) was originally discovered in *Bombina bombina*, a European frog, and it is an aminated tetradecapeptide (14 amino acids), having the primary structure: pGlu-Gln-Arg-Leu-Gly-Asn-Gln-Trp-Ala-Val-Gly-His-Leu-MetNH_2 (where: pGlu is 5-oxo-proline and all amino acids are in L-conformation) [91]. The adsorption of this neuropeptide and its C-terminal fragments on different metal noble NPs was analyzed using SERS spectroscopy [92]. The SERS analysis allowed to determine a particular surface geometry and orientation of BN and its fragments on specific noble metal surface, and their specific interaction with the surface. SERS spectra acquired for BN and fragments onto Ag surface indicated that the orientation and conformation of BN was dictated by the interaction of a thioether atom and L-tryptophan with the Ag surface. The results also clearly showed that L-histidine residue and C=O did not interact with the Ag. By comparing the SERS spectra of BN-like peptides adsorbed on the platinum surface with those on the silver surface it became obvious that BN interacts with Pt and Ag NPs through the same molecular fragments, acquiring a conformation involving a BN backbone shortening from the N-terminal end [93]. However, the BN peptides modified their conformation by adsorbing on AuNPs. Specifically, the changes on SERS spectra can be related to (1) the elongation of the C-terminal peptide fragment, which caused movement of the L-methionine side chain in the direction of the Au surface and the weakening of the interaction between the amide bond and Au; (2) the L-histidine residue assisted in the peptides interaction with the Au NPs; and (3) L-tryptophan dramatically changed the SERS fingerprint on Au surface [94].

Neuropeptide Y or neuropeptide tyrosine (NPY) is one of the most abundant neuropeptides in the mammalian brain and it is an aminated peptide having a primary structure of Tyr^{1}-Pro-Ser-Lys-Pro-Asp-Asn-Pro-Gly-Glu-Asp-Ala-Pro-Ala-Glu-Asp-Leu-Ala-Arg^{19}-Tyr^{20}-Tyr^{21}-Ser^{22}-Ala-Leu^{24}-Arg-His^{26}-Tyr_{27}-Ile-Asn-Leu^{30}-Ile-Thr-Arg^{33}-Gln-Arg^{35}-Tyr^{36}-NH_2. The acetyl-[$Leu^{28,31}$]-$NPY^{24\text{-}36}$ fragment, the selective Y_2R agonist was immobilized onto different metal nanostructured surfaces and investigated by SERS [95]. The adsorption geometry was triggered mainly by tyrosine and arginine residues interaction with the Ag and Au surfaces, and evident spectral changes were observed. The tyrosine ring on the Au surface existed in the phenoxyl radical and the interaction between the ring and Au was weaker than those between the tyrosinate ring and Ag. Domin et al. [96] demonstrated that in neuropeptide Y (NPY) and its native NPY^{3-36}, NPY^{13-36}, and NPY^{22-36} and mutated acetyl-($Leu^{28,31}$)-NPY^{24-36} C-terminal fragments, the NPY^{32-36} C-terminal fragment (Thr^{32}—Arg^{33}—Gln^{34}—Arg^{35}—$Tyr^{36}NH_2$) was involved in the adsorption process onto Ag surface.

Neurotoxic peptides are key factors in the development of neurodegenerative diseases [90]. Alzheimer's and Huntington's diseases involve proteases mechanisms to convert precursors proteins into neurotoxic peptides, neurotoxic β-amyloid ($A\beta$) peptides, and mutant NH_2-terminal fragment of the huntingtin (htt) proteins, respectively, that accumulated as aggregates in the disease process. For this, these neurotoxic peptides and their aggregates are potential biomarker for monitoring of neurodegenerative diseases for diagnostic purposes. Chou et al. [97] demonstrated the feasibility of using SERS to detect $A\beta^{1\text{-}40}$, one of the two most prevalent $A\beta$ species in CSF. They fabricated a nanofluidic trapping device with the ability to encourage aggregation of Au NPs, improving the reproducibility and sensitivity. The system was able to distinguish between α-helices and β-sheet as well as other protein conformational changes, which facilitated the discrimination between harmless monomeric form of $A\beta$ and more toxic β-sheet oligomeric or protofibril/fibril forms. Demeritte et al. [98] designed an antibody-conjugated SERS substrate based on plasmonic nanoparticles-coated graphene oxide to capture and identify trace levels of β-amyloid and tau protein, biomarkers of Alzheimer's disease, selectively from whole blood sample. They reported that the SERS detection limit is 500 fg/mL and 0.15 ng/mL for β amyloid and tau protein, respectively. In the label-free SERS spectra the strongest bands are mainly amide I, II, and III due to the α-helical and β-sheet conformation of β-amyloid. Other bands are associated with histidine residue bands, phenylalanine, tyrosine band, and D, G band from graphene oxide. It is very interesting to note that SERS spectra from tau protein is much different that β-amyloid and that the strongest peaks are mainly due to tyrosine. Clearly, these data show that biomarkers for Alzheimer's disease can be identified using their SERS fingerprints. Recently, Huefner et al. [99] employed a holistic (i.e., application of both principal component (PCA) and linear discriminant analysis (LDA)) approach using spectral profiles generated using SERS of serum samples of healthy and Huntington's disease patients covering a wide spectrum of disease' stages. SERS revealed significant correlations with disease progression, in particular progression from premagnified through to advanced Huntington's disease was associated with serum molecules related to protein misfolding and nucleotide catabolism.

As our understanding of specific diseases increases, multiplexed biomarkers assay will result in disease-specific signatures. In this line, SERS is an optimal technique for multiplexing measurements. Monfared et al. [100] combined SERS with a partial least squares (PLS) analysis to achieve the simultaneous detection of both glutamate and γ-amino butyric acid in blood serum with an LOD of 8 μM. The PLS calibration model showed a high coefficient of determination ($R^2 > 0.98$) between the spectral data and the respective concentrations of glutamate and GABA.

The multiplexing capabilities of SERS was demonstrated by monitoring of neurotransmitters secretion near living neuron using a plasmonic nanosensor with addressable location. This nanosensor was fabricated from borosilicate nanopipettes analogous to the patch clamp and these were decorated with Au NPs. The monitoring of ATP, glutamate, acetylcholine, GABA, and DA, among other neurotransmitters, was achieved in a single experiment using this dynamic SERS nanosensor. The SERS spectra of these neurotransmitters were identified with a barcoding data processing method and time series of the neurotransmitter levels were constructed. This nanosensor was coupled to a combined Raman-fluorescence microscope to visualize and position the tip of the nanosensor with submicron spatial resolution above GFP-expressing dopaminergic neurons to locally detect the neurotransmitters [101,102]. Ryzhikova and coworkers [103] reported the identification of Alzheimer's disease through analysis of blood serum using SERS in combination with multivariate statistical analysis. SERS spectra of serum samples from Alzheimer's disease patients were further compared to spectra from patient with other neurodegenerative dementias (Lewy body dementia, Parkinson's disease, and frontotemporal demential) and healthy controls to develop a simple test for Alzheimer's disease. The classification of SERS spectra from serum samples was performed using artificial neural networks achieving a diagnostic sensitivity around 96% for differentiating Alzheimer's disease samples from healthy controls in a binary model and 98% for differentiating Alzheimer's disease, healthy controls, and other neurodegenerative dementias in a tertiary model. The results from this proof-of-concept study demonstrated the great potential of SERS blood serum analysis to be developed further into a novel clinical POC assay for the effective and accurate diagnosis.

Neurological issues may also appear when humans are in contact with toxic organic pollutants such as pesticides. Organophosphate and carbamates are particularly toxic to humans in connexion to their high toxicity to acetylcholinesterase (ACHE). The inhibition of ACHE activity can lead to a loss of memory and an impairment of neuromuscular functions. ACHE is one important enzyme in the brain, which hydrolyzes the acetylcholine (ATC) neurotransmitter to choline and acetic acid. Therefore, there has been developed a fast and simple SERS assay using naked Au NPs based on the inhibition of ACHE activity in presence of pesticides such as carbaryl and paraoxon [104]. Thus, the direct SERS detection of choline could be correlated with the pesticide' exposure: lower choline detections means lower ACHE activity, which is inhibited by these pesticides. LODs obtained using this strategy was better than the conventional SERS analysis of the pesticide: 2 nM for carbaryl and 0.04 Pm for paraoxon. Moreover, besides inhibitors quantification, the use of simple chemometric tools, such as PCA, allowed the identification of the inhibitors in a complex mixture.

Currently different alternative biomarkers for developing a schizophrenia and major depressive diagnostic and the early stages of Alzheimer and Parkinson are been investigated. An ideal candidate biomarker for this is the family of low density lipoprotein receptor—related proteins, from now referred to as LRPs. LRPs are multifunctional scavenger and signaling receptors capable to bind a variety of ligands [105]. These receptors are expressed by a broad range of cells such as neurons, astrocytes, oligodendrocytes, macrophages, and vascular endothelial cells that constitute the blood—brain barrier. Thus, they are involved in a wide range of different biological functions. In this respect, to maintain tissue and cellular integrity, LRP1 is known as a phagocytic receptor of myelin debris in the central nervous system. When LRP1 is degraded by matrix metalloproteinases (MMPs), cell debris is accumulated in the brain leading to neuronal death. Therefore, the detection of increasing amounts of LRP fragments in peripheral blood and cerebrospinal fluid (CSF) must be indicative to degeneration and inflammatory events in the brain (Fig. 2.6). There is recent experimental evidence of changes in the concentration of different soluble fragments of LRP receptors in patients with mental diseases and Alzheimer. C. Spuch et al. [106] described a new soluble form of LRP2-220 kDa secreted from choroid plexus epithelial cells localized in the CFS, being reduced in Alzheimer' disease patients. Now, they have evidence about the existing relationship between small fragments of LRP1 and LRP2 and mood disorders seem to reveal these species as promising biomarkers for major depression and early stages of Alzheimer' disease. Hence, a comprehensive identification of the presence of these fragments in blood (serum and plasma) may constitute an ideal starting point in order to develop simple and inexpensive strategies for diagnosis, prognosis,

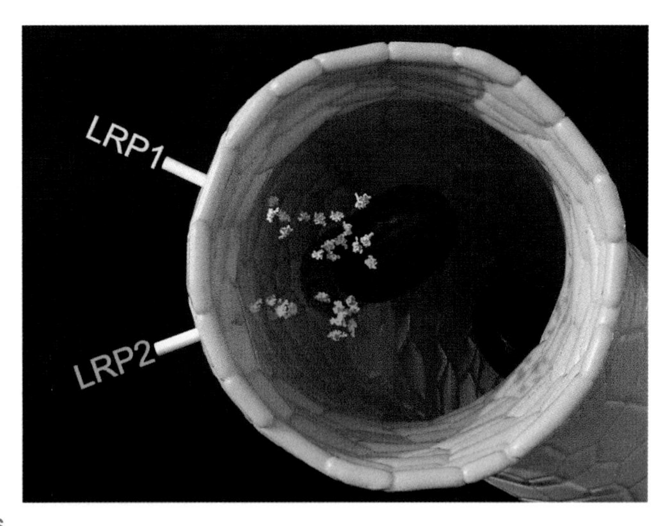

FIGURE 2.6

Illustrative scheme of LRP1 and LRP2, which are cell-surface receptors consisting of a large extracellular region, a single transmembrane domain, and a C-terminal cytoplasmic tail. Soluble fragments of the receptors LRP1 and LRP2 (yellow and green proteins in the scheme) release, through the blood—brain barrier to the extracellular space by a proteolytic mechanism and diffusion to blood vessels. It has been reported that the soluble fragments, which can be detected in CSF and blood, can be used such as potential biomarkers in neurodegenerative diseases and neurological diseases.

Credit: http://miguelspuch.wix.com/mspuchdesign.

and monitoring of different brain disease over time, such as Alzheimer's disease, Multiple Sclerosis, Schizophrenia, or depression. A future SERS—based sensing technology could be based on the functionalization of metal nanostructured surface with specific antibodies, since there are currently commercially available antibodies to recognize LRP1 (e.g., supplier Sigma—Aldrich L2170), and LRP2 (e.g., supplier Santa Cruz Biotechnology sc-H245). Thus, the strategy would be to capture these biomarkers by their specific antibodies and to detect them by monitoring changes in the SERS fingerprint of the antibody, as a result of a specific lock—key binding, as reported previously as a label-free indirect SERS strategy [38]. It is worth to point out that the applicability of this approach is closely bound to the fabrication of highly sensitive SERS-active plasmonic nanostructures.

Importantly, the ample availability of a large library of SERS fingerprints will allow the simultaneous detection of a panel of biomarkers with high selectivity enabling an accurate diagnosis of a complex disease. On the other hand, the development of the SERS multiplexed will expand the number of potential biomarkers to choose between, including new small peptides that which cannot be employed with the current technology because of sensitivity limitations. Moreover, the introduction of nanoenabling SERS sensing technology could enable rapid detection of multiple biomarkers at POC and may facilitate fast personalized healthcare delivery. The SERS analytical tools presented in this section may serve as POC tool for neurological diseases management program to obtain bioinformatics needed to optimize therapeutics for diagnosis management of these type of diseases.

Infectious diseases

Diagnosis is the key component in disease elimination to improve global health. However, there is a tremendous need for diagnostic innovation for infectious diseases because they are responsible for 15 million deaths each year, while more than 95% of these deaths take place in low-income countries [107]. Most infectious diseases are caused by pathogenic microorganisms including viruses, bacteria, parasites, and fungi. Compared to other diseases, infectious diseases can be exponentially transmitted among populations in a relatively short period of time thus threatening the general public health and economy. It is estimated that over half of the world population are at risk for infectious diseases, making them one of the most dangerous threats to humanity [108].

Adequate and prompt treatment to illnesses cannot be made properly without diagnosis in the first place. Sensitive, specific, and rapid diagnostic testing not only paves the way toward effective treatment but also plays a critical role in preventing the transmission of infectious diseases. For example, meningitis [109] caused by bacteria is usually very serious, since 5%–10% of patients die, typically within 24–48 h after the onset of symptoms, and requires a rapid detection method and urgent medical attention with appropriate antibiotic therapy. Malaria is a global health problem, an estimated 3.2 billion people in 97 countries and territories are at risk of being infected with malaria and developing the disease, and 1.2 billion are at high risk (>1 in 1000 chance of getting malaria in a year). According to the latest estimation, 214 million (range: 149–303 million) cases of malaria occurred globally in 2015 and the disease led to 438,000 deaths (range 236,000–635,000) [110]. Taking into account the high mortality rates, the rapid detection of these pathogenic microorganisms in body fluids and subsequent effective treatment is essential. Approximately 10.4 million new cases of tuberculosis were estimated in 2016 according to the WHO, with <64% cased diagnosed [111], preventing timely therapeutic interventions [112]. Therefore, even though tuberculosis has now become a largely treatable disease, it remains the worldwide leading infectious cause of death [111], claiming around 1.3 million deaths every year [113]. It would be impossible to achieve the goal of the End Tuberculosis strategy, with 90% reduction in incidence and 95% reduction in mortality by 2035 [114], without improved tuberculosis diagnostic tools to deliver timely therapeutic interventions.

Viruses are the most abundant biological entities on earth, more than archaea and bacteria combined [115]. Viruses can evolve rapidly and cause new epidemic, such as the zoonotic H5N1 pandemic [116,117], Zika [118], Ebola [119,120], and the most recent COVID-19 (SARS-CoV-2 virus) pandemic [121] outbreaks. For example, Influenza viruses infect 5%−15% of the global human population annually, and a seasonal epidemic occurs almost every years, mainly in the winter. Influenza virus infection often leads to severe illness, causing 250,000 to 500,000 deaths per year globally [122]. In addition to circulating seasonal influenza viruses, fatal new variants regularly emerge as threats. The recent 2009 flu pandemic caused by a new variant had persisted for longer than a year, eventually resulting in approximately 300,000 deaths [123]. Currently, 1.67 million unknown viruses are estimated to be circulating in animal reservoirs, a number of which with the potential for zoonotic transmission to humans [124]. Closely monitoring virus strains, as well as rapidly identifying unknown viruses, would aid in preparation for future outbreaks According to the WHO, early detection can halt virus spread by enabling the rapid deployment of proper countermeasures. Rapid diagnosis of infection by such new virus strain is critical for controlling viral spread in its early stage because patients can be treated before the onset of severe illness or effectively quarantined until the relevant vaccines or drugs are available.

Central clinical laboratories offer sensitive and specific assays, such as blood culture, high-throughput immunoassays (e.g., enzyme-linked immunosorbent assay, ELISA), real-time reverse transcription polymerase chain reaction (RT PCR), and mass spectrometry (MS) analysis [125,126]. However, they are often time- and labor-intensive, costly, and dependent on sophisticated instruments and well-trained operators. In addition, the requirement for predefined labels or probes such as a target pathogen-specific antibodies and DNA/RNA oligomers limits the use of these conventional methods for the rapid identification of newly emerging pathogens and the early and direct detection of known pathogens [127,128]. Timely diagnosis of bacterial infections can significantly reduce complications and lower fatalities. Therefore, there is a strong need to develop diagnostic technologies that can be used at the point-of-care to detect infectious diseases. POC tests provide rapid "on-site" results, and in resource-limited settings, supporting timely and proper treatment [129]. According to the WHO, POC tests that address infectious disease control needs, especially for the developing countries, should follow "ASSURED" criteria: Affordable, Sensitive, Specific, User-friendly, Rapid and robust, Equipment-free and Deliverable to end users [113].

Early diagnosis of these infectious diseases is critical but challenging because the biomarkers and whole pathogens are present at low concentrations, demanding bioanalytical techniques that can deliver high sensitivity with ensured specificity. Commercially, the biosensor market is expected to grow steadily with a compound annual growth rate of 8.84% between 2017 and 2022 and reach 27 billion USD by 2022 [130]. Owing to the high development of plasmonic nanostructures that allow high sensitivities, even up to single molecules, SERS has gained attention as an optical analytical method for early disease. Moreover, SERS offers ideal properties as readout modules for POC technologies, such as high sensitivity, label-free, and real-time monitoring. The integration of plasmonics and microfluidic technologies can potentially serve as an ideal platform for the development of POCs systems for the diagnosis of infectious diseases toward inexpensive, robust, and portable solutions [131].

Several reports have been focused on label-free SERS strategies for the identification of pathogens by differentiation of their proteins and lipids compositions on the membrane (Fig. 2.7). Pathogens are larger, e.g., micron size bacteria and around 100 nm size virus, as well as molecules, < 10 nm, conventionally analyzable by SERS. However, because pathogens possess unique surface protein and lipids profiles on their outer layer, they can generate distinctive Raman signals when their individual surface molecules are adequately in contact with noble metal nanostructures. Several reports highlighted that SERS can provide a powerful label-free strategy to quickly identify newly emerging and potentially fatal pathogens.

For example, a label-free influenza virus detection method using SERS has been developed [132]. Influenza viruses are enveloped viruses with seven- or eight-segmented genomes and are common human pathogens. The SERS fingerprints from the surface of a noninfluenza virus, two different influenza viruses, and a generically shuffled influenza virus were acquired. Importantly, nonreplicating pseudotyped viruses were constructed to ensure the safety for workers. The results demonstrated that influenza virus components produced enhanced Raman peaks, on Au NPs, that are easily distinguishable from those of a noninfluenza virus component, vesicular stomatitis virus G protein (V_{VSVG}). Moreover, the SERS fingerprint of virus with the surface components of a newly emerging influenza strain, A/California/04/2009 (H_1N_1) ($V_{CAL\ HA\ +\ NA}$) was different from those of viruses with components of the conventional laboratory-adapted influenza strain, A/WSN/33 (H_1N_1) ($V_{WSN\ HA\ +\ NA}$). Interestingly, the virus simultaneously displaying surface components of both influenza strains, a model mutant with genome reassortment ($V_{WSN\ HA\ +\ CAL\ NA}$), also produced a SERS signal pattern that was clearly distinguishable from those of each strain. Specifically, the SERS fingerprint from V_{VSVG} showed two peaks (1231 and 1587 cm^{-1}) matched with those from $V_{CAL\ HA\ +\ NA}$ and this $V_{CAL\ HA\ +\ NA}$ exhibited distinct two peaks (923 and 1356 cm^{-1}). In the case of $V_{WSN\ HA\ +\ NA}$, two peaks (1231 and 1587 cm^{-1}) that were common to all of the viruses, and two unique peaks were obtained (740 and 1107 cm^{-1}). This method may be used to detect other new viruses, such as the MERS viruses.

Using the same basis explained above, Kaminska et al. [133] reported the detection and identification of the most common meningitis pathogens *Neisseria meningitidis, Streptococcus pneumoniae,* and *Haemophilus influenza*e in cerebrospinal fluid (CSF) by acquiring their SERS spectra, which showed different characteristic peaks due to their unique cell membrane biomarkers such as lipids and polysaccharides (Fig. 2.8). The design of the SERS substrate involving Au−Ag-coated, nuclepore track-etched polycarbonate membranes allowed simultaneous filtration of CSF and immobilization of bacteria to enhance and detect their SERS spectra. The SERS fingerprint for bacteria was the result of the contribution of the molecular vibrations of nucleic acid, amides, n-alkanes, flavins, and other constituent compounds of bacterial cell. The multivariate PCA statistical method was applied (1) to extract the

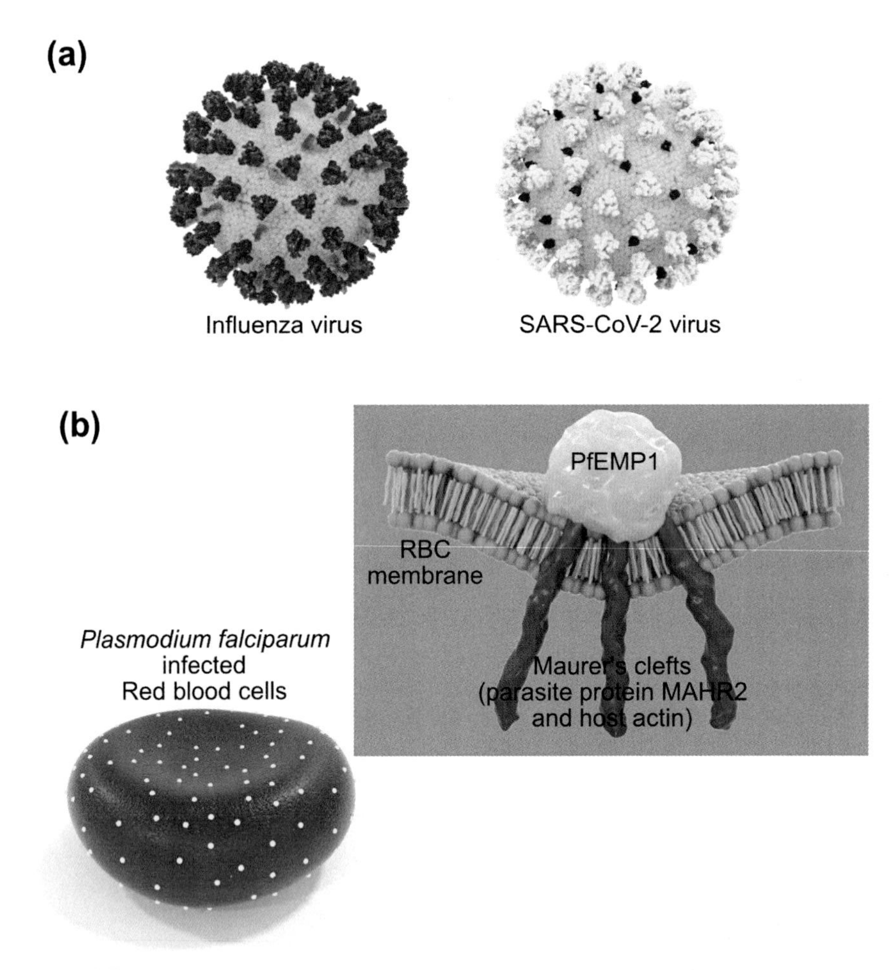

FIGURE 2.7

(A) Schematic illustration of the morphology of influenza H1N1 (red) and SARS-CoV-2 (yellow) viruses. Despite both being enveloped viruses, they possess unique surface protein and lipids profiles on their outer layer, allowing direct identification using their SERS fingerprint. (B) *Plasmodium falciparum*, the most virulent of the human malaria parasites, spends part of its life cycle inside the red blood cells (RBCs) of its host. The invasion of *P. falciparum* in the RBCs causes continuous changes in the RBC membrane, which can be correlated with different port invasion times. *P. falciparum* establishes an exomembrane system, known as Maurer's clefts in the host RBC cytoplasm, providing a pathway for trafficking proteins, including virulence determinants, to the RBC surface. The inset shows a simple schematic illustration of the RBC membrane with Maurer's clefts, which plays a role in trafficking of *P. falciparum* membrane protein 1 (PtEMP1) and other adhesins to the RBC surface.

Credit: http://miguelspuch.wix.com/mspuchdesign.

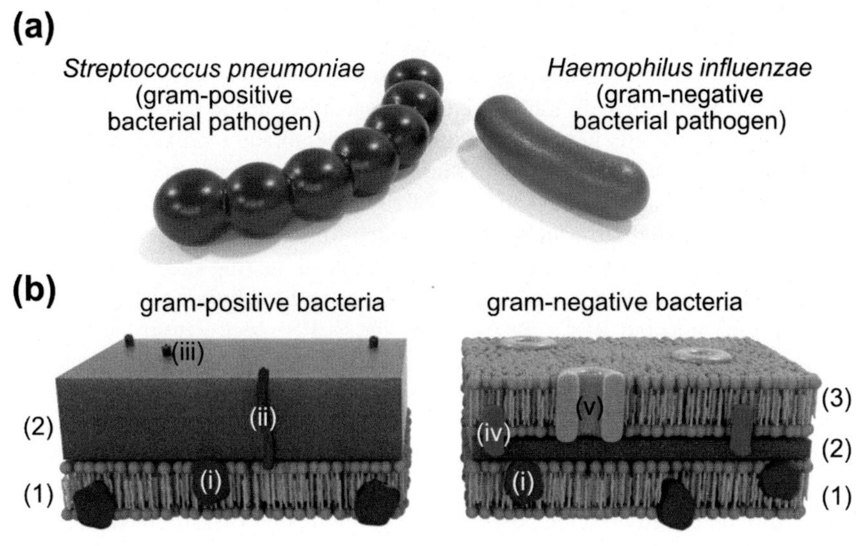

FIGURE 2.8

(A) Schematic illustration of the morphology of *Streptococcus pneumoniae*, a gram-positive pathogenic bacterium, and *Haemophilus influenza*, a gram-negative pathogenic bacterium. (B) There are clear structural differences between gram-positive and gram-negative bacteria: (1) both gram-positive and gram-negative bacteria have cytoplasmic membrane composed mainly of phospholipids and membrane proteins (i); (2) gram-positive bacteria have a smooth and single-layered cell wall of 20–80 nm thickness composed of multiple layers of peptidoglycan that forms a rigid and thick structure, while gram-negative bacteria have a wavy and double-layered cell wall of 8–10 nm, which is made up of an outer membrane (3) and several layers of peptidoglycan (2). The cell wall of gram-positive bacteria also has teichoic acids—(ii) lipoteichoic acid and (iii) teichoic wall acid—and phosphate. In the case of gram-negative bacteria, the peptidoglycan stays intact to lipoproteins (iv) of the outer membrane (3) that is located in the fluid-like periplasm between the cytoplasmic (1) and the outer membrane (3). The periplasm is contained with proteins and degrading enzymes, which assist in transporting molecules. In addition, the cell walls of the gram-negative bacteria lacks the teichoic acid, but it has porins (v) that make permeable the outer membrane to nutrition, water, food, iron, and among others.

Credit: http://miguelspuch.wix.com/mspuchdesign.

biochemical information from the recorded bacterial spectra, (2) to perform the statistical classification of analyzed pathogens, and, finally, (3) to identify the spectrum of an unknown sample by comparison to the library of known bacterial spectra. This approach allowed to evaluate the CSF neopterin level in patients with diagnosed meningococcal meningitis, which confirmed that bacterial meningitis caused by *N. meningitidis, H. influenza*, and *S. pneumoniae* was associated with elevated CSF fluid neopterin levels compared with control CSF samples. The neopterin concentration can be used to predict meningitis, but cannot be applied to qualify the species of bacteria inducing the meningitis infection.

In the case of malaria, Chen et al. [134] reported the applicability of free-label SERS in combination with multivariate PCA analysis in the detection of malaria infected red blood cells in blood due to the cell membrane modification of normal red blood cells (RBCs) during malaria infection. They used a reproducible and sensitive SERS substrate involving an array of silver nanorods fabricated by an oblique angle deposition method, which offered a batch to batch variation of $<15\%$. They demonstrated that the SERS fingerprints of RBCs and the ring stage of *Plasmodium falciparum*—infected RBCs (iRBCs) could be used to differentiate different stages of *P. falciparum* infected RBCs. In particular the SERS spectra from the ring stage iRBCs (had a characteristic Raman peak at 1599 cm^{-1}) were different from those of RBCs and late-stage iRBCs (with a characteristic peak at 723 cm^{-1}) which was consistent with ongoing modification of the iRBC membrane, especially in the membrane proteins expressed during the *Plasmodium* life cycle in RBC host (Fig. 2.8). Since the ring stage iRBCs of *P. falciparum* are found in blood circulation, such a difference provides the possibility to develop point-of-care for rapid malaria detection.

Hunter et al. [135] developed an optofluidic label-free SERS platform based on a microfluidic driven hollow-core photonic crystal fiber in combination with Ag NPs. This SERS in hollow crystal fiber provided bulk 400-fold enhancement of Raman signal compared with a conventional SERS substrate. By confining both light and cells within this fiber, spectral events generated by the flowing cells facilitate a novel method of cell counting to simultaneously quantify and qualify infections. The recyclability of the substrate was possible by regenerating repeatedly the fiber with appropriate flushing fluid, improving the applicability of the POC system in low-income countries. Moreover, the portable platform in combination with partial least squares discriminant analysis (PLS-DA) and genetic support vector machines allowed the online multiplexed detection and identification of planktonic bacteria, such as *Staphylococcus aureus, Escherichia coli,* and *Pseudomonas aeruginosa*, in fetal bovine serum to levels as low as 4 CFU/mL in 15 min. This compares favorably to more established methods such as qPCR and biosensing techniques. Further development of this device has promising potential as a rapid POC system for infection management in the clinic.

There are infectious diseases that modify the expression of some biomolecules, such as proteins, in physiological fluids and as consequence the composition of this fluids varies and these modification in the level of such biomolecules can be identified by the changes in the SERS fingerprint in comparison with the "healthy fluid." In this line, human tear fluid provides biological information about the eyes, and ocular diseases result in distinctive changes in biochemical composite fluids [136]. Thus, three indistinguishable infectious eye diseases were analyzed, adenoviral conjunctivitis (caused by adenovirus), herpes simplex keratitis (caused by herpes simplex virus), and herpes zoster ophthalmicus (caused by varicella-zoster virus), through the tear fluids using label-free SERS paper strip [137]. The healthy tear fluids produced a SERS fingerprints that was clearly distinguishable from those of diseased tear fluids. In order to discriminate between each type of infectious eye diseases, PCA analysis in combination with multisupport vector machine (SVM) classification algorithm with a linear kernel using the training data was applied to the SERS spectral data, showing a difference between normal and infectious eyes, with high sensitivity (100%) and high specificity (100%) for classifying normal and infectious eyes. SERS spectroscopy and PCA-SVM analysis not only distinguished between normal and infectious eyes but could also identify infection types based on the cellular biochemical and structural changes induced by viral and bacterial infections and allergic reaction.

Importantly, label-free SERS techniques facilitate inherent multiplexing due to intrinsic differences in SERS spectra from pathogenic microorganisms; however, this must be coupled with a multivariate analysis to decompose the spectrum if multiple pathogens are present. Thus, label-free SERS methodology in combination with machine learning or deep learning algorithms can be integrated in a POC system. This was demonstrated in the detection of infectious diseases using a POC system. A POC system named VIRRION was utilized to simultaneously enriched and optically detect virus strains in just a few minutes without using any labels. The platform was constructed with aligned nitrogen-doped carbon nanotubes arrays, decorated with AuNPs to separate, enrich, and detect fluidic human samples by direct SERS identification [138]. The setup involved a portable platform that captured viruses based on their size, coupled to a Raman spectrometer and a machine learning algorithm and database, resulting in a viral capture with a 70-fold enrichment enhancement and successful virus identification with 90% accuracy in real time directly from clinical samples. Interesting, after viral capture and detection on a chip, viruses remained viable and got purified in a microdevice that permitted subsequent in-depth characterizations by various conventional methods. The portable platform was validated using different subtypes of avian influenza A viruses and human samples with respiratory infections. VIRRION platform successfully enriched rhinovirus, influenza virus, and parainfluenza viruses, and maintained the stoichiometric viral proportions when the samples contained more than one type of virus, thus emulating coinfection. After enrichment using the device, the abundance of viral-specific reads was sequenced and a significant increase from 4.1% to 31.8% for parainfluenza and from 0.08% to 0.44% influenza virus was observed. In addition, the SERS database was expanded by collecting more spectra from different types of viruses. This POC system constituted an innovative system that could be used to quickly track and monitor viral outbreaks in real time.

Future challenges and perspectives

Despite the promising proof-of-concept works and preclinical demonstrations, all these approaches would be inconclusive until tests are performed in larger number of patient samples. One of the main reasons behind the gap between fundamental research and clinical applications of SERS is the batch to batch reproducibility of SERS substrates. This issue has become one of the main challenges to solve and the central criticisms of SERS from competing technologies such as fluorescence. Nevertheless, despite consecutive batches not being reproducible, a standardization and calibration step can be easily implemented within SERS-based sensors in the same way is for more established sensors such as electrochemical or fluorescent kits [139]. From the instrumentation point of view, most the reported works use Raman spectrometers which are not often found in clinical facilities. As a consequence, the translation of the technology encounters higher resistance from hospital administrators and medical doctors. In order to alleviate this, SERS researchers should act simultaneously in several fronts. For example, reproducible and simple SERS substrates with higher sensitivity will allow the use of portable or handheld Raman instruments. At the same time, Raman instrumentation developers should work in the miniaturization of their instruments, while compromising the sensitivity the minimum possible. Finally, a joint effort of the SERS community should be to spread the benefits of SERS and educate the clinical community on its advantages. Last but not least, the label-free SERS arena should follow closely advances in machine learning and artificial intelligence, as it will be the only way in which a high-throughput reading and interpretation will be possible from such rich data source as Raman spectroscopy.

References

[1] M. Fleischmann, P.J. Hendra, A.J. McQuillan, Raman spectra of pyridine adsorbed at a silver electrode, Chem. Phys. Lett. 26 (1974) 163–166, https://doi.org/10.1016/0009-2614(74)85388-1.

[2] D. Cialla-May, X.-S. Zheng, K. Weber, J. Popp, Recent progress in surface-enhanced Raman spectroscopy for biological and biomedical applications: from cells to clinics, Chem. Soc. Rev. 46 (2017) 3945–3961, https://doi.org/10.1039/C7CS00172J.

[3] S. Abalde-Cela, S. Carregal-Romero, J.P. Coelho, A. Guerrero-Martínez, Recent progress on colloidal metal nanoparticles as signal enhancers in nanosensing, Adv. Colloid Interface Sci. 233 (2016) 255–270, https://doi.org/10.1016/j.cis.2015.05.002.

[4] S. Abalde-Cela, P. Aldeanueva-Potel, C. Mateo-Mateo, L. Rodríguez-Lorenzo, R.A. Álvarez-Puebla, L.M. Liz-Marzán, Surface-enhanced Raman scattering biomedical applications of plasmonic colloidal particles, J. Royal Soc. Interface 7 (4) (2010), https://doi.org/10.1098/rsif.2010.0125.focus.

[5] L. Guerrini, N. Pazos-Perez, E. Garcia-Rico, R. Alvarez-Puebla, Cancer characterization and diagnosis with SERS-encoded particles, Cancer Nanotechnol. 8 (2017) 5, https://doi.org/10.1186/s12645-017-0031-3.

[6] L.A. Lane, X. Qian, S. Nie, SERS nanoparticles in medicine: from label-free detection to spectroscopic tagging, Chem. Rev. 115 (2015) 10489–10529, https://doi.org/10.1021/acs.chemrev.5b00265.

[7] X.-S. Zheng, I.J. Jahn, K. Weber, D. Cialla-May, J. Popp, Label-free SERS in biological and biomedical applications: recent progress, current challenges and opportunities, Spectrochim. Acta A Mol. Biomol. Spectrosc. 197 (2018) 56–77, https://doi.org/10.1016/j.saa.2018.01.063.

[8] A. Barucci, C. D'Andrea, E. Farnesi, M. Banchelli, C. Amicucci, M. de Angelis, et al., Label-free SERS detection of proteins based on machine learning classification of chemo-structural determinants, Analyst 146 (2021) 674–682, https://doi.org/10.1039/D0AN02137G.

[9] E. Farnesi, A. Barucci, C. D'Andrea, M. Banchelli, C. Amicucci, M. de Angelis, et al., Label-free SERS/machine learning procedures for protein classification, in: D. Contini, Y. Hoshi, T.D. O'Sullivan (Eds.), Diffus. Opt. Spectrosc. Imaging VIII, vol. 11920, SPIE, 2021, pp. 225–226, https://doi.org/10.1117/12.2615448.

[10] T. Rojalin, D. Antonio, A. Kulkarni, R.P. Carney, Machine learning-assisted sampling of SERS substrates improves data collection efficiency, Appl. Spectrosc. (2021), https://doi.org/10.1177/00037028211034543.

[11] Z.S. Ballard, H.-A. Joung, A. Goncharov, J. Liang, K. Nugroho, D. Di Carlo, et al., Deep learning-enabled point-of-care sensing using multiplexed paper-based sensors, NPJ Digit Med. 3 (2020) 66, https://doi.org/10.1038/s41746-020-0274-y.

[12] World Health Organization, Cancer, n.d.

[13] A.S.B. Primeau, Cancer Recurrence Statistics, 2018.

[14] D.R. Welch, D.R. Hurst, Defining the hallmarks of metastasis, Cancer Res. 79 (2019) 3011–3027, https://doi.org/10.1158/0008-5472.CAN-19-0458.

[15] C.L. Chaffer, R.A. Weinberg, A perspective on cancer cell metastasis, Science 331 (2011) 1559.

[16] C. Lopes, P. Piairo, A. Chícharo, S. Abalde-Cela, L. Pires, L. Muinelo-Romay, L. Costa, L. Diéguez, HER2 expression in circulating tumour cells isolated from metastatic breast cancer patients using a size-based microfluidic device, Cancers (2021), https://doi.org/10.3390/cancers13174446.

[17] S. Abalde-Cela, P. Piairo, L. Diéguez, The significance of circulating tumour cells in the clinic, Acta Cytol. 63 (2019) 466–478, https://doi.org/10.1159/000495417.

[18] P. Bheda, R. Schneider, Epigenetics reloaded: the single-cell revolution, Trends Cell Biol. 24 (2014) 712–723, https://doi.org/10.1016/j.tcb.2014.08.010.

[19] G.L. D'Adamo, J.T. Widdop, E.M. Giles, The future is now? Clinical and translational aspects of "Omics" technologies, Immunol. Cell Biol. 99 (2021) 168–176, https://doi.org/10.1111/imcb.12404.

[20] S. Abalde-Cela, L. Wu, A. Teixeira, K. Oliveira, L. Diéguez, Multiplexing liquid biopsy with surface-enhanced Raman scattering spectroscopy, Adv. Opt. Mater. 9 (2021) 2001171, https://doi.org/10.1002/adom.202001171.

[21] E.-A. Raiber, R. Hardisty, P. van Delft, S. Balasubramanian, Mapping and elucidating the function of modified bases in DNA, Nat. Rev. Chem. 1 (2017) 69, https://doi.org/10.1038/s41570-017-0069.

[22] R. Esteve-Puig, A. Bueno-Costa, M. Esteller, Writers, readers and erasers of RNA modifications in cancer, Cancer Lett. 474 (2020) 127–137, https://doi.org/10.1016/j.canlet.2020.01.021.

[23] A. Barhoumi, N.J. Halas, Detecting chemically modified DNA bases using surface-enhanced Raman spectroscopy, J. Phys. Chem. Lett. 2 (2011) 3118–3123, https://doi.org/10.1021/jz201423b.

[24] S.R. Panikkanvalappil, M.A. Mackey, M.A. El-Sayed, Probing the unique dehydration-induced structural modifications in cancer cell dna using surface enhanced Raman spectroscopy, J. Am. Chem. Soc. 135 (2013) 4815.

[25] L.-J. Xu, Z.-C. Lei, J. Li, C. Zong, C.J. Yang, B. Ren, Label-free surface-enhanced Raman spectroscopy detection of DNA with single-base sensitivity, J. Am. Chem. Soc. 137 (2015) 5149–5154, https://doi.org/10.1021/jacs.5b01426.

[26] F. Gao, J. Lei, H. Ju, Label-free surface-enhanced Raman spectroscopy for sensitive DNA detection by DNA-mediated silver nanoparticle growth, Anal. Chem. 85 (2013) 11788–11793, https://doi.org/10.1021/ac4032109.

[27] R.P. Johnson, J.A. Richardson, T. Brown, P.N. Bartlett, A label-free, electrochemical SERS-based assay for detection of DNA hybridization and discrimination of mutations, J. Am. Chem. Soc. 134 (2012) 14099–14107, https://doi.org/10.1021/ja304663t.

[28] J.D. Driskell, R.A. Tripp, Label-free SERS detection of microRNA based on affinity for an unmodified silver nanorod array substrate, Chem. Commun. 46 (2010) 3298–3300, https://doi.org/10.1039/C002059A.

[29] L. Wu, A. Garrido-Maestu, J.R.L. Guerreiro, S. Carvalho, S. Abalde-Cela, M. Prado, et al., Amplification-free SERS analysis of DNA mutation in cancer cells with single-base sensitivity, Nanoscale 11 (2019) 7781–7789, https://doi.org/10.1039/C9NR00501C.

[30] L. Wu, A. Teixeira, A. Garrido-Maestu, L. Muinelo-Romay, L. Lima, L.L. Santos, et al., Profiling DNA mutation patterns by SERS fingerprinting for supervised cancer classification, Biosens. Bioelectron. 165 (2020) 112392, https://doi.org/10.1016/j.bios.2020.112392.

[31] Z. Fan, R. Kanchanapally, P.C. Ray, Hybrid graphene oxide based ultrasensitive SERS probe for label-free biosensing, J. Phys. Chem. Lett. 4 (2013) 3813–3818, https://doi.org/10.1021/jz4020597.

[32] H.T. Ngo, H.-N. Wang, A.M. Fales, T. Vo-Dinh, Label-free DNA biosensor based on SERS molecular sentinel on nanowave chip, Anal. Chem. 85 (2013) 6378–6383, https://doi.org/10.1021/ac400763c.

[33] G. Qi, D. Wang, C. Li, K. Ma, Y. Zhang, Y. Jin, Plasmonic SERS Au nanosunflowers for sensitive and label-free diagnosis of DNA base damage in stimulus-induced cell apoptosis, Anal. Chem. 92 (2020) 11755–11762, https://doi.org/10.1021/acs.analchem.0c01799.

[34] L. Bi, Y. Rao, Q. Tao, J. Dong, T. Su, F. Liu, et al., Fabrication of large-scale gold nanoplate films as highly active SERS substrates for label-free DNA detection, Biosens. Bioelectron. 43 (2013) 193–199, https://doi.org/10.1016/j.bios.2012.11.029.

[35] A. Barhoumi, N.J. Halas, Label-free detection of DNA hybridization using surface enhanced Raman spectroscopy, J. Am. Chem. Soc. 132 (2010) 12792–12793, https://doi.org/10.1021/ja105678z.

[36] C.E. Talley, J.B. Jackson, C. Oubre, N.K. Grady, C.W. Hollars, S.M. Lane, et al., Surface-enhanced Raman scattering from individual Au nanoparticles and nanoparticle dimer substrates, Nano Lett. 5 (2005) 1569.

[37] J.B. Jackson, N.J. Halas, Surface-enhanced Raman scattering on tunable plasmonic nanoparticle substrates, Proc. Natl. Acad. Sci. 101 (2004) 17930–17935, https://doi.org/10.1073/pnas.0408319102.

[38] K. Kant, S. Abalde-Cela, Surface-enhanced Raman scattering spectroscopy and microfluidics: towards ultrasensitive label-free sensing, Biosensors 8 (2018) 62, https://doi.org/10.3390/bios8030062.

[39] Y. Chen, G. Chen, S. Feng, J. Pan, X. Zheng, Y. Su, et al., Label-free serum ribonucleic acid analysis for colorectal cancer detection by surface-enhanced Raman spectroscopy and multivariate analysis, J. Biomed. Opt. 17 (2012) 067003, https://doi.org/10.1117/1.jbo.17.6.067003.

[40] L. Guerrini, Ž. Krpetić, D. van Lierop, R.A. Alvarez-Puebla, D. Graham, Direct surface-enhanced Raman scattering analysis of DNA duplexes, Angew Chem. 127 (2015) 1160.

[41] J. Morla-Folch, H. Xie, P. Gisbert-Quilis, S.G. Pedro, N. Pazos-Perez, R.A. Alvarez-Puebla, et al., Ultrasensitive direct quantification of nucleobase modifications in DNA by surface-enhanced Raman scattering: the case of cytosine, Angew Chem. Int. Ed. 54 (2015) 13650–13654, https://doi.org/10.1002/anie.201507682.

[42] L. Li, S.F. Lim, A. Puretzky, R. Riehn, H.D. Hallen, DNA methylation detection using resonance and nanobowtie-antenna-enhanced Raman spectroscopy, Biophys. J. 114 (2018) 2498–2506, https://doi.org/10.1016/j.bpj.2018.04.021.

[43] X. Li, T. Yang, C.S. Li, Y. Song, D. Wang, L. Jin, et al., Polymerase chain reaction - surface-enhanced Raman spectroscopy (PCR-SERS) method for gene methylation level detection in plasma, Theranostics 10 (2020) 898–909, https://doi.org/10.7150/thno.30204.

[44] X. Luo, Y. Xing, D.D. Galvan, E. Zheng, P. Wu, C. Cai, et al., Plasmonic gold nanohole array for surface-enhanced Raman scattering detection of DNA methylation, ACS Sensors 4 (2019) 1534–1542, https://doi.org/10.1021/acssensors.9b00008.

[45] J.-A. Huang, M.Z. Mousavi, Y. Zhao, A. Hubarevich, F. Omeis, G. Giovannini, et al., SERS discrimination of single DNA bases in single oligonucleotides by electro-plasmonic trapping, Nat. Commun. 10 (2019) 5321, https://doi.org/10.1038/s41467-019-13242-x.

[46] J. Wang, D. Lin, J. Lin, Y. Yu, Z. Huang, Y. Chen, et al., Label-free detection of serum proteins using surface-enhanced Raman spectroscopy for colorectal cancer screening, J. Biomed. Opt. 19 (2014) 087003, https://doi.org/10.1117/1.jbo.19.8.087003.

[47] J. Lin, R. Chen, S. Feng, J. Pan, B. Li, G. Chen, et al., Surface-enhanced Raman scattering spectroscopy for potential noninvasive nasopharyngeal cancer detection, J. Raman Spectrosc. 43 (2012) 497.

[48] J. Lin, R. Chen, S. Feng, J. Pan, Y. Li, G. Chen, et al., A novel blood plasma analysis technique combining membrane electrophoresis with silver nanoparticle-based SERS spectroscopy for potential applications in noninvasive cancer detection, Nanomedicine 7 (2011) 655–663, https://doi.org/10.1016/J.NANO.2011.01.012.

[49] J. Wang, S. Feng, J. Lin, Y. Zeng, L. Li, Z. Huang, et al., Serum albumin and globulin analysis for hepatocellular carcinoma detection avoiding false-negative results from alpha-fetoprotein test negative subjects, Appl. Phys. Lett. 103 (2013) 204106, https://doi.org/10.1063/1.4830047.

[50] W. Wang, S. Feng, I. Tai, H. Zeng, Label-free surface-enhanced Raman spectroscopy for detection of colorectal cancer using blood plasma, Biomed. Opt. Exp. (2015), https://doi.org/10.1364/boe.6.003494.

[51] V. Shalabaeva, L. Lovato, R. La Rocca, G.C. Messina, M. Dipalo, E. Miele, et al., Time resolved and label free monitoring of extracellular metabolites by surface enhanced Raman spectroscopy, PLoS One 12 (2017) e0175581, https://doi.org/10.1371/journal.pone.0175581.

[52] C.M. Perlaki, Q. Liu, M. Lim, Raman spectroscopy based techniques in tissue engineering—an overview, Appl. Spectrosc. Rev 49 (2014) 513–532, https://doi.org/10.1080/05704928.2013.863205.

[53] S. Wachsmann-Hogiu, T. Weeks, T. Huser, Chemical analysis in vivo and in vitro by Raman spectroscopy—from single cells to humans, Curr. Opin. Biotechnol. 20 (2009) 63–73, https://doi.org/10.1016/j.copbio.2009.02.006.
[54] W.C. Zúñiga, V. Jones, S.M. Anderson, A. Echevarria, N.L. Miller, C. Stashko, et al., Raman spectroscopy for rapid evaluation of surgical margins during breast cancer lumpectomy, Sci. Rep. 9 (2019) 14639, https://doi.org/10.1038/s41598-019-51112-0.
[55] D. DePaoli, É. Lemoine, K. Ember, M. Parent, M. Prud'homme, L. Cantin, et al., Rise of Raman spectroscopy in neurosurgery: a review, J. Biomed. Opt. 25 (2020) 1–36, https://doi.org/10.1117/1.JBO.25.5.050901.
[56] H. Huang, H. Shi, S. Feng, J. Lin, W. Chen, Z. Huang, et al., Silver nanoparticle based surface enhanced Raman scattering spectroscopy of diabetic and normal rat pancreatic tissue under near-infrared laser excitation, Laser Phys. Lett. 10 (2013), https://doi.org/10.1088/1612-2011/10/4/045603.
[57] Z. Huang, Z. Li, R. Chen, G. Chen, D. Lin, G. Xi, et al., The application of Silver nanoparticle based SERS in diagnosing thyroid tissue, J. Phys. Conf. Ser. 277 (2011) 12014, https://doi.org/10.1088/1742-6596/277/1/012014.
[58] S. Feng, J. Lin, Z. Huang, G. Chen, W. Chen, Y. Wang, et al., Esophageal cancer detection based on tissue surface-enhanced Raman spectroscopy and multivariate analysis, Appl. Phys. Lett. 102 (2013) 43702, https://doi.org/10.1063/1.4789996.
[59] S. Yamazoe, M. Naya, M. Shiota, T. Morikawa, A. Kubo, T. Tani, et al., Large-area surface-enhanced Raman spectroscopy imaging of brain ischemia by gold nanoparticles grown on random nanoarrays of transparent boehmite, ACS Nano 8 (2014) 5622–5632, https://doi.org/10.1021/nn4065692.
[60] C.M. Girish, S. Iyer, K. Thankappan, V.V.D. Rani, G.S. Gowd, D. Menon, et al., Rapid detection of oral cancer using Ag–TiO_2 nanostructured surface-enhanced Raman spectroscopic substrates, J. Mater. Chem. B 2 (2014) 989–998, https://doi.org/10.1039/C3TB21398F.
[61] S.M. Asiala, N.C. Shand, K. Faulds, D. Graham, Surface-enhanced, spatially offset Raman spectroscopy (SESORS) in tissue analogues, ACS Appl. Mater. Interfaces 9 (2017) 25488–25494, https://doi.org/10.1021/acsami.7b09197.
[62] B. Gardner, P. Matousek, N. Stone, Direct monitoring of light mediated hyperthermia induced within mammalian tissues using surface enhanced spatially offset Raman spectroscopy (T-SESORS), Analyst 144 (2019) 3552–3555, https://doi.org/10.1039/C8AN02466A.
[63] R.A. Odion, P. Strobbia, B.M. Crawford, T. Vo-Dinh, Inverse surface-enhanced spatially offset Raman spectroscopy (SESORS) through a monkey skull, J. Raman Spectrosc. 49 (2018) 1452–1460, https://doi.org/10.1002/jrs.5402.
[64] A. Teixeira, J. Hernández-Rodríguez, L. Wu, K. Oliveira, K. Kant, P. Piairo, et al., Microfluidics-driven fabrication of a low cost and ultrasensitive SERS-based paper biosensor, Appl. Sci. 9 (2019) 1387, https://doi.org/10.3390/app9071387.
[65] P. Reokrungruang, I. Chatnuntawech, T. Dharakul, S. Bamrungsap, A simple paper-based surface enhanced Raman scattering (SERS) platform and magnetic separation for cancer screening, Sensors Actuators B Chem. 285 (2019) 462–469, https://doi.org/10.1016/j.snb.2019.01.090.
[66] K.V. Kong, W.K. Leong, Z. Lam, T. Gong, D. Goh, W.K.O. Lau, et al., A rapid and label-free SERS detection method for biomarkers in clinical biofluids, Small 10 (2014) 5030–5034, https://doi.org/10.1002/smll.201401713.

[67] T. Moore, A. Moody, T. Payne, G. Sarabia, A. Daniel, B. Sharma, In vitro and in vivo SERS biosensing for disease diagnosis, Biosensors 8 (2018) 46, https://doi.org/10.3390/bios8020046.

[68] WHO, Cancer, 2018. https://www.who.int/News-Room/Fact-Sheets/Detail/the-Top-10-Causes-of-Death. https://www.who.int/health-topics/cancer#tab=tab_1. (Accessed 6 April 2020).

[69] A. Kumar, A. Singh, Ekavali, A review on Alzheimer's disease pathophysiology and its management: an update, Pharmacol. Rep. 67 (2015) 195–203, https://doi.org/10.1016/j.pharep.2014.09.004.

[70] L.M. Bekris, C.-E. Yu, T.D. Bird, D.W. Tsuang, Review article: genetics of Alzheimer disease, J. Geriatr. Psychiatry Neurol. 23 (2010) 213–227, https://doi.org/10.1177/0891988710383571.

[71] C.J.L. Murray, T. Vos, R. Lozano, M. Naghavi, A.D. Flaxman, C. Michaud, et al., Disability-adjusted life years (DALYs) for 291 diseases and injuries in 21 regions, 1990-2010: a systematic analysis for the Global Burden of Disease Study 2010, Lancet 380 (2012) 2197–2223, https://doi.org/10.1016/S0140-6736(12)61689-4.

[72] J. Alonso, M.C. Angermeyer, S. Bernert, R. Bruffaerts, T.S. Brugha, H. Bryson, et al., Prevalence of mental disorders in Europe: results from the European study of the epidemiology of mental disorders (ESEMeD) project, Acta Psychiatr. Scand. 109 (2004) 21–27, https://doi.org/10.1111/j.1600-0047.2004.00327.x.

[73] H.J. Caruncho, T. Rivera-Baltanas, R. Romay-Tallon, L.E. Kalynchuk, J.M. Olivares, Patterns of membrane protein clustering in peripheral lymphocytes as predictors of therapeutic outcomes in major depressive disorder, Front. Pharmacol. 10 (2019) 190, https://doi.org/10.3389/fphar.2019.00190.

[74] R. Aroca, Surface-enhanced Vibrational Spectroscopy, John Wiley & Sons, Ltd., 2006.

[75] N.D. Israelsen, C. Hanson, E. Vargis, Nanoparticle properties and synthesis effects on surface-enhanced Raman scattering enhancement factor: an introduction, Sci. World J. 2015 (2015) 124582, https://doi.org/10.1155/2015/124582.

[76] B. Pergolese, M. Muniz-Miranda, A. Bigotto, Study of the adsorption of 1,2,3-triazole on silver and gold colloidal nanoparticles by means of surface enhanced Raman scattering, J. Phys. Chem. B 108 (2004) 5698–5702, https://doi.org/10.1021/jp0377228.

[77] E.M. Sundin, J.D. Ciubuc, K.E. Bennet, K. Ochoa, F.S. Manciu, Comparative computational and experimental detection of adenosine using ultrasensitive surface-enhanced Raman spectroscopy, Sensors 18 (2018) 2696, https://doi.org/10.3390/s18082696.

[78] F. Manciu, M. Manciu, J. Ciubuc, E. Sundin, K. Ochoa, M. Eastman, et al., Simultaneous detection of dopamine and serotonin—a comparative experimental and theoretical study of neurotransmitter interactions, Biosensors 9 (2018) 3, https://doi.org/10.3390/bios9010003.

[79] P. Song, X. Guo, Y. Pan, Y. Wen, Z. Zhang, H. Yang, SERS and in situ SERS spectroelectrochemical investigations of serotonin monolayers at a silver electrode, J. Electroanal. Chem. 688 (2013) 384–391, https://doi.org/10.1016/j.jelechem.2012.09.008.

[80] J. Ciubuc, K. Bennet, C. Qiu, M. Alonzo, W. Durrer, F. Manciu, Raman computational and experimental studies of dopamine detection, Biosensors 7 (2017) 43, https://doi.org/10.3390/bios7040043.

[81] G.D. Fleming, R. Koch, J.M. Perez, J.L. Cabrera, Raman and SERS study of N-acetyl-5-methoxytryptamine, melatonin - the influence of the different molecular fragments

on the SERS effect Dedicated to the memory of Dennis P, Strommen. Vib Spectrosc. 80 (2015) 70–78, https://doi.org/10.1016/j.vibspec.2015.08.002.

[82] M.R. Bailey, R.S. Martin, Z.D. Schultz, Role of surface adsorption in the surface-enhanced Raman scattering and electrochemical detection of neurotransmitters, J. Phys. Chem. C 120 (2016) 20624–20633, https://doi.org/10.1021/acs.jpcc.6b01196.

[83] A.S. Moody, B. Sharma, Multi-metal, multi-wavelength surface-enhanced Raman spectroscopy detection of neurotransmitters, ACS Chem. Neurosci. 9 (2018) 1380–1387, https://doi.org/10.1021/acschemneuro.8b00020.

[84] A. Choudhury, T. Sahu, P.L. Ramanujam, A.K. Banerjee, I. Chakraborty, R.A. Kumar, et al., Neurochemicals, behaviours and psychiatric perspectives of neurological diseases, Neuropsychiatry 08 (2018) 395–424, https://doi.org/10.4172/neuropsychiatry.1000361.

[85] V.S. Tiwari, A. Khetani, A.M.T. Monfared, B. Smith, H. Anis, V.L. Trudeau, Detection of amino acid neurotransmitters by surface enhanced Raman scattering and hollow core photonic crystal fiber, in: S. Achilefu, R. Raghavachari (Eds.), Reporters, Markers, Dye. Nanoparticles, Mol. Probes Biomed. Appl. IV, vol. 8233, SPIE, 2012, p. 82330Q, https://doi.org/10.1117/12.907754.

[86] M. Siek, A. Kaminska, A. Kelm, T. Rolinski, R. Holyst, M. Opallo, et al., Electrodeposition for preparation of efficient surface-enhanced Raman scattering-active silver nanoparticle substrates for neurotransmitter detection, Electrochim. Acta 89 (2013) 284–291, https://doi.org/10.1016/j.electacta.2012.11.037.

[87] W.A. El-Said, J.W. Choi, In-situ detection of neurotransmitter release from PC12 cells using Surface Enhanced Raman Spectroscopy, Biotechnol. Bioprocess. Eng. 19 (2014) 1069–1076, https://doi.org/10.1007/s12257-014-0092-7.

[88] P. Wang, M. Xia, O. Liang, K. Sun, A.F. Cipriano, T. Schroeder, et al., Label-free SERS selective detection of dopamine and serotonin using graphene-Au nanopyramid heterostructure, Anal. Chem. 87 (2015) 10255–10261, https://doi.org/10.1021/acs.analchem.5b01560.

[89] C.X. Shi, Z.P. Chen, Y. Chen, Q. Liu, R.Q. Yu, Quantification of dopamine in biological samples by surface-enhanced Raman spectroscopy: comparison of different calibration models, Chemom. Intell. Lab. Syst. 169 (2017) 87–93, https://doi.org/10.1016/j.chemolab.2017.09.006.

[90] V.Y.H. Hook, Protease pathways in peptide neurotransmission and neurodegenerative diseases, Cell Mol. Neurobiol. 26 (2006) 447–467, https://doi.org/10.1007/s10571-006-9047-7.

[91] J. Battey, E. Wada, Two distinct receptor subtypes for mammalian bombesin-like peptides, Trends Neurosci. 14 (1991) 524–528, https://doi.org/10.1016/0166-2236(91)90005-F.

[92] E. Podstawka, Y. Ozaki, L.M. Pronicwicz, Structures and bonding on a colloidal silver surface of the various length carboxyl terminal fragments of bombesin, Langmuir 24 (2008) 10807–10816, https://doi.org/10.1021/la8012415.

[93] A. Tąta, B. Gralec, E. Proniewicz, Unsupported platinum nanoparticles as effective sensors of neurotransmitters and possible drug curriers, Appl. Surf. Sci. 435 (2018) 256–264, https://doi.org/10.1016/j.apsusc.2017.11.100.

[94] A. Tąta, A. Szkudlarek, Y. Kim, E. Proniewicz, Adsorption of bombesin and its carboxyl terminal fragments onto the colloidal gold nanoparticles: SERS studies, Vib. Spectrosc. 84 (2016) 1–6, https://doi.org/10.1016/j.vibspec.2016.02.006.

[95] H. Domin, D. Świch, N. Piergies, E. Pita, Y. Kim, E. Proniewicz, Characterization of the surface geometry of acetyl-[Leu28,31]-NPY(24-36), a selective Y2 receptor agonist, onto the Ag and Au surfaces, Vib. Spectrosc. 85 (2016) 1−6, https://doi.org/10.1016/j.vibspec.2016.03.018.

[96] H. Domin, E. Pieta, N. Piergies, D. Świech, Y. Kim, L.M. Proniewicz, et al., Neuropeptide Y and its C-terminal fragments acting on Y2 receptor: Raman and SERS spectroscopy studies, J. Colloid Interface Sci. 437 (2015) 111−118, https://doi.org/10.1016/j.jcis.2014.09.053.

[97] I.H. Chou, M. Benford, H.T. Beier, G.L. Coté, M. Wang, N. Jing, et al., Nanofluidic biosensing for β-amyloid detection using surface enhanced Raman spectroscopy, Nano Lett. 8 (2008) 1729−1735, https://doi.org/10.1021/nl0808132.

[98] T. Demeritte, B.P. Viraka Nellore, R. Kanchanapally, S.S. Sinha, A. Pramanik, S.R. Chavva, et al., Hybrid graphene oxide based plasmonic-magnetic multifunctional nanoplatform for selective separation and label-free identification of alzheimer's disease biomarkers, ACS Appl. Mater. Interfaces 7 (2015) 13693−13700, https://doi.org/10.1021/acsami.5b03619.

[99] A. Huefner, W.-L. Kuan, S.L. Mason, S. Mahajan, R.A. Barker, Serum Raman spectroscopy as a diagnostic tool in patients with Huntington's disease, Chem. Sci. (2020), https://doi.org/10.1039/c9sc03711j.

[100] A.M.T. Monfared, V.S. Tiwari, V.L. Trudeau, H. Anis, Surface-enhanced Raman scattering spectroscopy for the detection of glutamate and $\\gamma$-Aminobutyric acid in serum by partial least squares analysis, IEEE Photonics J. 7 (2015) 1−16, https://doi.org/10.1109/JPHOT.2015.2423284.

[101] F. Lussier, Brulé, T. Brulé, M. Vishwakarma, T. Das, J.P. Spatz, J.-F.O. Masson, Dynamic-SERS optophysiology: a nanosensor for monitoring cell secretion events, Nano Lett. (2016), https://doi.org/10.1021/acs.nanolett.6b01371.

[102] F. Lussier, T. Brulé, M.J. Bourque, C. Ducrot, L.É. Trudeau, J.F. Masson, Dynamic SERS nanosensor for neurotransmitter sensing near neurons, Faraday Discuss 205 (2017) 387−407, https://doi.org/10.1039/c7fd00131b.

[103] E. Ryzhikova, N.M. Ralbovsky, L. Halámková, D. Celmins, P. Malone, E. Molho, et al., Multivariate statistical analysis of surface enhanced Raman spectra of human serum for alzheimer's disease diagnosis, Appl. Sci. 9 (2019) 3256, https://doi.org/10.3390/app9163256.

[104] A. El Alami, F. Lagarde, U. Tamer, M. Baitoul, P. Daniel, Enhanced Raman spectroscopy coupled to chemometrics for identification and quantification of acetylcholinesterase inhibitors, Vib. Spectrosc. 87 (2016) 27−33, https://doi.org/10.1016/j.vibspec.2016.09.005.

[105] C. Spuch, S. Ortolano, C. Navarro, LRP-1 and LRP-2 receptors function in the membrane neuron. Trafficking mechanisms and proteolytic processing in Alzheimer's disease, Front. Physiol. 3 (JUL) (2012) 269, https://doi.org/10.3389/fphys.2012.00269.

[106] C. Spuch, D. Antequera, C. Pascual, S. Abilleira, M. Blanco, M.J. Moreno-Carretero, et al., Soluble megalin is reduced in cerebrospinal fluid samples of Alzheimer's disease patients, Front. Cell Neurosci. 9 (2015) 134, https://doi.org/10.3389/fncel.2015.00134.

[107] P. Charlier, G. Héry-Arnaud, Y. Coppens, J. Malaurie, V. Hoang-Oppermann, P. Deps, et al., Global warming and planetary health: an open letter to the WHO from scientific and indigenous people urging for paleo-microbiology studies, Infect. Genet. Evol. 82 (2020) 104284, https://doi.org/10.1016/j.meegid.2020.104284.

[108] H. Hwang, B.-Y. Hwang, J. Bueno, Biomarkers in infectious diseases, Dis. Markers (2018), https://doi.org/10.1155/2018/8509127. ID 8509127.
[109] S.A.E. Logan, E. MacMahon, Viral meningitis, BMJ 336 (2008) 36–40, https://doi.org/10.1136/bmj.39409.673657.AE.
[110] WHO, World Malaria Report 2015, WHO Press, 2015.
[111] A.L. García-Basteiro, A. DiNardo, B. Saavedra, D.R. Silva, D. Palmero, M. Gegia, et al., Point of care diagnostics for tuberculosis, Rev. Port. Pneumol. 24 (2018) 73–85, https://doi.org/10.1016/j.rppnen.2017.12.002.
[112] R. McNerney, P. Daley, Towards a point-of-care test for active tuberculosis: obstacles and opportunities, Nat. Rev. Microbiol. 9 (2011) 204–213, https://doi.org/10.1038/nrmicro2521.
[113] A. Tay, A. Pavesi, S.R. Yazdi, C.T. Lim, M.E. Warkiani, Advances in microfluidics in combating infectious diseases, Biotechnol. Adv. 34 (2016) 404–421, https://doi.org/10.1016/j.biotechadv.2016.02.002.
[114] G. de Vries, R. Riesmeijer R, National Tuberculosis Control Plan 2016–2020: Towards Elimination 2016. https://rivm.openrepository.com/handle/10029/603648. (Accessed 7 April 2020).
[115] C.A. Suttle, Viruses in the sea, Nature 437 (2005) 356–361, https://doi.org/10.1038/nature04160.
[116] J.S.M. Peiris, M.D. De Jong, Y. Guan, Avian influenza virus (H5N1): a threat to human health, Clin. Microbiol. Rev. 20 (2007) 243–267, https://doi.org/10.1128/CMR.00037-06.
[117] J.H. Beigel, J. Farrar, A.M. Han, F.G. Hayden, R. Hyer, M.D. De Jong, et al., Avian influenza A (H5N1) infection in humans, N. Engl. J. Med. 353 (2005) 1374–1385, https://doi.org/10.1056/NEJMra052211.
[118] M.T. Aliota, L. Bassit, S.S. Bradrick, B. Cox, M.A. Garcia-Blanco, C. Gavegnano, et al., Zika in the Americas, year 2: what have we learned? What gaps remain? A report from the Global Virus Network, Antiviral Res. 144 (2017) 223–246, https://doi.org/10.1016/j.antiviral.2017.06.001.
[119] E.O. Saphire, S.L. Schendel, B.M. Gunn, J.C. Milligan, G. Alter, Antibody-mediated protection against Ebola virus, Nat. Immunol. 19 (2018) 1169–1178, https://doi.org/10.1038/s41590-018-0233-9.
[120] H. Feldmann, F. Feldmann, A. Marzi, Ebola: lessons on vaccine development, Annu. Rev. Microbiol. 72 (2018) 423–446, https://doi.org/10.1146/annurev-micro-090817-062414.
[121] J. Bedford, D. Enria, J. Giesecke, D.L. Heymann, C. Ihekweazu, G. Kobinger, et al., COVID-19: towards controlling of a pandemic, Lancet 395 (2020) 1015–1018, https://doi.org/10.1016/S0140-6736(20)30673-5.
[122] R. Gasparini, D. Amicizia, P.L. Lai, D. Panatto, Influenza vaccination: from epidemiological aspects and advances in research to dissent and vaccination policies, J. Prev. Med. Hyg. 57 (2016) E1–E4, https://doi.org/10.15167/2421-4248/jpmh2016.57.1.605.
[123] K.I. Lim, R. Klimczak, J.H. Yu, D.V. Schaffer, Specific insertions of zinc finger domains into Gag-Pol yield engineered retroviral vectors with selective integration properties, Proc. Natl. Acad. Sci. U. S. A. 107 (2010) 12475–12480, https://doi.org/10.1073/pnas.1001402107.

[124] D. Carroll, P. Daszak, N.D. Wolfe, G.F. Gao, C.M. Morel, S. Morzaria, et al., The global virome project, Science 359 (2018) 872–874, https://doi.org/10.1126/science.aap7463.

[125] B. Ge, Q. Li, G. Liu, M. Lu, S. Li, H. Wang, Simultaneous detection and identification of four viruses infecting pepino by multiplex RT-PCR, Arch. Virol. 158 (2013) 1181–1187, https://doi.org/10.1007/s00705-013-1604-z.

[126] F.S. Dawood, A.D. Iuliano, C. Reed, M.I. Meltzer, D.K. Shay, P.Y. Cheng, et al., Estimated global mortality associated with the first 12 months of 2009 pandemic influenza A H1N1 virus circulation: a modelling study, Lancet Infect. Dis. 12 (2012) 687–695, https://doi.org/10.1016/S1473-3099(12)70121-4.

[127] A. Ahmed, J.V. Rushworth, N.A. Hirst, P.A. Millner, Biosensors for whole-cell bacterial detection, Clin. Microbiol. Rev 27 (2014) 631–646, https://doi.org/10.1128/CMR.00120-13.

[128] B. Li, H. Tan, S. Anastasova, M. Power, F. Seichepine, G.Z. Yang, A bio-inspired 3D micro-structure for graphene-based bacteria sensing, Biosens. Bioelectron. 123 (2019) 77–84, https://doi.org/10.1016/j.bios.2018.09.087.

[129] H. Chen, A.E.V. Hagström, J. Kim, G. Garvey, A. Paterson, F. Ruiz-Ruiz, et al., Flotation immunoassay: masking the signal from free reporters in sandwich immunoassays, Sci. Rep. 6 (2016) 1–8, https://doi.org/10.1038/srep24297.

[130] Biosensors Market Worth $31.5 Billion by 2024 Growing With a CAGR of 8.3%, n.d. https://www.marketsandmarkets.com/PressReleases/biosensors.asp. (Accessed 7 April 2020).

[131] H. Chen, K. Liu, Z. Li, P. Wang, Point of care testing for infectious diseases, Clin. Chim. Acta 493 (2019) 138–147, https://doi.org/10.1016/j.cca.2019.03.008.

[132] J.Y. Lim, J.S. Nam, S.E. Yang, H. Shin, Y.H. Jang, G.U. Bae, et al., Identification of newly emerging influenza viruses by surface-enhanced Raman spectroscopy, Anal. Chem. 87 (2015) 11652–11659, https://doi.org/10.1021/acs.analchem.5b02661.

[133] A. Kamińska, E. Witkowska, A. Kowalska, A. Skoczyńska, P. Ronkiewicz, T. Szymborski, et al., Rapid detection and identification of bacterial meningitis pathogens in: ex vivo clinical samples by SERS method and principal component analysis, Anal. Methods 8 (2016) 4521–4529, https://doi.org/10.1039/c6ay01018k.

[134] F. Chen, B.R. Flaherty, C.E. Cohen, D.S. Peterson, Y. Zhao, Direct detection of malaria infected red blood cells by surface enhanced Raman spectroscopy, Nanomed. Nanotechnol. Biol. Med. 12 (2016) 1445–1451, https://doi.org/10.1016/j.nano.2016.03.001.

[135] R. Hunter, A.N. Sohi, Z. Khatoon, V.R. Berthiaume, E.I. Alarcon, M. Godin, et al., Optofluidic label-free SERS platform for rapid bacteria detection in serum, Sensors Actuators, B Chem. 300 (2019) 126907, https://doi.org/10.1016/j.snb.2019.126907.

[136] C. Camerlingo, M. Lisitskiy, M. Lepore, M. Portaccio, D. Montorio, S. Prete, et al., Characterization of human tear fluid by means of surface-enhanced Raman spectroscopy, Sensors 19 (2019) 1177, https://doi.org/10.3390/s19051177.

[137] W. Kim, J.C. Lee, J.H. Shin, K.H. Jin, H.K. Park, S. Choi, Instrument-free synthesizable fabrication of label-free optical biosensing paper strips for the early detection of infectious keratoconjunctivitides, Anal. Chem. 88 (2016) 5531–5537, https://doi.org/10.1021/acs.analchem.6b01123.

[138] Y.T. Yeh, K. Gulino, Y.H. Zhang, A. Sabestien, T.W. Chou, B. Zhou, et al., A rapid and label-free platform for virus capture and identification from clinical samples, Proc. Natl. Acad. Sci. U. S. A. 117 (2020) 895–901, https://doi.org/10.1073/pnas.1910113117.

[139] S. Guo, C. Beleites, U. Neugebauer, S. Abalde-Cela, N.K. Afseth, F. Alsamad, et al., Comparability of Raman spectroscopic configurations: a large scale cross-laboratory study, Anal. Chem. (2020), https://doi.org/10.1021/acs.analchem.0c02696.

CHAPTER

SERS probes and tags for biomedical applications

3

Pietro Strobbia[1], Andrew Fales[2]

[1]*Department of Chemistry, University of Cincinnati, Cincinnati, OH, United States;*
[2]*Center for Devices and Radiological Health, U.S. Food and Drug Administration, Silver Spring, MD, United States*

Introduction

Imaging and diagnostic techniques based on Surface-enhanced Raman scattering (SERS) are emerging as a potential alternative to the current medical standard practices. The latter suffer of applicability issues, such as complex instrumentation and/or procedures as well as low accuracy and sensitivity. In general, these techniques are characterized by a tradeoff between accuracy and complexity. In this landscape, SERS-based techniques can offer a cheap and reliable alternative to these current standard methods. The translation of SERS into clinical practice will provide clinicians with additional tools that can be used to complement their current toolbox and to expand the reach of standard imaging and diagnostic methods.

Established biomedical imaging methods currently used to assist surgeons and clinicians are magnetic resonance imaging, computed tomography, and positron emission tomography. These imaging techniques are invasive and expensive and are therefore currently used predominantly for neurosurgery at specialized medical centers [1]. Ultrasound imaging has additional drawbacks such as limited sensitivity and signal specificity [2]. The shortcomings of these techniques have fueled research into intraoperative optical imaging to address this unmet clinical need. To this end, fluorescence techniques have been widely explored for their use with contrast agents [1,3,4]. While being noninvasive, fluorescent techniques suffer from multiple disadvantages such as photobleaching and limited depth penetration. Additionally, autofluorescence from biological structures in the tumor area can affect the signal and lead to false positives [4].

Current biomedical diagnostic methods routinely used in clinical settings predominantly utilize colorimetry and fluorescence as optical readouts. These readouts are used because they require simple detection instrumentation, are easy to interpret, and can be coupled with imaging techniques to add diagnostic information. Colorimetry is used in both stained tissues (histopathology) and in common lateral flow assays. In both cases, a correct reading of the test depends on the test operator. Histopathology results are the standard for diagnosis of many diseases, ranging from cancer to malaria [4–6]. This technique requires a trained pathologist to read the

SERS for Point-of-care and Clinical Applications. https://doi.org/10.1016/B978-0-12-820548-8.00006-0

prepared slides, which involve a complex and long process. This type of diagnostic method can take multiple days in a well-equipped hospital but is almost inaccessible in the developing world, due to the low number of pathologist per capita. Rapid diagnostic tests (e.g., lateral flow assays) that uses calorimetry detection are prone to false negatives, due the low sensitivity arising from the tests or potentially from difficulties in the reading [7]. In a commonly used test for influenza, lateral flow assays are approved for positive diagnosis but not to rule out the infection, due to the lack of sensitivity.

In general, fluorescence techniques are more sensitive than colorimetry but require specific optical instrumentation to excite the fluorophores (i.e., fluorescence reporters/tags) and detect the signal readout. In fluorescence diagnostic applications, the excitation source is selected at a specific wavelength, which limits the background signal from endogenous species, improving sensitivity and specificity. Fluorescence microscopy techniques are used in many biomedical applications. Fluorescence microscopy is routinely used in clinical labs for advanced immunohistochemistry and other cell-identification procedures. Fluorescence in situ hybridization can be used to detect the expression of specific genes in tissues and can be used for molecular diagnostics [8]. Fluorescence is also used in flow-cytometry to detect the number of cells expressing specific antigens. This technique has been used to identify circulating tumoral cells [9]. Furthermore, fluorescence is also the typical readout of polymerase chain reaction (PCR), which is the current standard for genomic research and is routinely used as a confirmatory technique in many diagnosis [10].

While fluorescence is a powerful optical readout, the broad spectrum of fluorophores limits the number of reporters that can be detected simultaneously (i.e., multiplexed) in an assay. Common multiplexed fluorescence assays involve the detection of three or four different reporters; however, advances in diagnostic research have demonstrated that panels of biomarkers, rather than single biomarkers, are key for accurate diagnosis. Panels are especially important in genomic research and liquid biopsies for presymptomatic detection. These advances demonstrate the need for largely multiplexed assays. Current technology uses parallel assays to achieve multiplexing using fluorescence. An example are microarrays through which different targets are detected in different wells geometrically organized on a substrate. While microarrays meet the multiplexing requirement, they are not suitable for point-of-care (POC) diagnostic applications. The latter need low-complexity and low-cost solutions to be implemented successfully.

Raman spectroscopy is emerging as noninvasive and accurate alternative to standard biomedical imaging and diagnostic techniques. Raman scattering is an optical phenomenon that rely vibrational information relative to the sample under analysis. Its vibrational nature confers to this spectroscopy many advantages. Raman provides chemical information regarding the sample under analysis due to the unique vibrational spectrum of a molecule (vibrational fingerprint). The narrow bandwidth of vibrational (Raman) bands permits to detect a large number of peaks simultaneously. This advantage is the most important advance over a major limitation of the previously discussed optical readout methods, as a large number of biomarkers can be detected via Raman simultaneously. Additionally, Raman, unlike infrared spectroscopy, is performed using common silicon-based optical setups and detectors and can be performed on water-containing samples, making it more versatile than completing vibrational spectroscopies.

The use of Raman spectroscopy has been explored in biomedical applications to take advantage of the chemical information in Raman data for tissue identification and diagnostic purposes. Certain tumor tissues have been found to have specific vibrational signature and can be identified using Raman [11–14]. A prominent example is the detection of microcalcification, used toward the diagnosis of breast cancer. Specifically, transmission Raman spectroscopy is used for this application to detect microcalcification through the breast tissue and performing the test noninvasively [15]. Progress has also been made for the direct detection of glucose in blood using Raman [16,17]. These progresses are paving the way toward optical real-time monitoring of glucose, which would revolutionize the life of diabetes patients. Finally, the Raman signature of tumor tissue was used to categorize unlabeled pathology slides from a tumor resection directly in the operating room [18]. This application used stimulated Raman scattering (SRS) to quickly scan and image the tissue sections. SRS permits to only detect the vibrational signature specific for healthy and tumor tissue, increasing sensitivity and speed for Raman imaging.

While tumor cells have been identified using Raman in liquid biopsies [19], the detection of biomarkers only present in traces remain extremely challenging for spontaneous Raman. This type of trace analysis is key for early diagnosis and POC diagnostics. In all the applications discussed above, the Raman signatures detected are caused by main components in the tissue or by biomarkers present in large concentrations. This shortcoming is due to the inherently low Raman signal that reduces the sensitivity for the detection of biomarkers. Raman cross sections (i.e., the probability of observing a Raman photon) are in the order of 10^{-29}. In contrast, fluorescence cross sections are in the order of 10^{-16} (calculated as quantum efficiency times extinction coefficient). This difference demonstrates why is challenging to perform trace analysis using Raman spectroscopy. Without the possibility of improving the sensitivity of Raman scattering this technique would have had limited analytical applications. To this end, researchers started utilizing SERS to exploit the advantages of Raman scattering in bioanalytical and biomedical applications.

In summary, current optical methods (i.e., colorimetry and fluorescence) offer many advantages over other types of readout. However, there is a trade-off between the complexity of the assay and its accuracy and/or multiplexing capabilities. This trade-off has hindered the development of advanced POC diagnostic assays. Raman spectroscopy can potentially fill the need for highly multiplexed assay, but the inherently low sensitivity of this method makes it a challenge to be implemented in diagnostic applications. This landscape has made SERS-based diagnostic technology emerging as a potentially revolutionary tool for POC diagnostics.

Surface-enhanced Raman scattering

SERS was discovered in 1977 when molecules adsorbed on nanoscale roughened silver electrodes were observed to exhibit significantly amplified Raman scattering [20,21]. In the following years, many theories were developed for the SERS phenomenon [22–25]. The leading theory is that there are two processes producing the observed surface-enhancement, an electromagnetic and a chemical enhancement. The electromagnetic enhancement is by far the largest contribution to the observed signal amplification [26]. This enhancement is due to the effect of localized surface plasmons (LSPs) on the surface of a metallic nanostructure [27]. When a metallic nanostructure (e.g., nanoparticle or nano-array substrate) is illuminated with a resonant wavelength, it can sustain a collective excitation of the conductive band electrons of the metal (i.e., the LSP). The resonant wavelength capable of exciting an LSP depends on the size, geometry, and shape of the nanostructure. An LSP function as a nano-antenna, effectively concentrating the far-field light in the near-field of the structure and amplifying all the scattering phenomena around the surface of the nanoparticle. Thereby, the Raman scattering observed for a molecule in proximity of the metallic surface will experience the amplification due to the LSP. In addition to the electromagnetic enhancement, the surface enhancement was observed to be dependent on the molecule under analysis, contributing to the idea of a chemical enhancement. The theory behind this form of enhancement is less univocal. This enhancement has been attributed both to changes in the molecule cross section due to the surface interaction, as well as to a charge-transfer mechanism with the LSP on the nanostructure [28,29].

While the theory behind SERS is fascinating, biomedical application of SERS relies on the total observed enhancement, which influences the sensitivity. The total observed enhancement is a combination of the chemical and electromagnetic effects. Enhancements of 11–14 orders of magnitudes have been reported, as well as the SERS detection of single molecules [30–32]. Common SERS enhancements observed in single nanoparticles range between five and eight orders of magnitude. The enhancement strongly depends on the material and geometry of the plasmonic nanostructure (i.e., the source of the LSP) used for SERS. How these characteristics influence the SERS enhancement will be discussed in the following sections of this chapter.

Due to this observed enhancement, SERS effectively permits to exploit the benefits of Raman and vibrational spectroscopies in analytical and bioanalytical applications. SERS enables to identify and detects biomarkers with high sensitivity, accuracy, and multiplexing. The sensitivity in SERS is given by the enhancement that makes, in some cases, Raman reporters as bright as fluorescence ones (i.e., comparable peak-signal output). Accuracy is due to the narrow bandwidth of Raman peaks, which allows for their readily extraction from background signal avoiding false positives caused by a varying background. The Raman bandwidth also enables the simultaneous detection of multiple Raman peaks that can be associated with different reporters/biomarkers or from a single reporter, which, respectively, produce multiplexing and further accuracy by connecting a diagnosis to multiple peaks in the spectrum. Representative SERS and fluorescence spectra are shown in

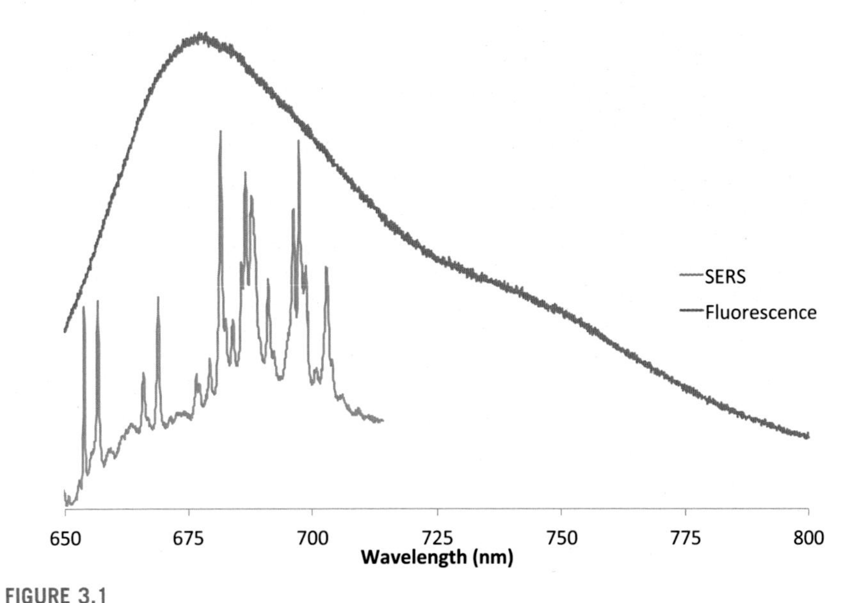

FIGURE 3.1

Representative SERS and fluorescence spectra collected at an excitation wavelength of 633 nm.

Fig. 3.1, demonstrating the multiple, narrow Raman peaks over a short spectral range compared to the broad, featureless fluorescence emission. Additionally, different fluorophores generally require the use of multiple excitation light sources or filters to match the absorption band, whereas a single laser can be used to generate Raman scattering from any Raman-active molecule.

These advantages are particularly interesting for noninvasive and POC biomedical applications, in which fluorescence and colorimetry are commonly used. The use of SERS tags could improve these technologies by making them more accurate and comprehensive. However, the successful development of these tags would need careful design of this multifaceted phenomenon, as SERS applications are influenced by multiple factors across different scientific disciplines (i.e., plasmonics, spectroscopy, analytical chemistry). Critical aspects for these technologies are the plasmonic structure used for SERS, the Raman reporter selection for the probe, as well as the sensing mechanism used to detect a specific target/biomarker. These critical topics will be the focus of this chapter.

Design considerations

Particle type

In selecting the optimal SERS probe for a specific application, the choice of nanoparticle(s) to serve as the source of SERS enhancement is a significant factor. One

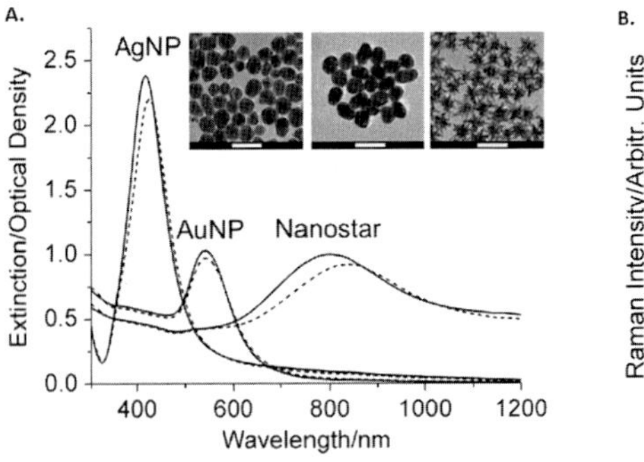

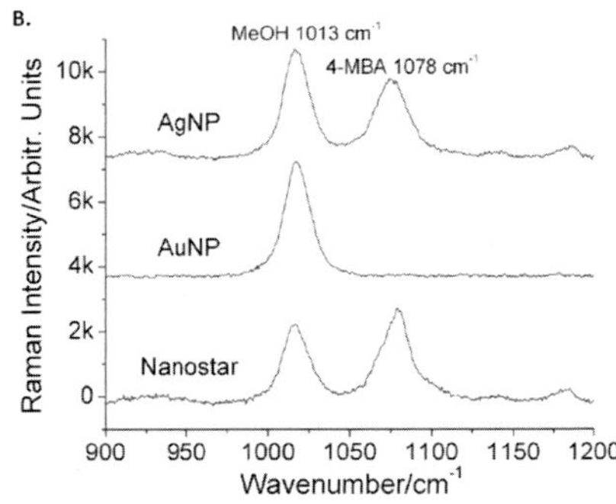

FIGURE 3.2

(A) Extinction spectra of 0.1 nM AgNP, AuNP, and nanostars before (*solid lines*) and after (*dashed lines*) the addition of Raman reporter p-MBA (1 μM final concentration) in MeOH (10% v/v final concentration). A slight red shift is observed in the plasmon band due to the change in refractive index at the particle surface and/or partial aggregation. Inset: TEM images of AgNP, AuNP, and nanostars (left to right; scale bars are 100 nm). (B) SERS spectra (baseline subtracted) of 1 μM 4-MBA in 0.1 nM AgNP, AuNP, and nanostar solutions examined through a Raman microscope under 785 nm excitation. Methanol (10% v/v) was used as an internal reference. Wavenumber 1013 and 1078 represent Raman signals from MeOH and 4-MBA, respectively.

aspect of particle design is the plasmonic material out of which the particle is made. For most biomedical applications, gold and/or silver will be the optimal choice. Particles made of these materials will have an LSP resonance in the visible to near-infrared ranges, which are compatible with commonly used laser excitation wavelengths. As shown in Fig. 3.2, the plasmon band of silver nanoparticles (AgNPs), gold nanoparticles (AuNPs), and gold nanostars can span from ~400 to 800 nm. It should be noted that aluminum plasmonic materials have grown in popularity in recent years as a low-cost alternative to gold and silver [33,34]. As described in the previous section, strong SERS enhancement occurs when the incident excitation light overlaps with the resonant wavelength of the LSP. Another way to generate a strong SERS signal is through aggregation or agglomeration, which creates small gaps between particles where the electromagnetic field is highly concentrated [35,36]. This typically involves the addition of salt (e.g., $MgCl_2$) to disrupt the electrostatic repulsion between particles, causing them to cluster [37]. Indeed, aggregation was heavily relied upon in early SERS biosensing and even led to the development of single-molecule detection using SERS [38,39].

While aggregation can be useful for obtaining high SERS enhancement factors, it generally results in issues with reproducibility due to the random nature of the aggregation process [40]. When monitoring the SERS signal over time, one will observe an increase in intensity up to a certain point, which will then rapidly decrease once the nanoparticle clusters become large enough to settle out of solution. The reliance on aggregation to obtain a measurable SERS signal was significantly reduced upon the development of wet chemical methods to synthesize anisotropic gold and silver nanoparticles in the early 2000s [41–43]. Whereas single spherical particles provide low enhancement (10–100 s of times), shapes with sharp points or corners ("hotspots"), such as nanorods and nanocubes, can offer average enhancement factors of $\sim 10^6$ without the need for aggregation. Much effort has been focused on the development of particles with higher enhancement in recent years. Of all the various shapes that have been developed to date, nanostars have emerged as one of the most promising for SERS in a nonaggregated state due to the high number of hotspots per particle.

Another advantage of anisotropic particles is the ability to tune the LSP wavelength. By modifying the size and shape of a nanoparticle, one can readily obtain any desired LSP wavelength within the visible to near-infrared range. For biomedical applications, an LSP wavelength in the near infrared, typically around 800 nm, is desirable. This wavelength not only falls within the "tissue optical window" where optical absorption and scattering are at a minimum but is also compatible for use with common Raman spectroscopy instrumentation operating at a 785-nm excitation wavelength. Thus, the particles can be designed to take advantage of both an optimal wavelength for tissue penetration and Raman instrumentation. Fig. 3.3 shows how the nanostar plasmon resonance can be tuned by altering the size and extent of branching. As the branches become longer and more numerous, the plasmon band exhibits a red-shift (Fig. 3.3B). In theory, the optimal SERS enhancement is obtained at a plasmon peak slightly red-shifted from the excitation wavelength to provide the highest enhancement of both the incident excitation light and generated Raman scattering. However, depending on the experimental conditions, self-absorption of the Raman scatter by the nanoparticles can significantly reduce the measured signal. In this case, with nanoparticles in solution at an optical density of ~ 1, the plasmon peak that is blue-shifted from the excitation results in the highest measured SERS signal, as the particle plasmon absorbs less of the Raman scattering generated at wavelengths above 785 nm (Fig. 3.3C).

While anisotropic nanoparticles offer favorable SERS enhancement and reproducibility properties, they are inherently unstable. Over time, anisotropic particles will coalesce toward a more thermodynamically stable, spherical shape [44]. This occurs more quickly with higher aspect ratio particles, as well as at elevated temperatures [44–46]. Therefore, it is recommended that all particles be stored at low temperature (4°C) when not in active use. There are also a variety of surface functionalization approaches that can help to stabilize the shape of an anisotropic nanoparticle [47]. A detailed discussion of these approaches is described in a later section of this chapter.

Raman reporter

A key aspect in the design of a SERS probe is the Raman reporter used. Raman reporters affect the Raman signal output due to their inherent cross sections, via the

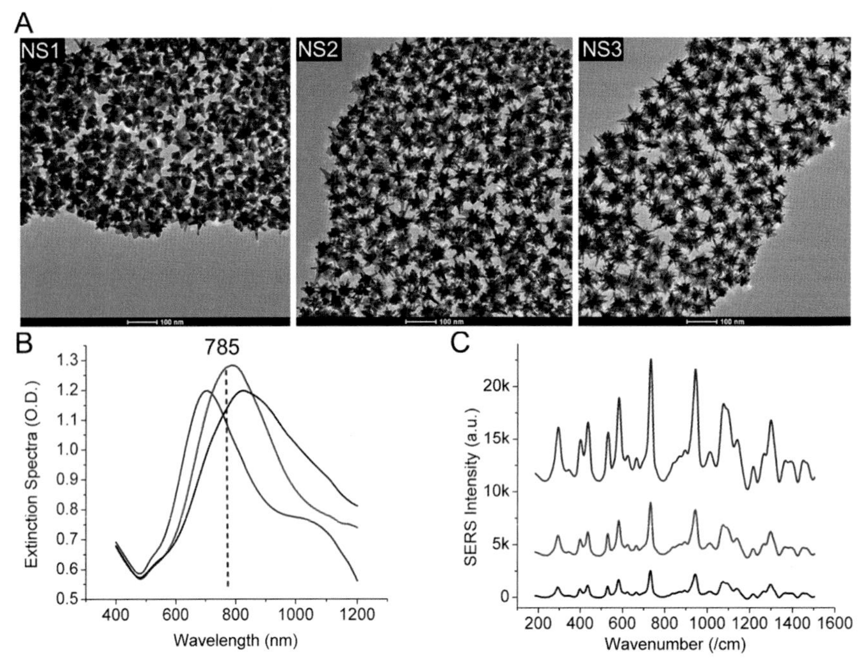

FIGURE 3.3

(A) TEM images of three different nanostar samples. (B, C) Extinction spectra and background-subtracted SERS spectra of the three nanostar samples labeled with IR-780 (NS1—blue, NS2—red, NS3—black). The dashed line at 785 nm in (B) indicates the laser excitation used to collect the SERS spectra.

Reprinted with permission from Yuan et al. Anal. Chem. 85 (1) (2013) 208–212. Copyright 2012 American Chemical Society.

resonant Raman phenomenon. Additional important consideration in the design of a probe are the functionalization method for the reporters and the complexity of their spectra. These aspects will be discussed in this section.

The selection rules for Raman scattering state that, during the specific molecular vibration, the polarizability has to change. Polarizability is a tensor representing the possible displacement of electrons in a molecule. The selection rules will influence the cross section of specific Raman bands, as large changes in polarizability will generate strong Raman peaks. Usually, aromatic rings and large structures where electrons are shared over multiple atoms have large polarizability and vibrations that displace such structures will have large Raman cross sections. Because of that, skeletal and ring-breathing modes commonly represent the strongest peaks in a Raman spectrum. As a practical rule, molecules with multiple aromatic ring and large conjugations are good SERS reporters.

Resonance Raman is a phenomenon that causes Raman scattering to be amplified when the energy of the excitation used for Raman is a close to the energy of an

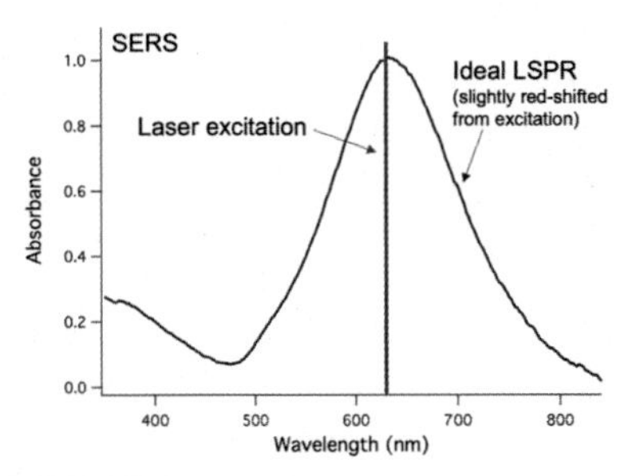

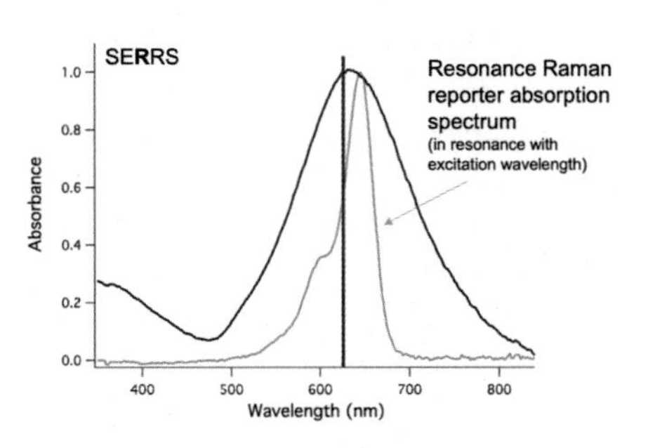

FIGURE 3.4

Spectra showing the ideal relative spectral position for laser excitation, LSPR, and absorption band. Left panel for SERS and right panel for SERRS. In SERS, optimal LSPR is resonant to the laser with the laser wavelength slightly blue-shifted with respect to the LSPR max wavelength, to permit enhancement of Stokes-shifted light as well as the laser. In SERRS, to this conformation is added the absorption band of the Raman reporter, also resonant to the laser to achieve resonance Raman.

electronic transition in the molecule. This phenomenon is due to the enhanced polarizability of molecule experienced during an electronic transition. We can take advantage of resonance Raman in SERS by selecting Raman reports for the SERS tags to be resonant to the excitation wavelength used. When resonant reporters are used in SERS, the scattering phenomenon is called surface-enhanced resonance Raman scattering (SERRS); however, often researchers will just refer to it as SERS. SERRS offers a mean to further amplify the Raman output of the SERS tags. Resonance Raman will further enhance the Raman signal of 10- to 1000-fold giving a significant additional improvement to the sensitivity. An issue with resonance Raman is photobleaching, the loss of resonance due to a photochemical transition. Once photobleached, the reporters will lose the ability to undergo the electronic transition, and therefore the resonance Raman enhancement. In practice, this causes SERRS tags to be less photo-stable than conventional SERS and it is important to balance the increased sensitivity with the photostability required for the specific application. Fig. 3.4 shows spectra describing this idea of resonance for SERS and SERRS, for SERS resonance between the LSPR of the nanostructure and laser and for SERRS the overlapping of LSPR, absorption band, and laser wavelength.

Another important aspect in Raman reporter selection is how these molecules are bound to the metallic surface. By definition, Raman reporters in SERS tags have to be on or in close proximity to the nanoparticle surface and there are different methods to accomplish this task. Raman reporters have been adsorbed on particle surface using electrostatic interaction, quasi-covalent bonds thiol functional groups, and embedded in the shell of core–shell particle. Several dyes have been shown to adsorb on the surface of metallic nanoparticles [48–50]. This process is driven by the affinity of the dye/reporter to the metallic surface and is strongly reporter-specific. A more reliable way to attach reporters to metallic surface is taking advantage of the strong binding characteristics of thiol groups to metallic surfaces, especially to gold [51]. Raman reporters with a thiol moiety can be easily attached to nanoparticles and displace previously ligand present from synthesis. These reporters are very stable on the nanoparticle surface (i.e., they are unlikely to detach) and often also increase the stability of the particles in solution via electrostatic or steric repulsion. Amine groups are also capable to bind to metallic surfaces likely due to a Lewis base mechanism. Amines bind preferably on silver surfaces and relatively stable. This binding process is commonly referred to as a chemisorption due to its stability. In core—shell nanoparticles, reporters can be added by embedding in the shell layer. For example, silica-coated nanoparticles can be synthesized including the reporter in solution during the silica coating steps [52]. This process produces particles with Raman reporters in proximity of the particle surface and stabilized inside the silica layer. More recently, DNA binding was also studied theoretically and experimentally, showing that poly-A chains strongly bind to gold nanoparticle surfaces and have been used to anchor multiple reporters on such surfaces.

Another aspect to account for in the selection of the reporter is the density of peaks in the Raman spectrum. This factor is particularly important for applications requiring multiplexing. Bulky reporters can have a large number of peaks that make it more challenging to isolate the contribution of single reporters due to overlapping. Recently, compendia of possible reporters for SERS-based magnetic capture and lateral flow assays were published showing, respectively, at least 6 and 10 nonoverlapping reporters. Additionally, simple machine learning algorithms, such as partial component analysis or partial least squares regression, can be used to isolate the contributions to the spectrum for each reporter during data analysis [53]. Using these statistical methods to analyze the data requires significant validation during the assay development and should be used with awareness of the process and results to avoid nontransferrable predictions.

Recent advances to improve the signal-to-background ratio in SERS tags include the use of triple-bond Raman reporters and a catalysis-controlled Raman reporter. The use of SERS tags functionalized with reporters of various synthesized triple bond gives unique single spectral emission in the Raman silent region ($\approx$1800–2600 cm^{-1}) [54]. These peaks can be easily identified and from the non-Raman background signal, offering a mean to easily resolved multiplexed spectra. An alternative approach used to increase the signal-to-background is to use a target-induced catalytic reaction to produce the Raman reporter [55]. This method is based on the catalytic (oxidizing) activity of gold nanosphere with respect to certain molecules. For example, tetramethylbenzidine oxidizes in the vicinity of gold nanospheres. The method uses a target nucleic acid sequence to capture the gold nanospheres on a silver cube functionalized with the molecule. In this way, the reporter molecule is catalytically oxidized only when the nanospheres are captured in the presence of the target nucleic acid. This strategy eliminates all interfering signal from the unbound SERS tags by only detecting the signal of the oxidized form of the Raman reporter molecule.

Surface coating

The surface coating on a nanoparticle serves multiple purposes. The first involves passivation of the surface to protect against degradation. This could be in the form of oxidation, which occurs rapidly in particles made of silver, and/or reshaping, which is a considerable problem for anisotropic nanoparticles. Both of these forms of degradation result in a significant reduction in SERS enhancement; oxidation by creating a non- or weakly SERS-active layer on the particle surface and reshaping by eliminating regions of highly concentrated EM field enhancement [56]. Nanoparticle reshaping can be prevented or reduced through steric confinement, such as by encapsulation in a rigid silica shell, or by stabilization with small thiol molecules that increase the activation energy for surface diffusion [47]. In the case of a silica shell, surface diffusion may still happen; however, the nanoparticle structure is physically confined so that any reshaping to occur is minimal. The second purpose of surface coatings is to impart colloidal stability within the intended use environment. Biomedical applications of SERS particles will often involve buffers with physiological salt concentrations, interferents such as proteins, or use in serum. Each of which presents challenges for the electrostatically stabilized particles that result from typical synthesis methods [57].

The most common method of surface functionalization involves use of dative bonds between thiols and the gold or silver surface. The strong Au–S and Ag–S bonds allow any thiolated molecule to be readily attached to the particle surface [58]. In some cases, in the presence of small-molecule thiols such as glutathione, a single thiol will not be sufficient to provide stability. Investigators have shown that multidentate thiol ligands can significantly increase bond strength and improve stability in cases where single thiol molecules would have detached [59]. To impart stability in most biological media, thiol-PEG can be used. When preparing SERS tags, the number of thiolated molecules attached to the surface should be optimized to provide a sufficient level of stability while allowing enough binding sites for the desired Raman reporter [60]. If there are too many stabilizing molecules bound to the particle surface, the number of Raman reporters will be suboptimal, which may be detrimental to use as a SERS tag.

There are some methods that allow for a complete self-assembled monolayer of Raman reporter molecules on the particle surface, while preserving the ability to add the desired surface functionality. This can be accomplished by chemically modifying the Raman reporter to include both binding affinity for the particle surface and an additional stabilizing moiety [61]. Alternatively, Raman reporter labeled particles can be encapsulated with polymer layers or a silica shell that can be further functionalized to impart colloidal stability. Such encapsulation strategies are also generally more resistant to degradation, as the Raman reporters bound to the nanoparticle surface are shielded from the external milieu [62].

The choice of surface functionality can also affect the cytotoxicity of nanoparticles. Silver particles are known to leach Ag^+ ions into solution, which can produce cytotoxic effects when used in vitro or in vivo. With appropriate surface coatings, the advantageous SERS properties of silver nanoparticles can be harnessed for use in biomedical applications [63]. Another well-studied issue is the toxicity of certain shape-directing surfactants used in the synthesis of certain anisotropic nanoparticles. Gold nanorod synthesis in particular relies on the use of high concentrations cetyltrimethylammonium bromide (CTAB), which is known to have toxic effects. A number of methods have been developed for surfactant exchange, to remove or reduce the CTAB concentration to a level that is no longer toxic to cells [64]. Again, thiolated molecules are a common strategy used to remove CTAB surfactant in a straightforward manner [65].

For in vivo applications, the surface coating is especially important in achieving delivery to the desired tissues. In general, a long circulation half-life will result in the highest number of particles delivered to the target area, such as a tumor [66]. This can occur passively through the enhanced permeation and retention (EPR) effect that results from the leaky vasculature in tumors, or though active targeting. Without the appropriate surface coating, particles may not have enough circulation time to reach the intended target location. Thiol-PEG is often used to impart stability and extended circulation time to nanoparticles in vivo [67]. While PEG cannot fully prevent the adsorption of biological components such as proteins to nanoparticles, it can help reduce the amount of adsorbed material that forms a protein corona around the PEG layer, impacting circulation and eventual fate within the body [68]. One issue that has arose with the use of PEGylated nanoparticles is the production of anti-PEG antibodies that result in accelerated blood clearance after repeat dosing. More recent studies have investigated alternatives to PEG that do not exhibit accelerated blood clearance. Zwitterionic coatings, consisting of both positively and negatively charged molecules, have been shown to resist the production of an immune response and retain favorable circulation characteristics after repeat administration [69].

Targeting

There are two potential strategies that can be used to target nanoparticles, active and passive targeting. The addition of nanoparticles to culture medium in vitro will often result in significant nanoparticle uptake without the need for targeted surface functionalities. Such methods have been used to study the intracellular composition and various biological processes with SERS. In 2002, Kneipp et al. demonstrated that colloidal gold nanoparticles uptaken from culture medium could be used as a SERS substrate for the chemical analysis of single living cells [70]. Further work showed that the SERS spectra collected from living cells change over time, indicative of the chemical environment within the endosomal environment [71]. Such label-free approaches rely on the intrinsic SERS signal of cellular components. Additional sensing capabilities can be achieved by using Raman reporters whose signal changes in response to the physicochemical environment. This strategy has been employed for intracellular pH sensing, using 4-mercaptobenzoic acid [72,73]. The protonation state of the carboxylic acid functional group varies with pH and results in changes of the associated Raman peaks [74].

Passive targeting in vivo involves modifying the physicochemical properties of a nanoparticle to improve circulation time. As described in the previous section, this can be accomplished with certain surface coatings that impart a "stealth" characteristic that reduces adsorption of biological molecules and clearance by the reticuloendothelial system. Certain disease states, such as cancer, result in rapid vascularization with tortuous and leaky structures. Prolonged circulation increases the number of nanoparticles that can extravasate into the tumor region and be used for detection or therapeutic purposes. This approach has been applied by many groups for the detection and treatment of tumors in animal models [75–77]. The SERS tags can be used to detect tumors and evaluate margins after treatment. When combined with a therapeutic modality, such as plasmonic photothermal therapy, a so-called theranostic SERS particle is produced.

Active targeting relies on the functionalization of nanoparticles with specific moieties to enhance delivery and uptake within a specific cell or tissue type. There are a number of different ligands that can be conjugated to nanoparticles for the purpose of active targeting. These can include small molecules, polysaccharides, carbohydrates, proteins, peptides, antibodies, and aptamers [78]. The ligands enhance nanoparticle uptake primarily by binding to cell surface receptors, causing internalization of both the receptor and any bound nanoparticles. These receptors are more highly expressed in the target cells/tissues than normal, leading to a higher relative nanoparticle distribution. Targeting ligands can be attached to nanoparticles with covalent linkages, streptavidin-biotin interaction, or by adsorption [58]. The choice of attachment strategy will depend on a number of factors, including ease of conjugation, compatibility with the nanoparticles or other reagents, and desired level of selectivity or tolerance for nonspecific binding [79].

Some common methods for attachment of targeting ligands include adsorption, bioaffinity (e.g., biotin/streptavidin), and N-hydroxysuccinimide/3-(3-dimethyl aminopropyl)carbodiimide (NHS/EDC). Adsorption of targeting moieties is the least robust of the above methods and may not be appropriate when directional functionalization is necessary for target binding (e.g., antibodies). While adsorption may be suitable for in vitro use of targeted SERS tags [80,81], such methods are generally not ideal for in vivo applications due to particle stability issues within the complex biological environment. The interaction between biotin-functionalized targeting molecules and streptavidin-functionalized nanoparticles can provide directional linkage with high affinity and has been extensively used for in vitro diagnostic assays, namely with fluorescent beads [82,83]. The use of NHS/EDC coupling is highly versatile as it has the ability to cross-link primary amines to carboxylic acid groups. Samanta et al. used NHS/EDC coupling to prepare antibody-conjugated SERS tags and demonstrated functionality both in vitro and in vivo [84]. A similar approach was used by Dinish et al. who prepared targeted SERS tags for in vivo multiplex detection [85].

Sensing mechanism

Numerous sensing mechanisms have been developed for sensing and diagnostic assays using SERS tags. In this section, we discuss the most used ones, highlighting first introduction and interesting advances (Fig. 3.5). We will also compare and contrast the methods, discussing advantages and disadvantages. The scope of this section is not to give a comprehensive list of every SERS assay but rather to inform the reader on the modalities of biomolecular sensing with SERS tags.

One of the first developed SERS sensing mechanisms is based on the aggregation of nanoparticles induced by a target analyte. This mechanism was pioneered in colorimetric plasmonic sensors in the 1990s for the detection of target genes and metallic ions [86,87]. While using aggregation to increase the SERS enhancement had been used in for various applications [31,70,88,89], this aggregation-based sensing mechanism (target-induced) was first used in SERS sensors to detect DNA target sequences [90,91]. In this work, the SERS tags are silver nanoparticles coated with a Raman reporter (resonant dye molecule) and functionalized with a DNA strand half-complementary to the target sequence, creating two batches of particles half-complementary to the opposite terminals of the target. The target DNA hybridize with the strands functionalized to the particles creating a network of the particles with a much stronger SERS enhancement than single nanoparticles and therefore increasing (turning on) the output signal. A variation over this mechanism has been developed to work as a "turn-off" sensor. These sensors are called plasmon coupling interference and are based on particles dimers being split by the target sequence and were applied to the detection of miR21, an important microRNA cancer biomarker [92]. In this "turn-off" sensing mechanism, the particles are connected in its initial state by a strand complementary to the target sequence. The connecting strand in the presence of the target is displaced, freeing the single nanoparticles from the dimer. The advantage of this method is its increased stability and accuracy. In fact, in aggregation sensing mechanisms stability is crucial for accurate sensing. A drawback of traditional ("turn-on") sensors is that they can undergo aggregation in the presence of certain contaminants, giving false-positive results.

The aggregation mechanism has also been used to detect other biotargets using binding affinity of specific enzymes. Lectin ConA was detected by exploiting the four glucose binding units included in this protein [93]. SERS tags were functionalized with a modified glucose molecule bound to the particle via a thiol group. The

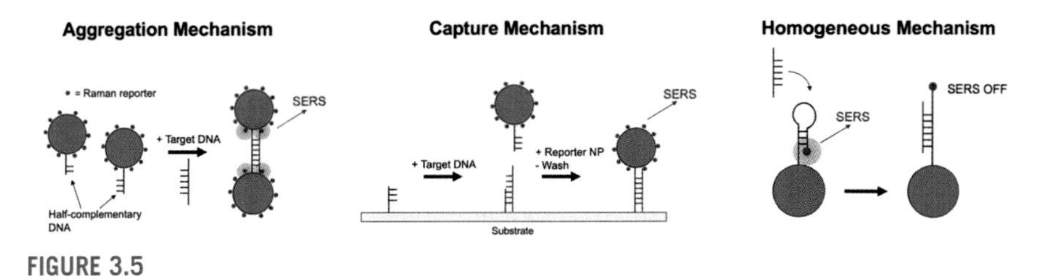

FIGURE 3.5

Schematic representation of different basic sensing mechanisms used in SERS diagnostic assays.

detection worked by having multiple nanoparticles aggregating on the protein and increasing the SERS signal observed for the tags. The requirements for the nanoparticle dimers to enhance SERS of a few nm gap/distance limits the number of target biomolecules detectable using this sensing mechanism. DNA detection remains the most common target for this aggregation sensing mechanism due to the gap requirement, easily achievable with DNA sequences.

A common strategy of SERS sensing is the capture of SERS tags on a substrate easily separated from the sample solution. This mechanism is very easy to accomplish and can work in tandem with fluidic or lab-on-chip systems; however, the common requirement of a washing steps in this mechanism limits the use of these methods for in situ analysis. Commonly used capture substrates are functionalized fixed surface or magnetic microparticle. These assays are commonly derived from standard molecular biology assays, such as ELISA, lateral flow assays, and magnetic capture. The advantage of using SERS tags in these assays is given by the chance for highly multiplexed detection. SERS tags can be used to detect numerous targets in a single pot/well by taking advantage of the narrow SERS peaks, whereas current state-of-the-art assays have to use microarrays or other form of performing parallel assays.

The first demonstration of this mechanism used in SERS was the detection of bacterial and HIV-related genes by capturing SERS nanoparticles on a substrate [94,95]. In the assays the target DNA is immobilized on a substrate and the probe, composed of a silver nanoparticle functionalized with a Raman reporter and a target-complementary DNA. The probe is captured on the substrate and the signal from the reporter identified. For HIV detection, this method was integrated with PCR by having the probe functionalized with a primer. In the presence of the target, the PCR process polymerize DNA after the primer generating the capture sequence. After polymerization, the probes are captured on the substrate surface by immobilized DNA complementary to the newly generated capture sequence on the probes.

An interesting example of this sensing strategy was demonstrated in an ELISA-like assay used to detect multiple nucleotide sequences associated with various viral diseases, showing the multiplexing advantage of SERS [96]. In this application, SERS tags based on gold nanoparticles are functionalized with a sequence half-complementary to the target and are captured in the presence of the target sequence on a microarray surface coated with DNA half-complementary to the other terminal of the target sequence. The SERS tags were then used as seeds to grow larger silver nanoparticles by addition of silver ions and hydroquinone, producing a strong SERS signal. It is noteworthy that the latter step is similar to the enzymatic transduction in ELISA assays and in this case is used to increase the observed SERS signal; however, this additional step is not necessary in SERS immunoassays that use SERS tags.

A similar mechanism is also used to detect important protein biomarkers (e.g., tumor necrosis factor α and prostate-specific antigen) [97,98]. Protein biomarkers are detected using primary and secondary antibodies as receptors, in place of complementary DNA sequences. One antibody is functionalized on the SERS tag and the other antibody is functionalized on the substrate surface. The sample is initially flown on the surface followed by the SERS tags, which are capture on the surface in the presence of the target protein. This mechanism was recently translated to paper substrates, producing deployable lateral flow assays. On paper, the flow is controlled by capillary force and the different elements necessary for the sensing steps are introduced on the flow, controlling their timing by varying the lateral position on the paper substrate. The capturing receptors are placed in the sensing region, which is where the SERS detection will focus and detect particles in the presence of the target. This assay was used to diagnose scrub typhus detecting biomarkers with capture antibodies in diluted blood samples [99]. A similar approach was used to detect miR29, a microRNA associated with cardiovascular diseases [100]. In the latter case, the assay is run on a 3D paper substrate, where the sensing region and the sensing elements are disposed on different paper sheets.

Magnetic microparticles are increasingly been used in sensing assays because they offer a mean for separation with magnetic forces applied externally to the system. In magnetic capture assays, all the sensing elements can be mixed simultaneously with the sample and then separated by collecting the microparticles on a magnet. In these assays the capture elements (e.g., DNA or antibody) are functionalized on the surface of the magnetic microparticle. One example of this method, using nanorattles as the SERS tag, is shown in Fig. 3.6. The assay commonly works

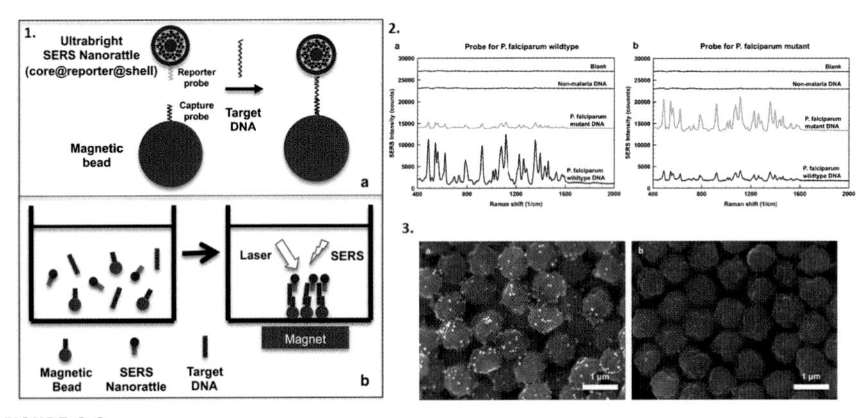

FIGURE 3.6

(1A) Nanorattle-based DNA detection using sandwich hybridization of magnetic beads that are loaded with capture probes, target sequency, and SERS nanorattles loaded with reporter probes. (1B) A magnet is applied to concentrate the hybridization sandwiches at a detection spot for SERS measurements. (2) Detection of wild-type *P. falciparum* and mutant *P. falciparum* with a single nucleotide difference using the nanorattle-based method (vertically shifted for clarity). Two probes, one for wild-type *P. falciparum* (2A) and one for mutant *P. falciparum* (2B), were designed and tested against *P. falciparum* wild-type DNA, *P. falciparum* mutant DNA, nonmalaria DNA, and blank. The wild-type DNA and the mutant DNA have a single base difference. (3A) SEM image of nanorattles bound onto magnetic beads' surface in the presence of complementary target sequences. (3B) Almost no nanorattle found on magnetic beads in the absence of complementary target sequences.

Adapted with permission from Ngo et al. Biosens. Bioelectron. 81 (2016) 8–14. Copyright 2016 Elsevier.

by incubating the sample with the magnetic particles and the SERS tags, both functionalized with a capture element (Fig. 3.6(1A)). After the required incubation time, the magnetic microparticles are separated using the magnet and the sample solution with the unused elements is discarded (Fig. 3.6(1B)). The number of SERS tags captured on the magnetic particles depends on the target concentration and can be probed by collecting a spectrum on the magnetically concentrated pellet (Fig. 3.6(2)). The attachment of SERS tags on the magnetic particle surface can be verified by scanning electron microscopy, as seen in Fig. 3.6(3). The magnetic concentration can be used to collect all the particles in a small spot, which if fully illuminated by the laser focus offer a mean to maximize the SERS response. This mechanism was used for the detection of RNA associated with a specificity down to a single nucleotide polymorphism [101]. In this application, the capture receptors on the surface of the magnetic microparticles and on the SERS tags are half-complementary DNA sequences, similarly as for the aggregation DNA-sensing mechanism. Because of the magnetic feature of this mechanism, this assay was integrated in a portable device capable of performing the necessary separation and washing steps in a single capillary [50]. Alternatively, magnetic particles can be used in microfluidics systems to separate/wash the sample after the assay, sorting particles in different channels using magnetic force. This integration was demonstrated in a microfluidic device developed to detect PSA, using this method [102].

Homogenous sensing mechanism are a specific type of sensing mechanism that works autonomously. This definition means that the sensor response is generated by the SERS tag itself rather than by aggregation or by its capture. The homogeneous nature of this sensing strategy is ideal for certain applications, such as in vivo and in situ sensing. One example of homogenous sensors is offered by a class of SERS sensors called "molecular sentinels" (MS) that can detect nucleic acid biotargets (e.g., DNA, mRNA, and miRNA) [103]. The MS sensing is composed of a nucleic acid probe strand bound to a SERS active surface and terminated with a Raman reporter molecule. The SERS signal from the Raman reporter is highest when in contact with or in close proximity to the metallic surface of the nanoparticle. The MS sequence is designed to naturally form a stem-loop, which controls the distance between the reporter and the surface of the nanoparticle. In the absence of target strand, the stem loop configuration of the MS will have the Raman reporter close to the nanoparticle and thus give an "on" SERS signal. Once the target sequence is introduced, it will partially bind to the complementary sequence on the MS and subsequently straighten out the stem loop. This process moves the Raman reporter farther away from the plasmonics-active surface which turns "off" the SERS signal. Inverse molecular sentinels (iMS) were developed to improve over the initial design by changing the mechanism to a "turn-on" signal, which reduce the shot noise [104]. The iMS works by binding a placeholder (i.e., a DNA strand complementary to the target sequence) to the molecular sentinel strand to lock the probe in an open position ("off") (Fig. 3.7). The stem-loop closes in the presence of the target sequence because of displacement of the placeholder stand from the placeholder-probe hybrid. The diagnostic capabilities of these sensors have been demonstrated by detecting differential expression of miRNA in

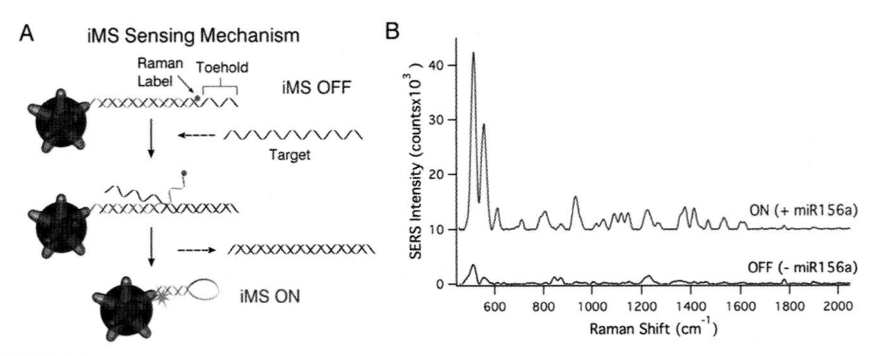

FIGURE 3.7

(A) Schematic representation of the iMS sensing mechanism. (B) Detection of miR156 target using iMS nanoprobes labeled with Cy7. SERS response of iMS nanoprobes in absence of target (OFF, blue spectrum) and in the presence of 2 μM synthetic miR156 target DNA (ON, red spectrum).

Reprinted with permission from Crawford et al. ACS Appl. Mater. Interfaces 11 (8) (2019) 7743–7754.

different breast cancer cell lines [105]. The iMS detection scheme has been further adapted for in situ analysis, by using on a fiber sensor with "dip-and-detect" capabilities [106]. This mechanism was also used on a "smart tattoo" biosensor that uses iMS nanoprobes for the detection of DNA, demonstrated in a large animal model in vivo [107]. Schematic representation of these sensors is shown in Fig. 3.8.

This type of homogenous sensing mechanism has also been adapted to detect biomolecules by taking advantage of aptamer binding sequences. Aptamer sequences are DNA or RNA sequences selected for their affinity to a specific molecules through a process known as systematic evolution of ligands by exponential enrichment [108]. Aptamer homogenous sensors use the binding competition between the complementary sequence and the target biomolecule. In these sensors the placeholder strand is an aptamer sequence and has the ability to bind to the probe sequence or to the target biomolecule. It is important to design a sequence that has a stronger binding affinity for the biomolecule to favor the sensing reaction. Thereby, in the presence of the target molecule, the placeholder will be displaced due to its binding affinity. The first example for this sensing mechanism was developed to sense adenosine, with the aptamer sequence producing the formation of a nanoparticles dimer [109]. Recent development have extended this mechanism to the detection of endocrine disrupting chemicals (i.e., BPA) directly in water samples [110].

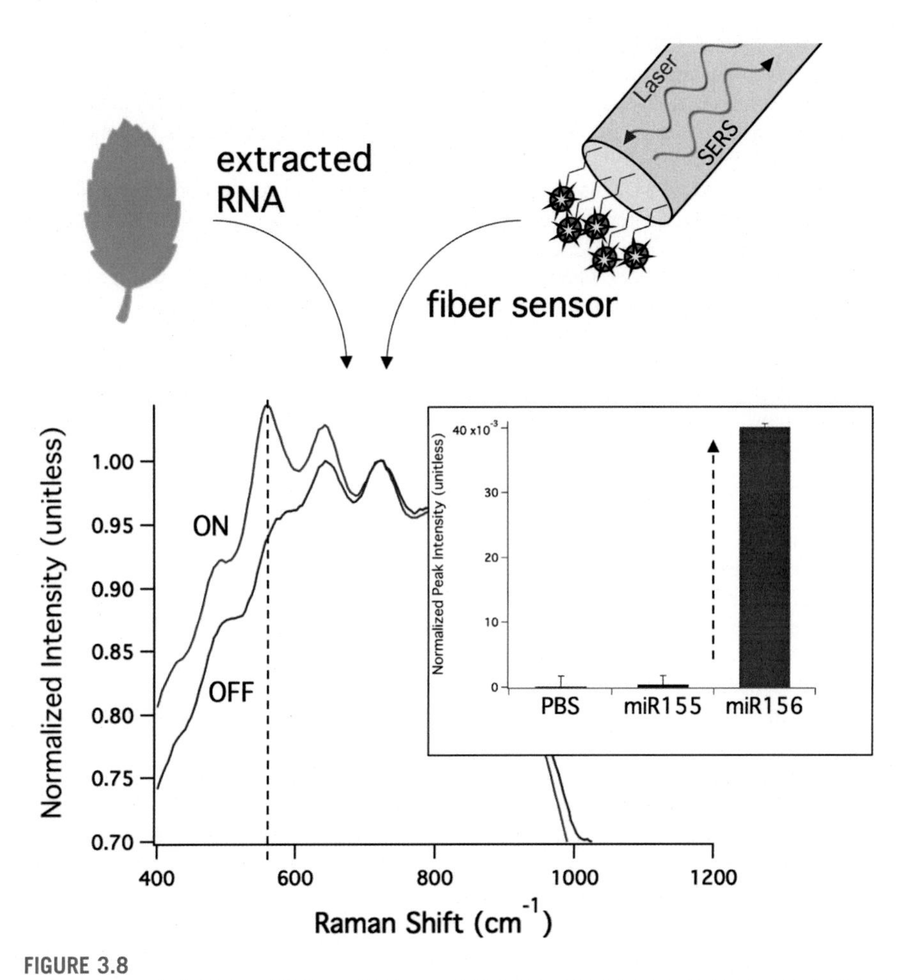

FIGURE 3.8

Schematic representation of "dip-and-detect" (left) and "smart tattoo" (right) SERS sensors.

References

[1] A.L. Vahrmeijer, M. Hutteman, J.R. Van Der Vorst, C.J. Van De Velde, J.V. Frangioni, Image-guided cancer surgery using near-infrared fluorescence, Nat. Rev. Clin. Oncol. 10 (9) (2013) 507.

[2] S. Li, P.-H. Wu, Magnetic resonance image-guided versus ultrasound-guided high-intensity focused ultrasound in the treatment of breast cancer, Chin. J. Cancer 32 (8) (2013) 441.

[3] W.C. Chan, D.J. Maxwell, X. Gao, R.E. Bailey, M. Han, S. Nie, Luminescent quantum dots for multiplexed biological detection and imaging, Curr. Opin. Biotechnol. 13 (1) (2002) 40–46.

[4] P. Demétrio de Souza França, N. Guru, S. Roberts, S. Kossatz, C. Mason, M. Abrahão, R.A. Ghossein, S.G. Patel, T. Reiner, Fluorescence-guided resection of tumors in mouse models of oral cancer, Sci. Rep. 10 (1) (2020) 11175.

[5] M.T. Makler, C.J. Palmer, A.L. Ager, A review of practical techniques for the diagnosis of malaria, Ann. Trop. Med. Parasitol. 92 (4) (1998) 419–434.
[6] W. Yasui, H. Yokozaki, F. Shimamoto, H. Tahara, E. Tahara, Molecular–pathological diagnosis of gastrointestinal tissues and its contribution to cancer histopathology, Pathol. Int. 49 (9) (1999) 763–774.
[7] G.P. Leonardi, A.M. Wilson, A.R. Zuretti, Comparison of conventional lateral-flow assays and a new fluorescent immunoassay to detect influenza viruses, J. Virol. Methods 189 (2) (2013) 379–382.
[8] A. Moter, U.B. Göbel, Fluorescence in situ hybridization (FISH) for direct visualization of microorganisms, J. Microbiol. Methods 41 (2) (2000) 85–112.
[9] C. Alix-Panabières, K. Pantel, Technologies for detection of circulating tumor cells: facts and vision, Lab Chip 14 (1) (2014) 57–62.
[10] P.M. Holland, R.D. Abramson, R. Watson, D.H. Gelfand, Detection of specific polymerase chain reaction product by utilizing the 5'——3' exonuclease activity of Thermus aquaticus DNA polymerase, Proc. Natl. Acad. Sci. U.S.A. 88 (16) (1991) 7276–7280.
[11] C. Kendall, M. Isabelle, F. Bazant-Hegemark, J. Hutchings, L. Orr, J. Babrah, R. Baker, N. Stone, Vibrational spectroscopy: a clinical tool for cancer diagnostics, Analyst 134 (6) (2009) 1029–1045.
[12] M. Jermyn, K. Mok, J. Mercier, J. Desroches, J. Pichette, K. Saint-Arnaud, L. Bernstein, M.-C. Guiot, K. Petrecca, F. Leblond, Intraoperative brain cancer detection with Raman spectroscopy in humans, Sci. Trans. Med. 7 (274) (2015), 274ra19-274ra19.
[13] T. Hollon, S. Lewis, C.W. Freudiger, X. Sunney Xie, D.A. Orringer, Improving the accuracy of brain tumor surgery via Raman-based technology, Neurosurg. Focus 40 (3) (2016). E9-E9.
[14] E.M. Barroso, R.W.H. Smits, T.C. Bakker Schut, I. ten Hove, J.A. Hardillo, E.B. Wolvius, R.J. Baatenburg de Jong, S. Koljenović, G.J. Puppels, Discrimination between oral cancer and healthy tissue based on water content determined by Raman spectroscopy, Anal. Chem. 87 (4) (2015) 2419–2426.
[15] N. Stone, P. Matousek, Advanced transmission Raman spectroscopy: a promising tool for breast disease diagnosis, Cancer Res. 68 (11) (2008) 4424–4430.
[16] S.M. Lundsgaard-Nielsen, A. Pors, S.O. Banke, J.E. Henriksen, D.K. Hepp, A. Weber, Critical-depth Raman spectroscopy enables home-use non-invasive glucose monitoring, PLoS One 13 (5) (2018) e0197134.
[17] J.W. Kang, Y.S. Park, H. Chang, W. Lee, S.P. Singh, W. Choi, L.H. Galindo, R.R. Dasari, S.H. Nam, J. Park, P.T.C. So, Direct observation of glucose fingerprint using in vivo Raman spectroscopy, Sci. Adv. 6 (4) (2020) eaay5206.
[18] D.A. Orringer, B. Pandian, Y.S. Niknafs, T.C. Hollon, J. Boyle, S. Lewis, M. Garrard, S.L. Hervey-Jumper, H.J.L. Garton, C.O. Maher, J.A. Heth, O. Sagher, D.A. Wilkinson, M. Snuderl, S. Venneti, S.H. Ramkissoon, K.A. McFadden, A. Fisher-Hubbard, A.P. Lieberman, T.D. Johnson, X.S. Xie, J.K. Trautman, C.W. Freudiger, S. Camelo-Piragua, Rapid intraoperative histology of unprocessed surgical specimens via fibre-laser-based stimulated Raman scattering microscopy, Nat. Biomed. Eng. 1 (2) (2017) 0027.
[19] E. Canetta, M. Mazilu, A. De Luca, A. Carruthers, K. Dholakia, S. Neilson, H. Sargeant, T. Briscoe, C.S. Herrington, A. Riches, Modulated Raman spectroscopy for enhanced identification of bladder tumor cells in urine samples, J. Biomed. Opt. 16 (3) (2011) 037002.

[20] D.L. Jeanmaire, R.P. Van Duyne, Surface Raman spectroelectrochemistry: Part I. Heterocyclic, aromatic, and aliphatic amines adsorbed on the anodized silver electrode, J. Electroanal. Chem. Inter. Electrochem. 84 (1) (1977) 1–20.
[21] M.G. Albrecht, J.A. Creighton, Anomalously intense Raman spectra of pyridine at a silver electrode, J. Am. Chem. Soc. 99 (15) (1977) 5215–5217.
[22] F.W. King, R.P. Van Duyne, G.C. Schatz, Theory of Raman scattering by molecules adsorbed on electrode surfaces, J. Chem. Phys. 69 (10) (1978) 4472–4481.
[23] M. Moskovits, Surface roughness and the enhanced intensity of Raman scattering by molecules adsorbed on metals, J. Chem. Phys. 69 (9) (1978) 4159–4161.
[24] J. Gersten, A. Nitzan, Electromagnetic theory of enhanced Raman scattering by molecules adsorbed on rough surfaces, J. Chem. Phys. 73 (7) (1980) 3023–3037.
[25] G.C. Schatz, Theoretical studies of surface enhanced Raman scattering, Acc. Chem. Res. 17 (10) (1984) 370–376.
[26] M. Moskovits, Surface-enhanced spectroscopy, Rev. Mod. Phys. 57 (3) (1985) 783–826.
[27] M. Moskovits, How the localized surface plasmon became linked with surface-enhanced Raman spectroscopy, Notes Rec. R. Soc. 66 (2) (2012) 195–203.
[28] A. Otto, On the contribution of charge transfer excitations to SERS, J Electr. Spectrosc. Rel. Phenom. 29 (1) (1983) 329–342.
[29] M. Moskovits, Persistent misconceptions regarding SERS, Phys. Chem. Chem. Phys. 15 (15) (2013) 5301–5311.
[30] H. Xu, J. Aizpurua, M. Käll, P. Apell, Electromagnetic contributions to single-molecule sensitivity in surface-enhanced Raman scattering, Phys. Rev. E 62 (3) (2000) 4318–4324.
[31] K. Kneipp, Y. Wang, H. Kneipp, L.T. Perelman, I. Itzkan, R.R. Dasari, M.S. Feld, Single molecule detection using surface-enhanced Raman scattering (SERS), Phys. Rev. Lett. 78 (9) (1997) 1667–1670.
[32] S. Nie, S.R. Emory, Probing single molecules and single nanoparticles by surface-enhanced Raman scattering, Science 275 (5303) (1997) 1102–1106.
[33] M.W. Knight, N.S. King, L. Liu, H.O. Everitt, P. Nordlander, N.J. Halas, Aluminum for plasmonics, ACS Nano 8 (1) (2014) 834–840.
[34] S. Ambardar, D. Nguyen, G. Binder, Z.W. Withers, D.V. Voronine, Quantum leap from gold and silver to aluminum nanoplasmonics for enhanced biomedical applications, Appl. Sci. 10 (12) (2020) 4210.
[35] H. Xu, M. Käll, Surface-plasmon-enhanced optical forces in silver nanoaggregates, Phys. Rev. Lett. 89 (24) (2002) 246802.
[36] F. Svedberg, Z. Li, H. Xu, M. Käll, Creating hot nanoparticle Pairs for surface-enhanced Raman spectroscopy through optical manipulation, Nano Lett. 6 (12) (2006) 2639–2641.
[37] K.A. Huynh, K.L. Chen, Aggregation kinetics of citrate and polyvinylpyrrolidone coated silver nanoparticles in monovalent and divalent electrolyte solutions, Environ. Sci. Technol. 45 (13) (2011) 5564–5571.
[38] K. Kneipp, H. Kneipp, Single molecule Raman scattering, Appl. Spectrosc. 60 (12) (2006) 322A–334A.
[39] A.M. Schwartzberg, C.D. Grant, A. Wolcott, C.E. Talley, T.R. Huser, R. Bogomolni, J.Z. Zhang, Unique gold nanoparticle aggregates as a highly active surface-enhanced Raman scattering substrate, J. Phys. Chem. B 108 (50) (2004) 19191–19197.
[40] R. Tantra, R.J.C. Brown, M.J.T. Milton, Strategy to improve the reproducibility of colloidal SERS, J. Raman Spectrosc. 38 (11) (2007) 1469–1479.

[41] N.R. Jana, L. Gearheart, C.J. Murphy, Wet chemical synthesis of high aspect ratio cylindrical gold nanorods, J. Phys. Chem. B 105 (19) (2001) 4065–4067.
[42] B. Nikoobakht, M.A. El-Sayed, Preparation and growth mechanism of gold nanorods (NRs) using seed-mediated growth method, Chem. Mater. 15 (10) (2003) 1957–1962.
[43] Y. Sun, Y. Xia, Shape-controlled synthesis of gold and silver nanoparticles, Science 298 (5601) (2002) 2176–2179.
[44] A.B. Taylor, A.M. Siddiquee, J.W.M. Chon, Below melting point photothermal reshaping of single gold nanorods driven by surface diffusion, ACS Nano 8 (12) (2014) 12071–12079.
[45] W.J. Kennedy, S. Izor, B.D. Anderson, G. Frank, V. Varshney, G.J. Ehlert, Thermal reshaping dynamics of gold nanorods: influence of size, shape, and local environment, ACS Appl. Mater. Interf. 10 (50) (2018) 43865–43873.
[46] R. Zou, Q. Zhang, Q. Zhao, F. Peng, H. Wang, H. Yu, J. Yang, Thermal stability of gold nanorods in an aqueous solution, Coll. Surf. A Physicochem. Eng. Aspec. 372 (1) (2010) 177–181.
[47] S. Centi, L. Cavigli, C. Borri, A. Milanesi, M. Banchelli, S. Chioccioli, B.N. Khlebtsov, N.G. Khlebtsov, P. Matteini, P. Bogani, F. Ratto, R. Pini, Small thiols stabilize the shape of gold nanorods, J. Phys. Chem. C 124 (20) (2020) 11132–11140.
[48] P. Lee, D. Meisel, Adsorption and surface-enhanced Raman of dyes on silver and gold sols, J. Phys. Chem. 86 (17) (1982) 3391–3395.
[49] T. Vo-Dinh, M.Y.K. Hiromoto, G.M. Begun, R.L. Moody, Surface-enhanced Raman spectrometry for trace organic analysis, Anal. Chem. 56 (9) (1984) 1667–1670.
[50] H.T. Ngo, E. Freedman, R.A. Odion, P. Strobbia, A.S.D.S. Indrasekara, P. Vohra, S.M. Taylor, T. Vo-Dinh, Direct detection of unamplified pathogen RNA in blood lysate using an integrated lab-in-a-stick device and ultrabright SERS nanorattles, Sci. Rep. 8 (1) (2018) 4075.
[51] H. Yuan, A.M. Fales, C.G. Khoury, J. Liu, T. Vo-Dinh, Spectral characterization and intracellular detection of Surface-Enhanced Raman Scattering (SERS)-encoded plasmonic gold nanostars, J. Raman Spectrosc. 44 (2) (2013) 234–239.
[52] S. Harmsen, M.A. Wall, R. Huang, M.F. Kircher, Cancer imaging using surface-enhanced resonance Raman scattering nanoparticles, Nat. Protoc. 12 (7) (2017) 1400–1414.
[53] K. Faulds, R.E. Littleford, D. Graham, G. Dent, W.E. Smith, Comparison of surface-enhanced resonance Raman scattering from unaggregated and aggregated nanoparticles, Anal. Chem. 76 (3) (2004) 592–598.
[54] Y. Zeng, K.M. Koo, A.-G. Shen, J.-M. Hu, M. Trau, Nucleic acid hybridization-based noise suppression for ultraselective multiplexed amplification of mutant variants, Small 17 (2) (2021) 2006370.
[55] J. Li, K.M. Koo, Y. Wang, M. Trau, Native MicroRNA targets trigger self-assembly of nanozyme-patterned hollowed nanocuboids with optimal interparticle gaps for plasmonic-activated cancer detection, Small 15 (50) (2019) 1904689.
[56] Y. Han, R. Lupitskyy, T.-M. Chou, C.M. Stafford, H. Du, S. Sukhishvili, Effect of oxidation on surface-enhanced Raman scattering activity of silver nanoparticles: a quantitative correlation, Anal. Chem. 83 (15) (2011) 5873–5880.
[57] L. Guerrini, R.A. Alvarez-Puebla, N. Pazos-Perez, Surface modifications of nanoparticles for stability in biological fluids, Materials 11 (7) (2018) 1154.

[58] M.H. Jazayeri, H. Amani, A.A. Pourfatollah, H. Pazoki-Toroudi, B. Sedighimoghaddam, Various methods of gold nanoparticles (GNPs) conjugation to antibodies, Sens. Bio-Sens. Res. 9 (2016) 17–22.

[59] E. Oh, K. Susumu, A.J. Mäkinen, J.R. Deschamps, A.L. Huston, I.L. Medintz, Colloidal stability of gold nanoparticles coated with multithiol-poly(ethylene glycol) ligands: importance of structural constraints of the sulfur anchoring groups, J. Phys. Chem. C 117 (37) (2013) 18947–18956.

[60] X. Qian, X.-H. Peng, D.O. Ansari, Q. Yin-Goen, G.Z. Chen, D.M. Shin, L. Yang, A.N. Young, M.D. Wang, S. Nie, In vivo tumor targeting and spectroscopic detection with surface-enhanced Raman nanoparticle tags, Nat. Biotechnol. 26 (1) (2008) 83–90.

[61] Y. Wang, S. Schlücker, Rational design and synthesis of SERS labels, Analyst 138 (8) (2013) 2224–2238.

[62] J. Langer, D. Jimenez de Aberasturi, J. Aizpurua, R.A. Alvarez-Puebla, B. Auguié, J.J. Baumberg, G.C. Bazan, S.E.J. Bell, A. Boisen, A.G. Brolo, J. Choo, D. Cialla-May, V. Deckert, L. Fabris, K. Faulds, F.J. García de Abajo, R. Goodacre, D. Graham, A.J. Haes, C.L. Haynes, C. Huck, T. Itoh, M. Käll, J. Kneipp, N.A. Kotov, H. Kuang, E.C. Le Ru, H.K. Lee, J.-F. Li, X.Y. Ling, S.A. Maier, T. Mayerhöfer, M. Moskovits, K. Murakoshi, J.-M. Nam, S. Nie, Y. Ozaki, I. Pastoriza-Santos, J. Perez-Juste, J. Popp, A. Pucci, S. Reich, B. Ren, G.C. Schatz, T. Shegai, S. Schlücker, L.-L. Tay, K.G. Thomas, Z.-Q. Tian, R.P. Van Duyne, T. Vo-Dinh, Y. Wang, K.A. Willets, C. Xu, H. Xu, Y. Xu, Y.S. Yamamoto, B. Zhao, L.M. Liz-Marzán, Present and future of surface-enhanced Raman scattering, ACS Nano 14 (1) (2020) 28–117.

[63] H.M. Fahmy, A.M. Mosleh, A.A. Elghany, E. Shams-Eldin, E.S. Abu Serea, S.A. Ali, A.E. Shalan, Coated silver nanoparticles: synthesis, cytotoxicity, and optical properties, RSC Adv. 9 (35) (2019) 20118–20136.

[64] A.M. Alkilany, P.K. Nagaria, M.D. Wyatt, C.J. Murphy, Cation exchange on the surface of gold nanorods with a polymerizable surfactant: polymerization, stability, and toxicity evaluation, Langmuir 26 (12) (2010) 9328–9333.

[65] L. Vigderman, P. Manna, E.R. Zubarev, Quantitative replacement of cetyl trimethylammonium bromide by cationic thiol ligands on the surface of gold nanorods and their extremely large uptake by cancer cells, Angewandte Chemie Int. Ed. 51 (3) (2012) 636–641.

[66] J.S. Suk, Q. Xu, N. Kim, J. Hanes, L.M. Ensign, PEGylation as a strategy for improving nanoparticle-based drug and gene delivery, Adv. Drug Deliv. Rev. 99 (2016) 28–51.

[67] G. Zhang, Z. Yang, W. Lu, R. Zhang, Q. Huang, M. Tian, L. Li, D. Liang, C. Li, Influence of anchoring ligands and particle size on the colloidal stability and in vivo biodistribution of polyethylene glycol-coated gold nanoparticles in tumor-xenografted mice, Biomaterials 30 (10) (2009) 1928–1936.

[68] R. García-Álvarez, M. Vallet-Regí, Hard and soft protein corona of nanomaterials: analysis and relevance, Nanomaterials 11 (4) (2021) 888.

[69] J. Zhao, Z. Qin, J. Wu, L. Li, Q. Jin, J. Ji, Zwitterionic stealth peptide-protected gold nanoparticles enable long circulation without the accelerated blood clearance phenomenon, Biomater. Sci. 6 (1) (2018) 200–206.

[70] K. Kneipp, A.S. Haka, H. Kneipp, K. Badizadegan, N. Yoshizawa, C. Boone, K.E. Shafer-Peltier, J.T. Motz, R.R. Dasari, M.S. Feld, Surface-enhanced Raman spectroscopy in single living cells using gold nanoparticles, Appl. Spectrosc. 56 (2) (2002) 150–154.
[71] J. Kneipp, H. Kneipp, M. McLaughlin, D. Brown, K. Kneipp, *In* vivo molecular probing of cellular compartments with gold nanoparticles and nanoaggregates, Nano Lett. 6 (10) (2006) 2225–2231.
[72] C.E. Talley, L. Jusinski, C.W. Hollars, S.M. Lane, T. Huser, Intracellular pH sensors based on surface-enhanced Raman scattering, Anal. Chem. 76 (23) (2004) 7064–7068.
[73] R. Luo, Y. Li, Q. Zhou, J. Zheng, D. Ma, P. Tang, S. Yang, Z. Qing, R. Yang, SERS monitoring the dynamics of local pH in lysosome of living cells during photothermal therapy, Analyst 141 (11) (2016) 3224–3227.
[74] Y. Liu, H. Yuan, A.M. Fales, T. Vo-Dinh, pH-sensing nanostar probe using surface-enhanced Raman scattering (SERS): theoretical and experimental studies, J. Raman Spectrosc. 44 (7) (2013) 980–986.
[75] G. von Maltzahn, A. Centrone, J.-H. Park, R. Ramanathan, M.J. Sailor, T.A. Hatton, S.N. Bhatia, SERS-coded gold nanorods as a multifunctional platform for densely multiplexed near-infrared imaging and photothermal heating, Adv. Mater. 21 (31) (2009) 3175–3180.
[76] C.L. Zavaleta, B.R. Smith, I. Walton, W. Doering, G. Davis, B. Shojaei, M.J. Natan, S.S. Gambhir, Multiplexed imaging of surface enhanced Raman scattering nanotags in living mice using noninvasive Raman spectroscopy, Proc. Natl. Acad. Sci. U.S.A. 106 (32) (2009) 13511–13516.
[77] Y. Liu, J.R. Ashton, E.J. Moding, H. Yuan, J.K. Register, A.M. Fales, J. Choi, M.J. Whitley, X. Zhao, Y. Qi, Y. Ma, G. Vaidyanathan, M.R. Zalutsky, D.G. Kirsch, C.T. Badea, T. Vo-Dinh, A plasmonic gold nanostar theranostic probe for in vivo tumor Imaging and photothermal therapy, Theranostics 5 (9) (2015) 946–960.
[78] J. Yoo, C. Park, G. Yi, D. Lee, H. Koo, Active targeting strategies using biological ligands for nanoparticle drug delivery systems, Cancers 11 (5) (2019) 640.
[79] M. Oliverio, S. Perotto, G.C. Messina, L. Lovato, F. De Angelis, Chemical functionalization of plasmonic surface biosensors: a tutorial review on issues, strategies, and costs, ACS Appl. Mater. Interf. 9 (35) (2017) 29394–29411.
[80] K.L. Nowak-Lovato, K.D. Rector, Targeted surface-enhanced Raman scattering nanosensors for whole-cell pH imagery, Appl Spectrosc 63 (4) (2009) 387–395.
[81] A.M. Fales, H. Yuan, T. Vo-Dinh, Cell-penetrating peptide enhanced intracellular Raman imaging and photodynamic therapy, Mol. Pharm. 10 (6) (2013) 2291–2298.
[82] Y.H. Ngo, W.L. Then, W. Shen, G. Garnier, Gold nanoparticles paper as a SERS bio-diagnostic platform, J Coll Interf Sci 409 (2013) 59–65.
[83] K.L. Kellar, R.R. Kalwar, K.A. Dubois, D. Crouse, W.D. Chafin, B.-E. Kane, Multiplexed fluorescent bead-based immunoassays for quantitation of human cytokines in serum and culture supernatants, Cytometry 45 (1) (2001) 27–36.
[84] A. Samanta, K.K. Maiti, K.-S. Soh, X. Liao, M. Vendrell, U.S. Dinish, S.-W. Yun, R. Bhuvaneswari, H. Kim, S. Rautela, J. Chung, M. Olivo, Y.-T. Chang, Ultrasensitive near-infrared Raman reporters for SERS-based in vivo cancer detection, Angewandte Chemie Int. Ed. 50 (27) (2011) 6089–6092.
[85] U.S. Dinish, G. Balasundaram, Y.-T. Chang, M. Olivo, Actively targeted in vivo multiplex detection of intrinsic cancer biomarkers using biocompatible SERS nanotags, Sci. Rep. 4 (1) (2014) 4075.

[86] R. Elghanian, J.J. Storhoff, R.C. Mucic, R.L. Letsinger, C.A. Mirkin, Selective colorimetric detection of polynucleotides based on the distance-dependent optical properties of gold nanoparticles, Science 277 (5329) (1997) 1078–1081.

[87] Y. Kim, R.C. Johnson, J.T. Hupp, Gold nanoparticle-based sensing of "spectroscopically silent" heavy metal ions, Nano Lett. 1 (4) (2001) 165–167.

[88] K. Kneipp, H. Kneipp, R. Manoharan, E.B. Hanlon, I. Itzkan, R.R. Dasari, M.S. Feld, Extremely large enhancement factors in surface-enhanced Raman scattering for molecules on colloidal gold clusters, Appl. Spectrosc. 52 (12) (1998) 1493–1497.

[89] K. Kneipp, H. Kneipp, G. Deinum, I. Itzkan, R.R. Dasari, M.S. Feld, Single-molecule detection of a cyanine dye in silver colloidal solution using near-infrared surface-enhanced Raman scattering, Appl. Spectrosc. 52 (2) (1998) 175–178.

[90] D. Graham, D.G. Thompson, W.E. Smith, K. Faulds, Control of enhanced Raman scattering using a DNA-based assembly process of dye-coded nanoparticles, Nat. Nanotechnol. 3 (9) (2008) 548–551.

[91] K. Faulds, R. Jarvis, W.E. Smith, D. Graham, R. Goodacre, Multiplexed detection of six labelled oligonucleotides using surface enhanced resonance Raman scattering (SERRS), Analyst 133 (11) (2008) 1505–1512.

[92] H.-N. Wang, T. Vo-Dinh, Plasmonic coupling interference (PCI) nanoprobes for nucleic acid detection, Small 7 (21) (2011) 3067–3074.

[93] D. Craig, J. Simpson, K. Faulds, D. Graham, Formation of SERS active nanoparticle assemblies via specific carbohydrate–protein interactions, Chem. Commun. 49 (1) (2013) 30–32.

[94] T. Vo-Dinh, K. Houck, D.L. Stokes, Surface-enhanced Raman gene probes, Anal. Chem. 66 (20) (1994) 3379–3383.

[95] N.R. Isola, D.L. Stokes, T. Vo-Dinh, Surface-enhanced Raman gene probe for HIV detection, Anal. Chem. 70 (7) (1998) 1352–1356.

[96] Y.C. Cao, R. Jin, C.A. Mirkin, Nanoparticles with Raman spectroscopic fingerprints for DNA and RNA detection, Science 297 (5586) (2002) 1536–1540.

[97] S. Laing, A. Hernandez-Santana, J. Sassmannshausen, D.L. Asquith, I.B. McInnes, K. Faulds, D. Graham, Quantitative detection of human tumor necrosis factor α by a resonance Raman enzyme-linked immunosorbent assay, Anal. Chem. 83 (1) (2011) 297–302.

[98] S. Laing, E.J. Irvine, A. Hernandez-Santana, W.E. Smith, K. Faulds, D. Graham, Immunoassay arrays fabricated by dip-pen nanolithography with resonance Raman detection, Anal. Chem. 85 (12) (2013) 5617–5621.

[99] S.H. Lee, J. Hwang, K. Kim, J. Jeon, S. Lee, J. Ko, J. Lee, M. Kang, D.R. Chung, J. Choo, Quantitative serodiagnosis of scrub typhus using surface-enhanced Raman scattering-based lateral flow assay platforms, Anal. Chem. 91 (19) (2019) 12275–12282.

[100] S. Mabbott, S.C. Fernandes, M. Schechinger, G.L. Cote, K. Faulds, C.R. Mace, D. Graham, Detection of cardiovascular disease associated miR-29a using paper-based microfluidics and surface enhanced Raman scattering, Analyst 145 (3) (2020) 983–991.

[101] H.T. Ngo, N. Gandra, A.M. Fales, S.M. Taylor, T. Vo-Dinh, Sensitive DNA detection and SNP discrimination using ultrabright SERS nanorattles and magnetic beads for malaria diagnostics, Biosens. Bioelectron. 81 (2016) 8–14.

[102] R. Gao, Z. Cheng, A.J. deMello, J. Choo, Wash-free magnetic immunoassay of the PSA cancer marker using SERS and droplet microfluidics, Lab Chip 16 (6) (2016) 1022–1029.

[103] H.-N. Wang, T. Vo-Dinh, Multiplex detection of breast cancer biomarkers using plasmonic molecular sentinel nanoprobes, Nanotechnology 20 (6) (2009) 065101.
[104] H.-N. Wang, A.M. Fales, T. Vo-Dinh, Plasmonics-based SERS nanobiosensor for homogeneous nucleic acid detection, Nanomed. Nanotechnol. Biol. Med. 11 (4) (2015) 811−814.
[105] H.-N. Wang, B.M. Crawford, A.M. Fales, M.L. Bowie, V.L. Seewaldt, T. Vo-Dinh, Multiplexed detection of MicroRNA biomarkers using SERS-based inverse molecular sentinel (iMS) nanoprobes, J. Phys. Chem. C 120 (37) (2016) 21047−21055.
[106] P. Strobbia, Y. Ran, B.M. Crawford, V. Cupil-Garcia, R. Zentella, H.-N. Wang, T.-P. Sun, T. Vo-Dinh, Inverse molecular sentinel-integrated fiberoptic sensor for direct and in situ detection of miRNA targets, Anal. Chem. 91 (9) (2019) 6345−6352.
[107] H.-N. Wang, J.K. Register, A.M. Fales, N. Gandra, E.H. Cho, A. Boico, G.M. Palmer, B. Klitzman, T. Vo-Dinh, SERS nanosensors for in vivo detection of nucleic acid targets in a large animal model, Nano Res. 11 (8) (2018) 4005−4016.
[108] C. Tuerk, L. Gold, Systematic evolution of ligands by exponential enrichment: RNA ligands to bacteriophage T4 DNA polymerase, Science 249 (4968) (1990) 505−510.
[109] N.H. Kim, S.J. Lee, M. Moskovits, Aptamer-mediated surface-enhanced Raman spectroscopy intensity amplification, Nano Lett. 10 (10) (2010) 4181−4185.
[110] E. Chung, J. Jeon, J. Yu, C. Lee, J. Choo, Surface-enhanced Raman scattering aptasensor for ultrasensitive trace analysis of bisphenol A, Biosens. Bioelectron. 64 (2015) 560−565.

CHAPTER

SERS biosensors for point-of-care infectious disease diagnostics

4

Hoan Thanh Ngo[1], Tuan Vo-Dinh[2]

[1]*School of Biomedical Engineering, International University, Vietnam National University, Ho Chi Minh City, Vietnam;* [2]*Fitzpatrick Institute for Photonics, Department of Biomedical Engineering, Department of Chemistry, Duke University, Durham, NC, United States*

Introduction

In the time of pandemics like the current ongoing COVID-19 pandemic, timely diagnostic of infectious diseases is critical to pandemic control and healthcare. Conventional methods for infectious disease diagnostics usually involve collecting and transporting samples to central laboratories inside clinics/hospitals for laboratory testing. Common types of laboratory tests for infectious diseases include staining and examination under microscopy, culturing, testing for host—response antibodies (antibodies produced by patient's immune system in response to pathogens), testing for pathogenic antigens (e.g., ELISA), and testing for pathogenic genetic materials (e.g., PCR). For patients living in rural and remote areas, these laboratory-based methods usually have long turn-around times of one to several days, since patients usually have to travel or samples have to be transported over long distances before reaching central laboratories. Moreover, the wait time at the central laboratories further prolongs the turn-around time. Since diagnostics will determine treatment strategies, long turn-around times cause delays in treatment. For highly contagious infectious diseases such as COVID-19, a long turn-around time in diagnosis also increases the risk of further spreading the disease to the surrounding communities.

Many efforts have been made to develop assays, biosensors, and systems that can rapidly detect infectious diseases at the point-of-care (POC) (e.g., patient home, at doctor's office, in the field ...) so that effective treatment can be started in a timely fashion. Generally, these efforts can be classified into two main directions. One direction is to integrate the existing assays that are currently being used in laboratories into portable, compact devices that are suitable for deployment at the POC. For example, efforts have been made to integrate nucleic acid extraction, PCR amplification, and PCR product detection into microfluidic chips for diagnosis of various diseases. Another direction is to develop novel assays for POC applications. In general, these assays can be categorized into classes based on (1) the type of target molecule (e.g., pathogenic antigen, pathogenic nucleic acid, host—response antibody, metabolites) or (2) the type of biorecognition element (e.g., antibody, aptamer,

SERS for Point-of-care and Clinical Applications. https://doi.org/10.1016/B978-0-12-820548-8.00004-7

nucleic acid, molecular imprint) or (3) the type of readout method (e.g., electrochemical, optical, colorimetric, plasmonic). Applications of these readout methods for POC infectious disease diagnostics have been discussed in a recent review [1].

Herein, we focus on POC infectious disease diagnostic assays and biosensors utilizing surface-enhanced Raman scattering (SERS) as the readout method. SERS is a spectrochemical process in which the Raman scattering of molecules adsorbed on metallic nanostructures is enhanced millions of times or more. It is a sensitive sensing method capable of achieving single molecule limits of detection [2]. In comparison to fluorescence, SERS exhibits narrower peaks, making it more suitable for multiplex detection in a single assay, a highly desirable feature for infectious disease detection of complex, real-life samples. In addition, SERS only requires a single excitation source for multiplexing and is more resistant to photobleaching. SERS can be excited in the near-IR region where light absorption of biological matrices is low. Therefore, its optical output is less affected by biological matrices, thus potentially simplifying the sample preparation process.

Application of SERS for infectious disease diagnostics at the POC has attracted a lot of research interest, as shown in several reviews relating to this topic [3—10]. In this work, we focus on SERS-based biosensors for POC infectious disease diagnostics that have been reported in the last 5 years. We categorize these biosensors based on their type of biorecognition element (i.e., bioreceptor). There are a total four categories: (1) antibody-based SERS biosensors, (2) aptamer-based SERS biosensors, (3) nucleic acid—based SERS biosensors, and finally (4) SERS biosensors without a bioreceptor. For each biosensor, we will discuss the following characteristics but not limited to: (1) working principle and analytical process, (2) limit of detection (LOD), (3) ability to work with real samples, and (4) other features when appropriate.

Antibody-based SERS biosensors

Antibody-based SERS biosensors are biosensors that utilize antibodies as biorecognition elements to capture target molecules and SERS as the readout method. Examples of target molecules that this kind of bioreceptor can capture include pathogens (virus, bacteria, toxins), pathogen-released proteins, host—response proteins, host—response antibodies, etc. Many antibody-based SERS biosensors have been developed and showed great promises because almost no sample preparation is required. It is noteworthy that the specificity of this kind of biosensor is strongly dependent on the antibody's specificity. In addition, antibody cost and stability are also issues that need to be taken into account.

One example of antibody-based SERS biosensor is the SERS nanosensor developed by Kearns et al. [11] for multiplex detection of three bacterial pathogens, including *Escherichia coli*, *Salmonella typhimurium*, and methicillin-resistant *Staphylococcus aureus* (Fig. 4.1). First, lectin-functionalized silver-coated magnetic nanoparticles and SERS-active silver nanoparticles functionalized with antibodies

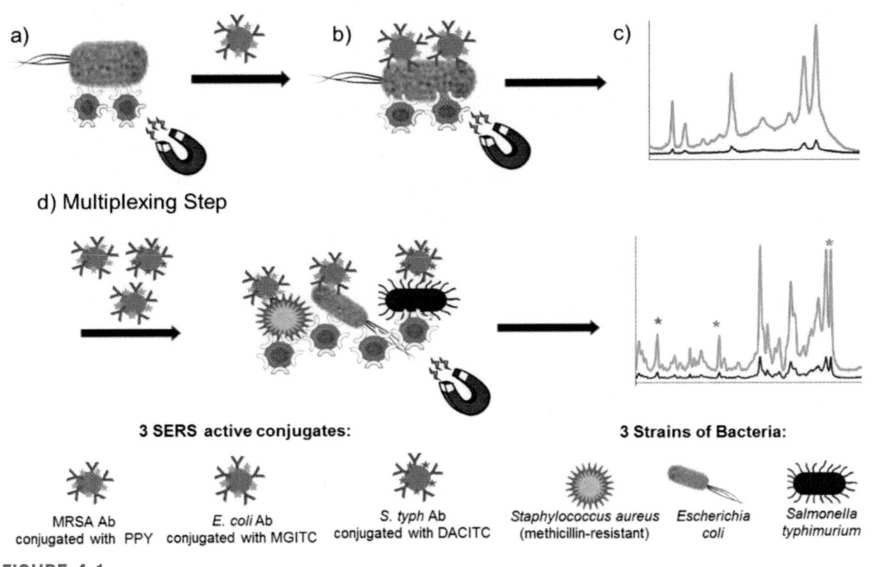

FIGURE 4.1

SERS detection of three bacterial pathogens including *Escherichia coli*, *Salmonella typhimurium*, and methicillin-resistant *Staphylococcus aureus* using nanosensors. Silver-coated magnetic nanoparticles functionalized with lectin were used for magnetic enrichment, and SERS-active silver nanoparticles functionalized with antibodies were used as SERS nanotags.

Reprinted with permission from H. Kearns, R. Goodacre, L. E. Jamieson, D. Graham, K. Faulds, SERS detection of multiple antimicrobial-resistant pathogens using nanosensors, Anal. Chem. 89 (23) (December 2017) 12666–12673. Copyright 2017 American Chemical Society.

were added to sample solutions and the mixture was thoroughly mixed for 30 min. Subsequently, the mixture was placed on a magnetic rack for 30 min for sample collection. Supernatant was then removed, and the remaining solution was resuspended in deionized water for analysis. The LOD for each bacteria strain was 10^1 colony-forming unit (CFU) per milliliter (mL). Multiplex detection of three bacteria in the same sample matrix was demonstrated as well. However, the possibility to analyze real clinical samples was not demonstrated in this work.

In another work, Chattopadhyay et al. [12] developed a SERS immunosensor for detection of *S. typhimurium*. Polymeric magnetic nanoparticles functionalized with antibodies against Salmonella's common structural antigen (capture probe) and Raman reporter–labeled gold nanoparticles functionalized with the same antibodies (signal probe) were prepared. For detection, capture probe solution and pathogen solution were first mixed and incubated for 1 h at 37°C. Then the immunocomplexes were isolated by magnetic force for 30 s before being mixed with a signal probe solution and incubated for another 1h at 37°C. The capture probe—*S. typhimurium*—signal probe complexes were washed twice with phosphate-buffered saline (PBS) buffer followed by resuspension in PBS solution, magnetic pulldown, and SERS measurement. The sensor could specifically detect *S. typhimurium* with LOD of 10 cells/mL. Furthermore, the authors also demonstrated that their method can detect the bacteria spiked in different real food matrices.

Liyan Bi et al. [13] developed SERS-active Au@Ag core–shell nanorods functionalized with biotinylated antibodies (SERS nanotags) for detection of *E. coli*. First, SERS nanotags were added to the sample solution. In the presence of *E. coli* in sample solution, SERS nanotags bound onto surface of *E. coli*, forming *E. coli*–SERS nanotag complexes. These complexes were then extracted using centrifugation and transferred into a glass capillary for SERS measurements. The method could specifically detect *E. coli* over the range of 10^7–10^2 CFU/mL with an LOD of 10^2 CFU/mL, which is lower than the standard diagnostic threshold for measurement of organisms in urine (10^4 CFU/mL). The performance of the method in blood samples was demonstrated by using mouse blood spiked with *E. coli* at 10^4 CFU/mL. Specifically, the authors demonstrated that by using principal component analysis (PCA), they could discriminate between SERS spectra of ampicillin-resistant *E. coli* and of ampicillin-susceptible *E. coli*. The antibiotic resistance testing took less than 3.5 h.

Shen et al. [14] developed a SERS-based lateral flow assay (LFA) that could differentially detect wild-type and vaccine-type *pseudorabies virus* (PRV), a pathogen that causes an acute infectious disease in pigs (Fig. 4.2). Gold–silver core–

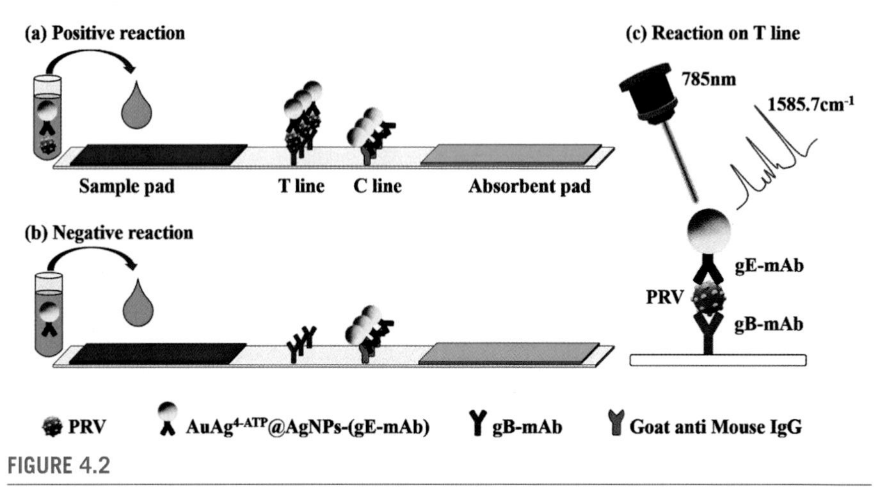

FIGURE 4.2

SERS-based LFA for differential diagnosis of wild-type *pseudorabies virus* and vaccine-type *pseudorabies virus*, a pathogen that causes an acute infectious disease in pigs. In the presence of wild-type *pseudorabies virus* (positive sample), both control line and test line exhibit SERS signal. Without the presence of wild-type *pseudorabies virus* (negative sample), only control line exhibits SERS signal.

Reprinted with permission from H. Shen et al., A novel SERS-based lateral flow assay for differential diagnosis of wild-type pseudorabies virus and gE-deleted vaccine, Sensors Actuators B Chem. 282 (June 2018) (2019) 152–157. Copyright 2018 Elsevier.

shell bimetallic nanoparticles embedded with 4-ATP and functionalized with anti-PRV gE monoclonal antibodies (gE-mAb) were used as SERS nanotags. The test line was spotted with anti-PRV gB monoclonal antibodies (gB-mAb) and the control line goat anti-mouse IgG. Only in the presence of wild-type PRV, the only PRV which expresses both protein gB and protein gE, sandwiches of gB-mAb—wild-type PRV—gE-mAb functionalized SERS nanotag were formed at the test line, and SERS signal could be measured by a portable Raman reader. On the other hand, SERS nanotags always bound to goat anti-mouse IgG at the control line. The results showed an LOD of 5 ng/mL and a linear detection range of 41–650 ng/mL. Particularly, experiments with clinical samples were conducted, and the results correlated with PCR. Although the method's sensitivity was inferior to PCR, it did not require nucleic acid extraction and target amplification.

Hwang et al. [15] developed a SERS-based LFA for detection of *Staphylococcal enterotoxin B* (SEB) using Raman reporter–labeled hollow gold nanospheres. The LFA strip was composed of the four sections: a sampling loading pad, a conjugate pad, a nitrocellulose membrane, and an absorbent pad. Analytical procedure started with dipping LFA strip into sample solution. Through capillary action, sample solution containing target antigens migrated through the sample loading pad to the conjugate pad where antigens were captured by hollow gold nanospheres functionalized with mouse anti-SEB (called SERS nanotags). These immunocomplexes then migrated to the test line in the nitrocellulose membrane where they were captured by rabbit anti-SEB antibodies immobilized at the test line. The excess SERS nanotags migrate past the test line to the control line (also in the nitrocellulose membrane) where they were captured by mouse anti-IgG immobilized at the control line. LOD was estimated to be 0.001 ng/mL (Fig. 4.3), which was three orders of magnitude more sensitive than that of a corresponding ELISA-based method. A follow-up work from this group demonstrated that their method can be adapted for detection of bacterial pathogens such as *Yersinia pestis*, *Francisella tularensis*, and *Bacillus anthracis* [16]. The assay time was 15 min, and only a small volume of pathogen sample (40 uL) was needed. The possibility to analyze real clinical samples was not demonstrated in these works.

Liu et al. [17] developed an LFA based on Fe_3O_4@Au SERS tags for multiplex detection of two inflammation biomarkers, the serum amyloid A (SAA), and the C-reactive protein (CRP), in unprocessed blood samples (Fig. 4.4). Fe_3O_4@Au SERS tags conjugated with SAA antibodies and Fe_3O_4@Au SERS tags conjugated with CRP antibodies were first added to sample solutions and incubated for 15 min. Following magnetic enrichment, Fe_3O_4@Au tags-SAA/CRP immune complexes were resuspended in running solution before being loaded onto the sample pad of LFA strip for immune binding and SERS measurement using a portable Raman system. Using this method, the authors could detect the two biomarkers, SAA and CRP, in unprocessed blood samples with an LOD of 0.1 and 0.01 ng/mL, respectively. This sensitivity is 100-fold–1000-fold more sensitive than that of a standard colloidal gold strip using the same antibodies. Besides SAA and CRP, this method could also detect influenza A H1N1 virus and human adenovirus as demonstrated in another work also reported by this research group [18].

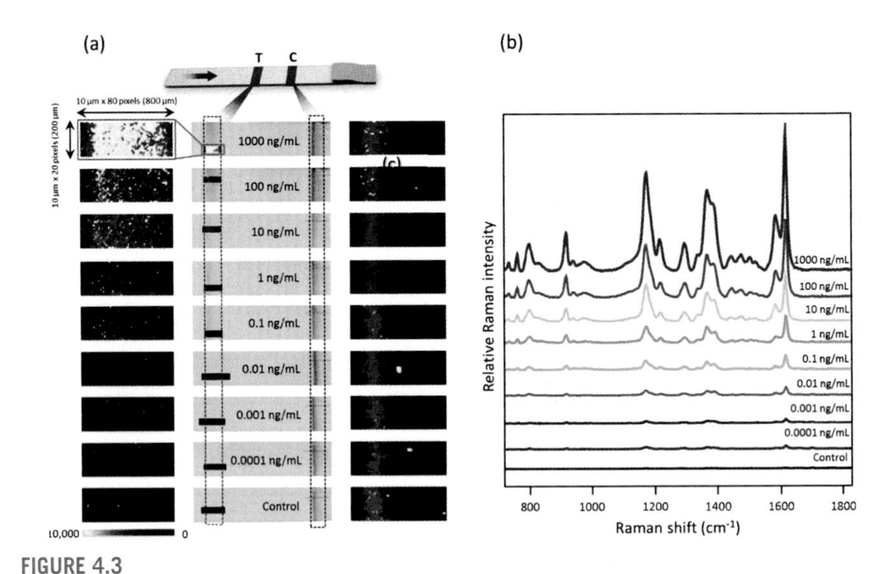

FIGURE 4.3

SERS-based LFA for detection of *Staphylococcal enterotoxin B* (SEB) using Raman reporter-labeled hollow gold nanospheres.

Reprinted with permission from J. Hwang, S. Lee, J. Choo, Application of a SERS-based lateral flow immunoassay strip for the rapid and sensitive detection of staphylococcal enterotoxin B, Nanoscale 8 (22) (2016) 11418–11425. Copyright 2016 Royal Society of Chemistry.

Aptamer-based SERS biosensors

Aptamer-based SERS biosensors are biosensors that utilize aptamers as biorecognition and SERS as the readout method. Target molecules of aptamer bioreceptors are the same as antibody bioreceptors. However, compared to antibody bioreceptors, aptamer bioreceptors exhibit several advantages including lower production time and cost, higher stability, and higher specificity. Therefore, application of aptamers in biosensing, including biosensing in which SERS is used as the readout method, has attracted much research interest.

Susana Díaz-Amaya et al. [19] developed an aptamer-based SERS biosensor for whole cell detection of *E. coli* O157:H7 (Fig. 4.5). SERS nanotags composed of 4-aminothiophenol gold nanoparticles functionalized with aptamers were added to sample solution. In the presence of *E. coli*, these SERS nanotags bind onto the surface of *E. coli* and sediment. This leads to a decrease in SERS nanotag content in the supernatant, thus leading to a reduction in the supernatant's SERS signal. In this method, the SERS signal showed negative correlation with *E. coli* concentration. Using this method, the authors could detect low concentrations of *E. coli* in both pure cultures ($\sim 10^1$ CFUmL^{-1}) and ground beef samples ($\sim 10^2$ CFUmL^{-1}). On the other hand, interferent microorganisms such as *E. coli* B1201, *E. coli* O55:H7, *Listeria monocytogenes*, *S. aureus*, and *Salmonella typhi* generated SERS responses similar to the background.

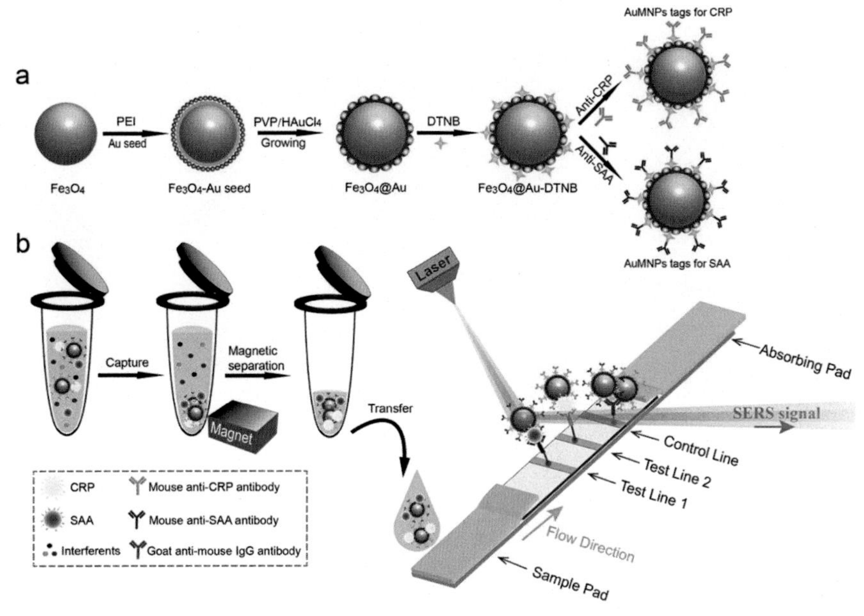

FIGURE 4.4

SERS-based LFA for simultaneous detection of serum amyloid A (SAA) and C-reactive protein (CRP) in unprocessed blood sample. SAA and CRP were tagged and separated from the sample solution by Fe_3O_4@Au SERS tags before being loaded onto LFA strip's sample pad for detection.

Reprinted with permission from X. Liu et al., Fe_3O_4@Au SERS tags-based lateral flow assay for simultaneous detection of serum amyloid A and C-reactive protein in unprocessed blood sample, Sensors Actuators B Chem. 320 (2020) 128350. Copyright 2020 Elsevier.

Ma et al. [20] developed a SERS aptasensor for *S. typhimurium* detection using (1) SERS nanotags composed of spiny gold nanoparticles decorated with 4-mercaptobenzoic acid and thiolated *S. typhimurium* aptamers, and (2) microtiter plate decorated with biotinylated *S. typhimurium* aptamers via streptavidin–biotin conjugation (Fig. 4.6). The analytical procedure started with adding sample solutions into each well of the microtiter plates and incubated at 37°C for 30 min followed by 3 times washing and air drying, SERS nanotags were then added, incubated at 37°C for 30 min, washed 3 times, and air dried prior to SERS measurements. Using this method, the authors achieved an LOD of 4 CFU/mL and a detection range from 10^1 to 10^5 CFU/mL. Notably, they demonstrated the ability to detect *S. typhimurium* inoculated into pork samples.

Nuo Duan et al. [21] developed a SERS aptasensor for detecting multiple pathogens by sandwich capturing of target pathogens between (1) a SERS-active substrate, which is a gold nanoparticle–decorated polydimethylsiloxane (PDMS) film functionalized with aptamers, and (2) SERS nanotags, which are gold nanoparticles

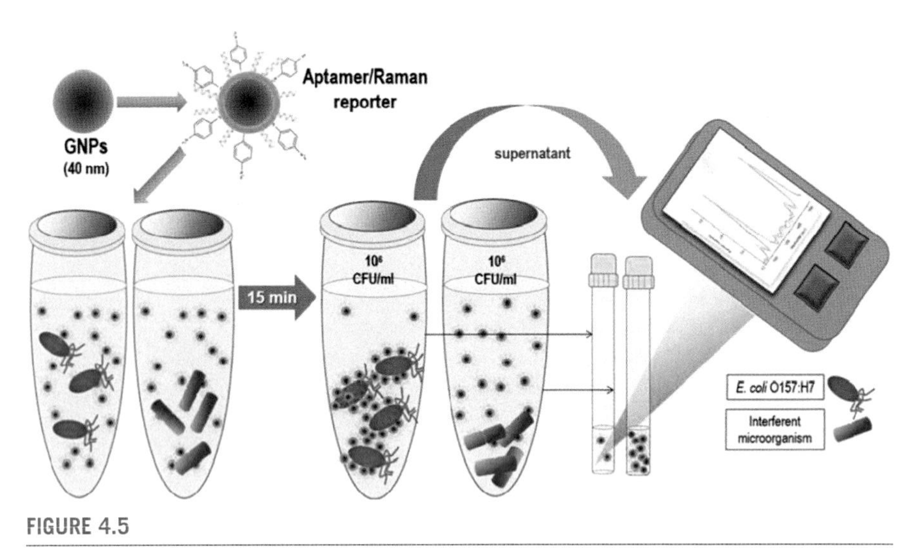

FIGURE 4.5

Aptamer-based SERS biosensor for whole cell analytical detection of *E. coli* O157:H7. In the presence of *E. coli* O157:H7 in sample solution, SERS nanotags bound onto *E. coli's* surface and sediment, resulting in less SERS nanotags in the supernatant, thus smaller SERS signal. The higher *E. coli* concentration, the smaller SERS signal.

Reprinted with permission from S. Díaz-Amaya, L. K. Lin, A. J. Deering, L. A. Stanciu, Aptamer-based SERS biosensor for whole cell analytical detection of E. coli *O157:H7, Anal. Chim. Acta 1081 (2019) 146–156. Copyright 2019 Elsevier.*

(AuNP) integrated with Raman reporters and functionalized with aptamers (Fig. 4.7). The analytical procedure included (1) dip the substrate into sample solution and incubate, (2) take the substrate out and wash twice with PBS buffer, (3) dip the washed substrate into SERS nanotags solution and incubate, and finally (4) take the substrate out, wash, and measure the SERS signal. Using this method, the authors could specifically detect *V. parahaemolyticus* and *S. typhimurium* at concentrations as low as 18 and 27 CFU/mL, respectively. Testing with real samples was also conducted. Seafood samples including salmon, oyster, and shrimp were spiked with the pathogens at concentrations from 10^2 to 10^4 CFU/mL before detection. In comparison to the classic plate count method, this method showed high accuracy with recoveries ranging from 82.9% to 95.1%.

Zhang et al. [22] developed a SERS-based method for detection of *E. coli* and *S. aureus*. Vancomycin-modified Fe_3O_4@Au magnetic nanoparticles (MNPs) were used as universal bacterial capturers and aptamer-modified AuNPs as specific SERS nanotags. Compared to the other works previously discussed, this is interesting because of the use of Vancomycin in combination with aptamer as biorecognition molecules. Vancomycin possesses broad-spectrum bacteria recognition capability, making Vancomycin-modified Fe_3O_4@Au MNPs universal bacterial capturers while aptamer-modified AuNPs serve as specific SERS tags.

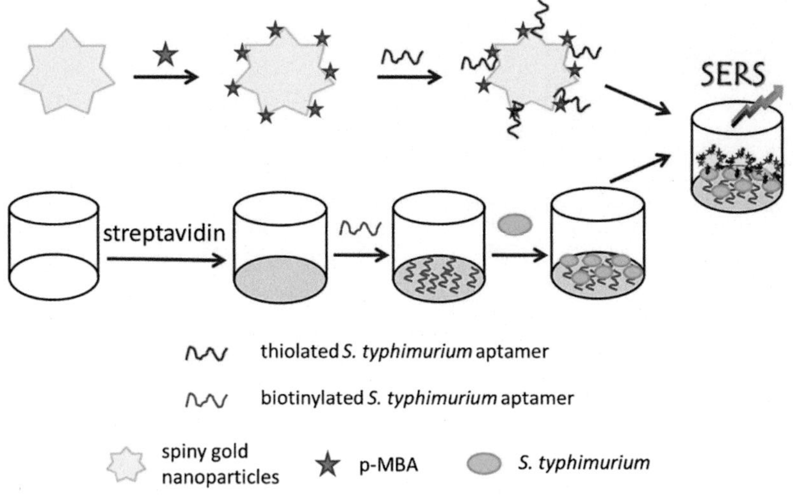

FIGURE 4.6

SERS aptasensor for *Salmonella typhimurium* (*S. typhimurium*) detection based on spiny gold nanoparticles. *S. typhimurium* are captured by biotinylated *S. typhimurium* aptamers at the bottom of microtiter plate's well before being labeled by SERS nanotags, spiny gold nanoparticles decorated with 4-mercaptobenzoic acid and thiolated *S. typhimurium* aptamers.

Reprinted with permission from X. Ma, X. Xu, Y. Xia, Z. Wang, SERS aptasensor for Salmonella typhimurium *detection based on spiny gold nanoparticles, Food Control 84 (2018) 232–237. Copyright 2017 Elsevier.*

The LODs of this method were 20 and 50 cells/mL for *S. aureus* and *E. coli*, respectively, and the possibility to analyze real urine clinical samples was demonstrated.

Nucleic acid–based SERS biosensors

Nucleic acid–based SERS biosensors are biosensors that utilize nucleic acid sequences as the biorecognition element and SERS as the readout method. Target molecules of these bioreceptors are DNA or RNA of pathogens. Since this type of biosensor can provide genetic information about the pathogen, it is highly specific. However, sample preparation is required because of the need to first expose a pathogen's DNA or RNA before capture and detection.

Our group reported the first analytical application of SERS [24] and developed a SERS chip–based method for detection of DNA through a single step, with no washing needed [23]. The SERS chip was composed of silver/gold coated nanosphere arrays, referred to as "Nanowave" [25,26]. For DNA detection, the chip was functionalized with Molecular Sentinel (MS) probes. Normally, these MS probes are in close state, keeping Raman labels in close proximity to SERS substrate's surface and inducing strong SERS signal. The DNA detection process starts

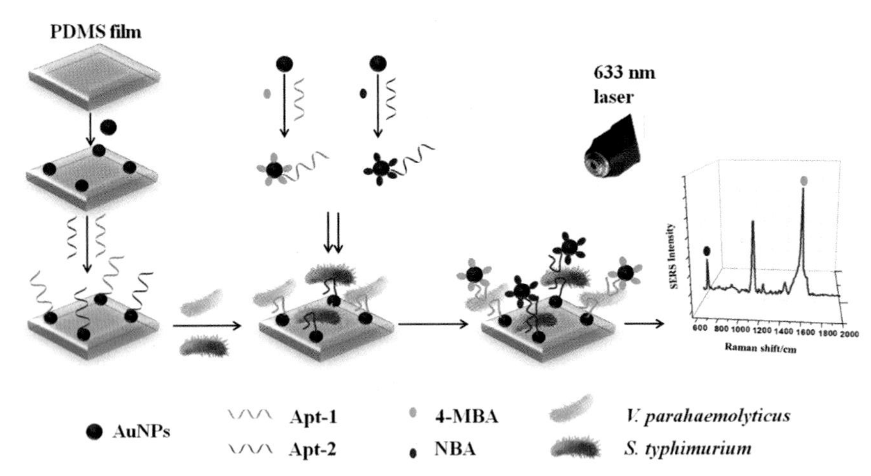

FIGURE 4.7

SERS aptasensor for simultaneous multiple pathogens detection using gold decorated PDMS substrate. *Vibrio parahaemolyticus* and *Salmonella typhimurium* were captured by aptamers immobilized on gold nanoparticle decorated PDMS film before being labeled by SERS nanotags, gold nanoparticles integrated with Raman reporters and aptamers.

Reprinted with permission from N. Duan, M. Shen, S. Qi, W. Wang, S. Wu, Z. Wang, A SERS aptasensor for simultaneous multiple pathogens detection using gold decorated PDMS substrate, Spectrochim. Acta - Part A Mol. Biomol. Spectrosc. 230 (2020) 118103. Copyright 2020 Elsevier.

with drop-casting sample solutions onto the SERS chip. In the presence of target DNA in the sample solution, target DNA hybridize with the MS probes, opening the MS probes. Raman labels are displaced away from the SERS substrate's surface, resulting in a decrease of SERS signal (Fig. 4.8). The method is single step and without any washing, making it easy-to-use. A follow-up study by our group showed that this method can be adapted for multiplex detection of two different target DNA sequences in a single assay [27]. Furthermore, we devised a variant of this method called the inverse molecular sentinel (iMS) method that provides off-to-on signal change in the presence of target DNA [28]. In later studies, we demonstrated the possibility of the iMS method for cancer diagnosis using clinical patients' samples [29] and for live plant assays [30] without the need of target amplification.

Donnelly et al. [31] developed a SERS-based DNA detection method using (1) silver nanoparticles conjugated with DNA probes and Raman dyes (SERS nanotags) and (2) silver-coated magnetic nanoparticles conjugated with DNA probes. In the presence of target DNA, DNA sandwich hybridization formed and created clusters of silver nanoparticles and silver-coated magnetic nanoparticles. A permanent magnet was then applied to concentrate the formed clusters for SERS measurement (Fig. 4.9). This method achieved 20 fmol LOD and was capable of multiplex detection of two different target DNA sequences. The method's ability to work with real clinical samples was not demonstrated in this work.

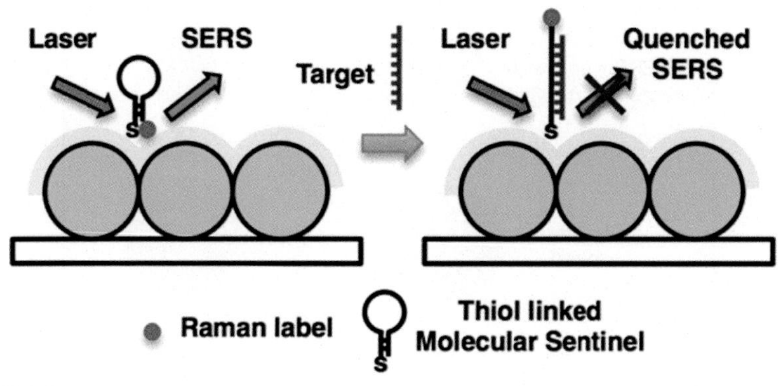

FIGURE 4.8

Label-free DNA biosensor based on SERS molecular sentinel on nanowave chip. Target DNA hybridizes with thiol-linked molecular sentinel probe, opening the probe and bringing Raman label away from nanowave chip surface, thus resulting in quenched SERS signal. The higher target DNA concentration, the lower SERS signal.

Reprinted with permission from H. T. Ngo, H.-N. Wang, A. M. Fales, T. Vo-Dinh, Label-free DNA biosensor based on SERS molecular sentinel on nanowave chip, Anal. Chem. 85 (13) (2013). Copyright 2013 American Chemical Society.

Our laboratory also developed a SERS-based DNA detection method using a sandwich hybridization-based method but utilized magnetic microbeads and ultrabright SERS nanorattles instead (Fig. 4.10) [32]. The method's novelty lies in the use of ultrabright SERS nanorattles, which are approximately three orders of magnitude brighter than 60 nm gold nanoparticles. With this method, we achieved an LOD

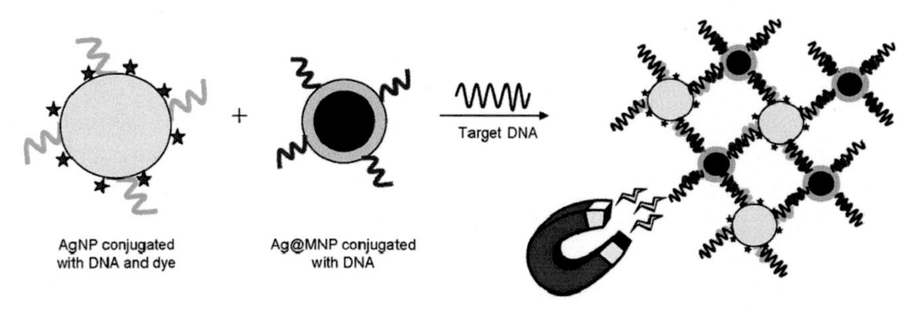

FIGURE 4.9

Silver and magnetic nanoparticles for sensitive DNA detection by SERS. Target DNA are sandwiched by silver-coated magnetic nanoparticles conjugated with DNA and SERS nanotags (silver nanoparticles conjugated with DNA and Raman dyes), forming clusters. A permanent magnet was then used to concentrate the clusters for SERS measurement.

Reprinted with permission from T. Donnelly, W. E. Smith, K. Faulds, D. Graham, Silver and magnetic nanoparticles for sensitive DNA detection by SERS, Chem. Commun. 50 (85) (2014) 12907–12910. Copyright 2014 Royal Society of Chemistry.

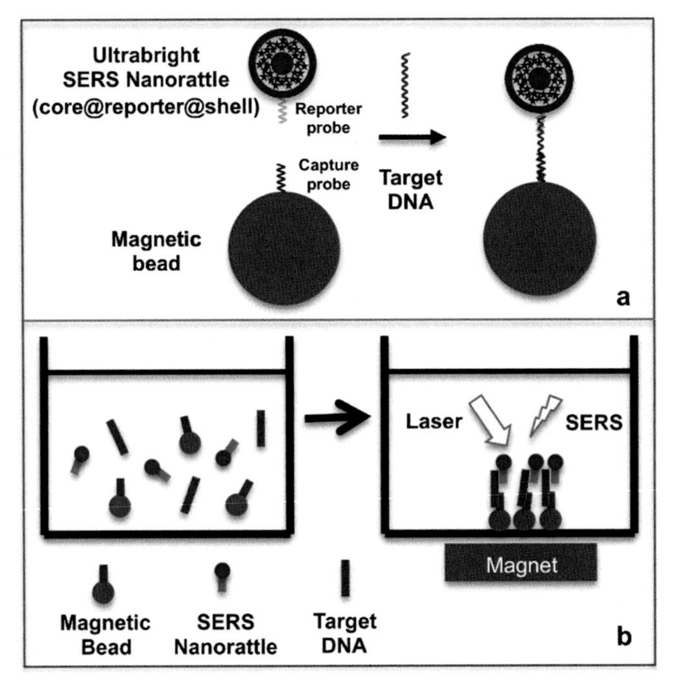

FIGURE 4.10

Sensitive DNA detection and SNP discrimination using ultrabright SERS nanorattles and magnetic microbeads. In the presence of target DNA, hybridization sandwiches are formed. A permanent magnet was used to concentrate the hybridization sandwiches at one spot for SERS measurement.

Reprinted with permission from H. T. Ngo, N. Gandra, A. M. Fales, S. M. Taylor, T. Vo-Dinh, Sensitive DNA detection and SNP discrimination using ultrabright SERS nanorattles and magnetic beads for malaria diagnostics, Biosens. Bioelectron. 81 (2016). Copyright 2016 Elsevier.

of ~3 p.m. or below 100 amol given the sample volume of 30 μL. Furthermore, we demonstrated that this method could differentiate wild-type and mutant malaria DNA with a single-nucleotide difference (single-nucleotide polymorphism-SNP). Real clinical samples were not used in this work.

In a follow-up work [33], we integrated the method into a "lab-in-a-stick" portable system that supports automatic washing of magnetic beads (Fig. 4.11). More importantly, we demonstrated that the method could directly detect malaria nucleic acid in blood lysate without nucleic acid extraction or target amplification. To the best of our knowledge, this was the first SERS-based method capable of directly detecting pathogenic RNA in blood lysate without RNA extraction and amplification. Recently, we tested the method with 20 real clinical tissue samples of head and neck squamous cell carcinoma and found 100% sensitivity and 97% specificity in comparison to pathology review [34]. Although RNA was extracted from clinical tissue samples prior to detection, target amplification was not needed.

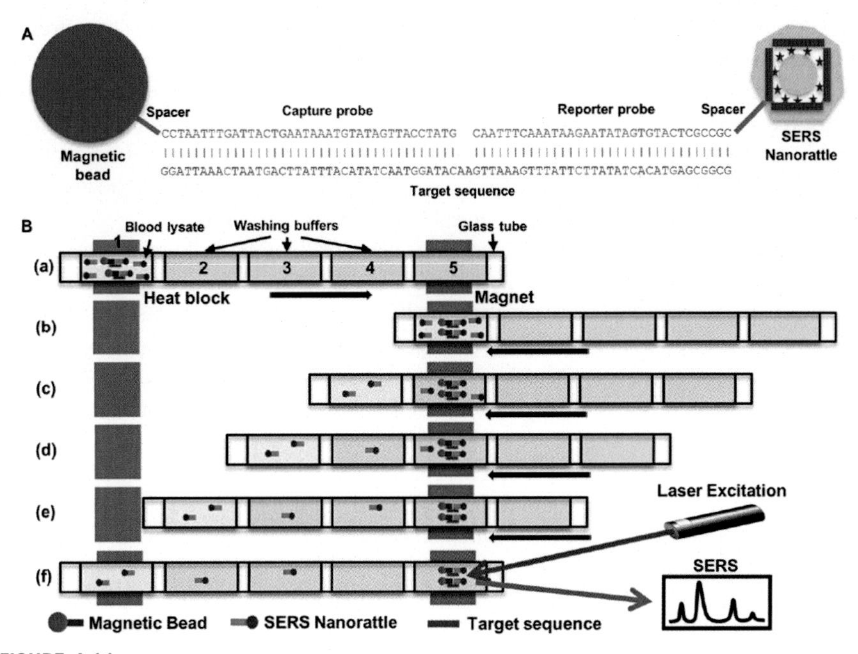

FIGURE 4.11

Direct detection of unamplified pathogen RNA in blood lysate using an integrated lab-in-a-stick device and ultrabright SERS nanorattles. Hybridization sandwiches of magnetic bead with capture probe—target DNA—SERS nanorattle with reporter probe were transported through three washing solutions by pulling the glass tube forward and backward relatively to a permanent magnet. After washing, the hybridization sandwiches are concentrated by the same permanent magnet for SERS measurement.

Reprinted from H. T. Ngo et al., Direct detection of unamplified pathogen RNA in blood lysate using an integrated lab-in-a-stick device and ultrabright SERS nanorattles, Sci. Rep. 8 (1) (2018).

It is noteworthy that the nucleic acid biosensors described above are based on signal amplification. Another promising direction for nucleic acid testing at the POC is to employ isothermal techniques such as loop-mediated isothermal amplification (LAMP) or recombinase polymerase amplification (RPA) for target amplification before detection. Especially, recent works combining LAMP/RPA isothermal amplification and a method based on clustered regularly interspaced short palindromic repeats (CRISPR) showed great potential for infectious disease diagnostics, particularly COVID-19 diagnosis, at the POC [35,36].

SERS biosensors without bioreceptor

This type of SERS biosensors does not utilize bioreceptors for target capturing. Instead, the targets of interest are detected through their intrinsic SERS fingerprints.

Target molecules are brought to close proximity of SERS-active nanostructures by different ways such as drop casting samples on SERS substrates or mixing samples with nanoparticles. The intrinsic SERS fingerprints of target molecules are then measured. The biggest advantage of this kind of SERS biosensors is its simple analytical procedure with almost no sample preparation required and no washing needed. However, since the produced SERS spectra are complex, sophisticated spectral analysis or machine learning methods are usually required.

Lim et al. [37] developed a method that can differentiate influenza virus DNA-infected cells from uninfected cells using SERS and PCA (Fig. 4.12). First, cell solutions were centrifuged and the culture media on top of the cell pellets was removed by aspiration. The pellets were then resuspended in PBS, followed by drop casting onto a SERS substrate composed of 80 nm spherical gold nanoparticles on cover glass. Following complete drying through vaporization at room temperature, SERS spectra were recorded. To differentiate influenza virus DNA-infected cells from uninfected cells, PCA was used. The results showed that the method could differentiate three groups including uninfected cells, wild-type influenza virus DNA-infected cells, and mutant influenza virus DNA-infected cells. Capability to analyze real clinical samples was not reported in this study.

Durmanov et al. [38] developed a novel SERS substrate composed of a thin silver film with nanocavities deposited on mica substrate by electron beam physical vapor deposition method. Sample solutions were analyzed by pipetting an aliquot onto the SERS substrate's surface followed by laser focusing and SERS measurement. Using a statistical classification approach (PCA and linear discriminant analysis) to analyze the acquired SERS spectra, the authors could differentiate four virus species including rabbit myxomatosis virus, canine distemper virus, tobacco mosaic virus, and potato virus X. Although effective in differentiating virus species, method's limit of sensitivity was not tested. Also, detection of virus in the presence of real sample matrix such as blood, which could exhibit interferent Raman bands and make spectral analysis challenging, was not demonstrated.

Gahlaut et al. [39] employed silver nanorod arrays fabricated by a glancing angle deposition technique for dengue virus detection at the early stage (within 5 days of the onset of symptoms). NS1 protein, which is present in the blood and tissue samples up to 5–7 days after the onset of dengue fever, was used as the biomarker. The detection process started with a centrifugation step to obtain blood serum, followed by drop casting it onto silver nanorods arrays, SERS measurement, and subsequent PCA analysis (Fig. 4.13). Pure NS1 samples, NS1-spiked fetal bovine serum samples, and clinical blood samples were used to test the method. The results showed that the method could differentiate healthy samples, dengue-negative fever samples, and dengue-positive fever samples. Notably, handheld Raman readers were used for SERS measurements, making this method field deployable.

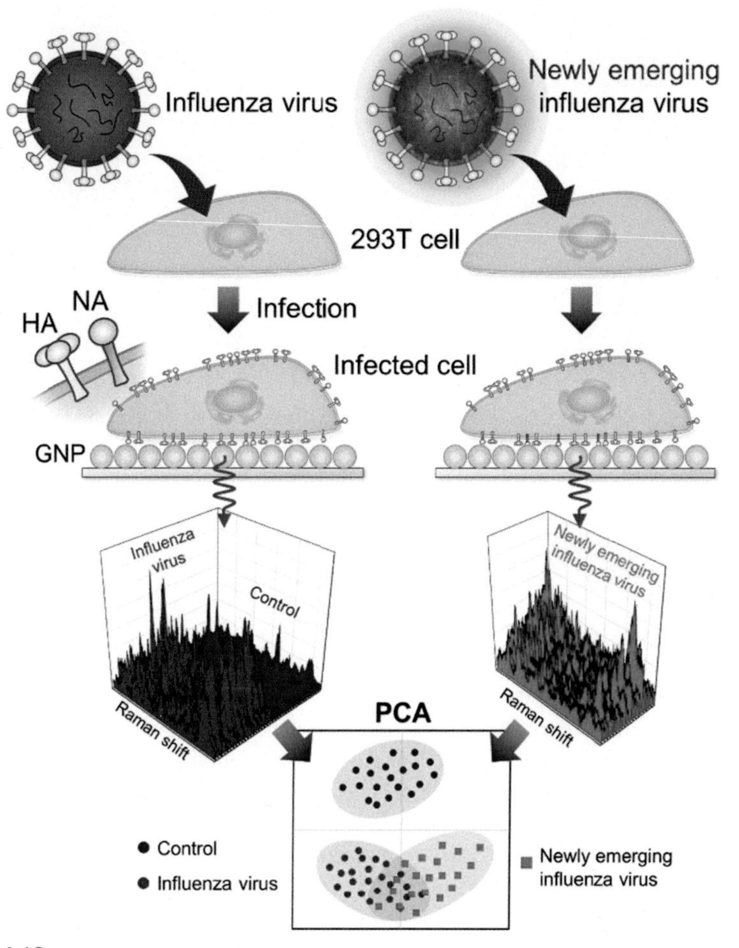

FIGURE 4.12

Identification of newly emerging influenza viruses by detecting the virally infected cells based on SERS and PCA. Cells infected by influenza virus and newly emerging influenza virus are drop casted onto SERS substrate followed by SERS measurement and PCA analysis. With this method, three groups of uninfected cells (control), wild-type influenza virus DNA-infected cells, and mutant influenza virus DNA-infected cells could be differentiated.

Reprinted with permission from J. Y. Lim et al., Identification of newly emerging influenza viruses by detecting the virally infected cells based on surface enhanced Raman spectroscopy and principal component analysis, Anal. Chem. 91 (9) (2019) 5677–5684. Copyright 2019 American Chemical Society.

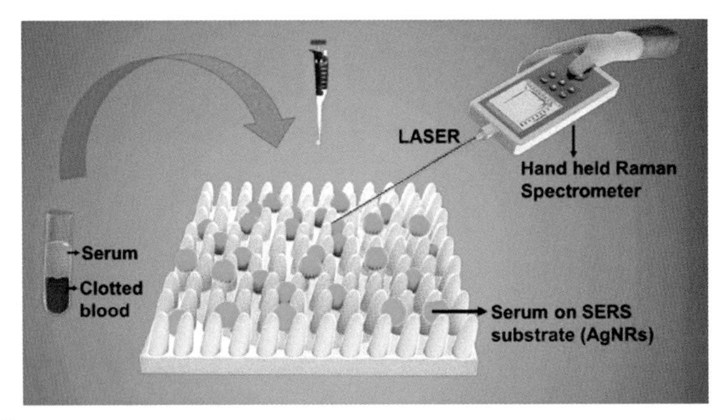

FIGURE 4.13

SERS platform for dengue diagnosis from clinical samples employing a hand held Raman spectrometer. Blood serum obtained through centrifugation was drop casted onto SERS substrate for SERS measurement by a handheld Raman spectrometer followed by PCA analysis.

Reprinted with permission from S. K. Gahlaut, D. Savargaonkar, C. Sharan, S. Yadav, P. Mishra, J. P. Singh, SERS platform for dengue diagnosis from clinical samples employing a hand held Raman spectrometer, Anal. Chem. 92 (3) (February 2020) 2527–2534. Copyright 2020 American Chemical Society.

Guo et al. [40] developed a "three-in-one" SERS adhesive tape that could capture bacteria from wound surfaces, release bacteria onto culture medium, and—after a few hours of culturing—detect bacterial presence (Fig. 4.14). To support bacteria capturing, the SERS adhesive tape had a layer of synthetic o-nitrobenzyl derivative molecule, forming an artificial biointerface that could capture pathogens via electrostatic interaction. The captured bacteria were then released onto culture medium via UV-based photocleavage of the o-nitrobenzyl moieties. After a few hours of culturing, SERS signal was measured. Using this method, the authors could detect the infection of *Pseudomonas aeruginosa* and *S. aureus* in a skin burn wound on mice.

Hunter et al. [41] developed an optofluidic system based on hollow-core photonic crystal fiber for detection of bacteria. Sample solutions were first mixed with silver nanoparticles before being injected into the optofluidic system. SERS signals were measured by focusing a laser beam onto one end of the photonic crystal fiber. Experiments with Rhodamine 6G (R6G) samples showed that the use of silver nanoparticles provided 200-fold enhancement of the R6G signal, while the use of photonic crystal fibers provided further 2-fold enhancement. Using genetic support vector machine for the analysis of SERS spectra of mono-culture bacteria in fetal bovine serum, the authors obtained 92% accuracy in classifying spectra into one of three classes (*P. aeruginosa*, *S. aureus*, and *E. coli*) and could count the number of Raman events. Using the same algorithm for the analysis of SERS spectra of mixed culture bacteria in fetal bovine serum, the authors could identify the density of each bacteria in mixed samples with an overall root mean squared error 13.27 CFU/mL. The LOD of bacteria spiked in FBS was 4 CFU/mL with a time requirement of 15 min.

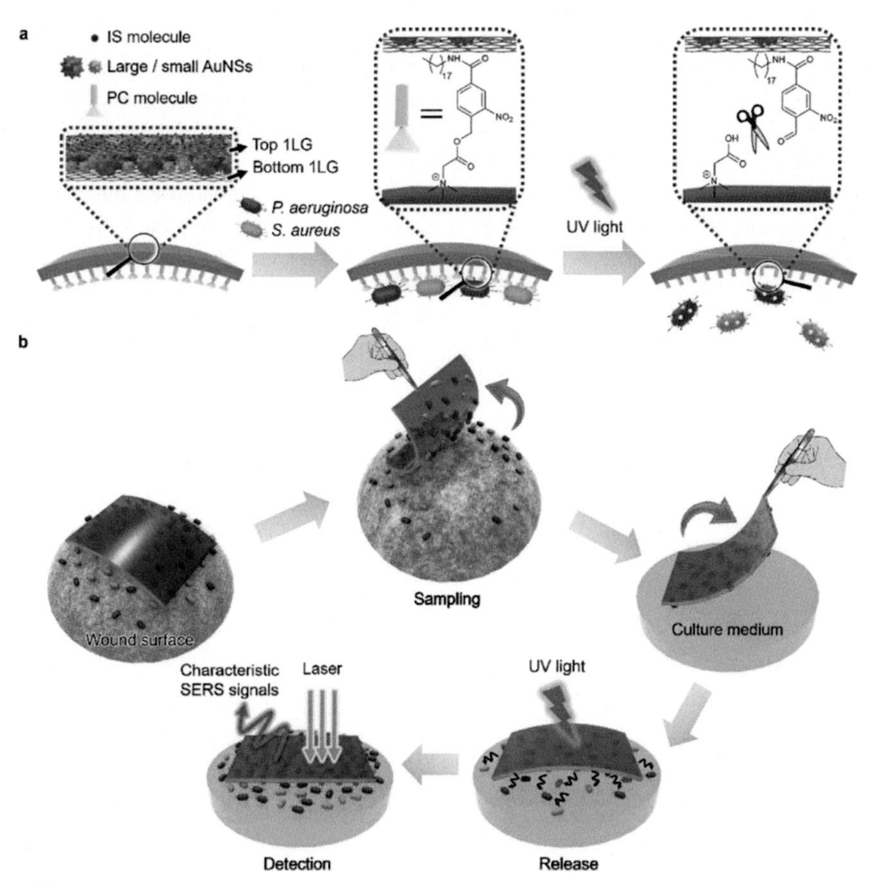

FIGURE 4.14

"Three-in-One" SERS adhesive tape for rapid sampling, release, and detection of wound infectious pathogens. Bacteria were captured by the SERS adhesive tape via electrostatic interaction followed by transferring onto culture medium via UV light photocleavage. After culturing, SERS signal was measured.

Reprinted with permission from J. Guo, Z. Zhong, Y. Li, Y. Liu, R. Wang, H. Ju, 'Three-in-One' SERS adhesive tape for rapid sampling, release, and detection of wound infectious pathogens, ACS Appl. Mater. Interfaces, 11 (40) (October 2019) 36399–36408. Copyright 2019 American Chemical Society.

Conclusion

SERS biosensors have great potential for infectious disease diagnostics at the POC. A wide variety of SERS biosensors have been successfully developed for this purpose. They can be grouped into four categories based on the type of biorecognition element: (1) antibody-based SERS biosensors, (2) aptamer-based SERS biosensors, (3) nucleic acid–based SERS biosensors, (4) and SERS biosensors with no bioreceptor. Each type has its own advantages and disadvantages. Many SERS biosensors showed excellent limits of detection, high specificity, capability to analyze real samples with minimum-to-no sample preparation, short turn-around time, and simplicity for operation. These features make SERS biosensors a potential candidate for diagnosis of infectious disease at the POC, which is in high demand, especially during the current pandemic. The current hurdle is likely the cost of Raman readers, which remain relatively high compared to other readers (absorption, colorimetric methods) for POC applications. However, with the accelerating rate of progress, miniaturization, and cost reduction of Raman spectrometer technology, we could expect see SERS biosensors being used in real-life diagnostic applications for infectious diseases right at the POC in the very near future.

Acknowledgments

The authors acknowledge the support of the National Institutes of Health (1R01GM135486-01) and the Department of Energy Office of Science (DE-SC0019393).

References

[1] M. Sharafeldin, J.J. Davis, Point of care sensors for infectious pathogens, Anal. Chem. (2021).

[2] R. Pilot, R. Signorini, C. Durante, L. Orian, M. Bhamidipati, L. Fabris, A review on surface-enhanced Raman scattering, Biosensors 9 (2) (2019).

[3] J.H. Granger, N.E. Schlotter, A.C. Crawford, M.D. Porter, Prospects for point-of-care pathogen diagnostics using surface-enhanced Raman scattering (SERS), Chem. Soc. Rev. 45 (14) (2016) 3865–3882.

[4] H. Marks, M. Schechinger, J. Garza, A. Locke, G. Coté, Surface enhanced Raman spectroscopy (SERS) for in vitro diagnostic testing at the point of care, Nanophotonics 6 (4) (2017) 681–701.

[5] M. Kahraman, E.R. Mullen, A. Korkmaz, S. Wachsmann-Hogiu, Fundamentals and applications of SERS-based bioanalytical sensing, Nanophotonics 6 (5) (2017) 831–852.

[6] S. Laing, K. Gracie, K. Faulds, Multiplex in vitro detection using SERS, Chem. Soc. Rev. 45 (7) (2016) 1901–1918.

[7] S. Pahlow, et al., Application of vibrational spectroscopy and imaging to point-of-care medicine: a review, Appl. Spectrosc. 72 (1) (September 2018) 52–84.

[8] L. Hamm, A. Gee, A.S. De Silva Indrasekara, Recent advancement in the surface-enhanced Raman spectroscopy-based biosensors for infectious disease diagnosis, Appl. Sci. 9 (7) (2019).

[9] L.F. Tadesse, et al., Toward rapid infectious disease diagnosis with advances in surface-enhanced Raman spectroscopy, J. Chem. Phys. 152 (24) (June 2020) 240902.

[10] K. Kim, et al., Recent advances in sensitive surface-enhanced Raman scattering-based lateral flow assay platforms for point-of-care diagnostics of infectious diseases, Sensors Actuators B Chem. (2020) 129214.

[11] H. Kearns, R. Goodacre, L.E. Jamieson, D. Graham, K. Faulds, SERS detection of multiple antimicrobial-resistant pathogens using nanosensors, Anal. Chem. 89 (23) (December 2017) 12666–12673.

[12] S. Chattopadhyay, P.K. Sabharwal, S. Jain, A. Kaur, H. Singh, Functionalized polymeric magnetic nanoparticle assisted SERS immunosensor for the sensitive detection of *S. typhimurium*, Anal. Chim. Acta 1067 (2019) 98–106.

[13] L. Bi, et al., SERS-active Au@Ag core-shell nanorod (Au@AgNR) tags for ultrasensitive bacteria detection and antibiotic-susceptibility testing, Talanta 220 (December 2020) 121397.

[14] H. Shen, et al., A novel SERS-based lateral flow assay for differential diagnosis of wild-type pseudorabies virus and gE-deleted vaccine, Sensors Actuators B Chem. 282 (June 2018) (2019) 152–157.

[15] J. Hwang, S. Lee, J. Choo, Application of a SERS-based lateral flow immunoassay strip for the rapid and sensitive detection of staphylococcal enterotoxin B, Nanoscale 8 (22) (2016) 11418–11425.

[16] R. Wang, et al., Highly sensitive detection of high-risk bacterial pathogens using SERS-based lateral flow assay strips, Sensors Actuators B Chem. 270 (2018) 72–79.

[17] X. Liu, et al., Fe3O4@Au SERS tags-based lateral flow assay for simultaneous detection of serum amyloid A and C-reactive protein in unprocessed blood sample, Sensors Actuators B Chem. 320 (2020) 128350.

[18] C. Wang, et al., Magnetic SERS strip for sensitive and simultaneous detection of respiratory viruses, ACS Appl. Mater. Interfaces 11 (21) (May 2019) 19495–19505.

[19] S. Díaz-Amaya, L.K. Lin, A.J. Deering, L.A. Stanciu, Aptamer-based SERS biosensor for whole cell analytical detection of *E. coli* O157:H7, Anal. Chim. Acta 1081 (2019) 146–156.

[20] X. Ma, X. Xu, Y. Xia, Z. Wang, SERS aptasensor for *Salmonella typhimurium* detection based on spiny gold nanoparticles, Food Control 84 (2018) 232–237.

[21] N. Duan, M. Shen, S. Qi, W. Wang, S. Wu, Z. Wang, A SERS aptasensor for simultaneous multiple pathogens detection using gold decorated PDMS substrate, Spectrochim. Acta Part A Mol. Biomol. Spectrosc. 230 (2020) 118103.

[22] C. Zhang, et al., Sensitive and specific detection of clinical bacteria: via vancomycin-modified Fe_3O_4@Au nanoparticles and aptamer-functionalized SERS tags, J. Mater. Chem. B 6 (22) (2018) 3751–3761.

[23] H.T. Ngo, H.-N. Wang, A.M. Fales, T. Vo-Dinh, Label-free DNA biosensor based on SERS molecular sentinel on nanowave chip, Anal. Chem. 85 (13) (2013).

[24] T. Vo-Dinh, M.Y.K. Hiromoto, G.M. Begun, R.L. Moody, Surface-enhanced Raman spectrometry for trace organic analysis, Anal. Chem. 56 (9) (August 1984) 1667–1670.

[25] T. Vo-Dinh, Surface-enhanced Raman spectroscopy using metallic nanostructures1The submitted manuscript has been authored by a contractor of the U.S Government under contract No. DE-AC05-96OR22464. Accordingly, the U.S. Government retains a nonexclusive, royalty-free license to publish or reproduce the published form of this contribution, or allow others to do so, for U.S. Government purposes.1, TrAC Trends Anal. Chem. 17 (8) (1998) 557–582.

[26] C.G. Khoury, T. Vo-Dinh, Plasmonic nanowave substrates for SERS: fabrication and numerical analysis, J. Phys. Chem. C 116 (13) (April 2012) 7534–7545.

[27] H.T. Ngo, H.-N. Wang, T. Burke, G.S. Ginsburg, T. Vo-Dinh, Multiplex detection of disease biomarkers using SERS molecular sentinel-on-chip Multiplex Platforms in Diagnostics and Bioanalytics, Anal. Bioanal. Chem. 406 (14) (2014).

[28] H.T. Ngo, H.-N. Wang, A.M. Fales, B.P. Nicholson, C.W. Woods, T. Vo-Dinh, DNA bioassay-on-chip using SERS detection for dengue diagnosis, Analyst 139 (22) (2014).

[29] B.M. Crawford, et al., Plasmonic nanobiosensors for detection of microRNA cancer biomarkers in clinical samples, Analyst 145 (13) (2020) 4587–4594.

[30] B.M. Crawford, et al., Plasmonic Nanobiosensing: from in situ plant monitoring to cancer diagnostics at the point of care, J. Phys. Photonics 2 (3) (2020) 34012.

[31] T. Donnelly, W.E. Smith, K. Faulds, D. Graham, Silver and magnetic nanoparticles for sensitive DNA detection by SERS, Chem. Commun. 50 (85) (2014) 12907–12910.

[32] H.T. Ngo, N. Gandra, A.M. Fales, S.M. Taylor, T. Vo-Dinh, Sensitive DNA detection and SNP discrimination using ultrabright SERS nanorattles and magnetic beads for malaria diagnostics, Biosens. Bioelectron. 81 (2016).

[33] H.T. Ngo, et al., Direct detection of unamplified pathogen RNA in blood lysate using an integrated lab-in-a-stick device and ultrabright SERS nanorattles, Sci. Rep. 8 (1) (2018).

[34] P.V. Dukes, et al., Plasmonic assay for amplification-free cancer biomarkers detection in clinical tissue samples, Anal. Chim. Acta 1139 (2020) 111–118.

[35] J. Joung, et al., Detection of SARS-CoV-2 with SHERLOCK one-pot testing, N. Engl. J. Med. 383 (15) (September 2020) 1492–1494.

[36] J.P. Broughton, et al., CRISPR–Cas12-based detection of SARS-CoV-2, Nat. Biotechnol. 38 (7) (2020) 870–874.

[37] J.Y. Lim, et al., Identification of newly emerging influenza viruses by detecting the virally infected cells based on surface enhanced Raman spectroscopy and principal component analysis, Anal. Chem. 91 (9) (2019) 5677–5684.

[38] N.N. Durmanov, et al., Non-labeled selective virus detection with novel SERS-active porous silver nanofilms fabricated by Electron Beam Physical Vapor Deposition, Sensors Actuators B Chem. 257 (2018) 37–47.

[39] S.K. Gahlaut, D. Savargaonkar, C. Sharan, S. Yadav, P. Mishra, J.P. Singh, SERS platform for dengue diagnosis from clinical samples employing a hand held Raman spectrometer, Anal. Chem. 92 (3) (February 2020) 2527–2534.

[40] J. Guo, Z. Zhong, Y. Li, Y. Liu, R. Wang, H. Ju, Three-in-One' SERS adhesive tape for rapid sampling, release, and detection of wound infectious pathogens, ACS Appl. Mater. Interfaces 11 (40) (October 2019) 36399–36408.

[41] R. Hunter, et al., Optofluidic label-free SERS platform for rapid bacteria detection in serum, Sensors Actuators, B Chem. 300 (May) (2019).

CHAPTER

5 SERS-based molecular sentinel nanoprobes for nucleic acid biomarker detection

Bridget Crawford[1,2], Hsin-Neng Wang[1,2], Tuan Vo-Dinh[1,2,3]

[1]*Department of Biomedical Engineering, Duke University, Durham, NC, United States;* [2]*Fitzpatrick Institute for Photonics, Duke University, Durham, NC, United States;* [3]*Department of Chemistry, Duke University, Durham, NC, United States*

Introduction

The development of sensitive and selective biosensing techniques is of great interest for various chemical and biological applications, ranging from environmental monitoring to medical diagnostics. In medical applications, sensitive diagnostic tools with the capability to detect multiple biomarkers simultaneously are highly desirable. Among potential diagnostic biomarkers, those composed of nucleic acids, such as DNA, Mrna, and microRNA (miRNA), have been considered extremely valuable in monitoring the presence and progression of various diseases. Northern blotting, microarrays, and quantitative reverse transcriptase PCR (qRT-PCR) are often employed as the conventional nucleic acid detection methods, which involve elaborate, time-consuming, and expensive processes that require special laboratory equipment [1,2]. MiRNA detection is of considerable difficulty as there are additional challenges that arise from the intrinsic characteristics of miRNAs, such as the short sequence lengths, low abundance, and high sequence similarity [3]. While northern blotting is the only current technique that allows for the quantitative visualization of miRNAs, it has low detection efficiency and requires complex methods that can introduce contamination [4]. Microarrays, which allow simultaneous detection of several hundred miRNAs, are only semiquantitative making them most suitable for comparing relative expression levels of miRNA between different cellular states [5]. Microarrays, therefore, require an additional form of validation, such as qRT-PCR, to quantify expression. qRT-PCR, the gold standard among miRNA detection techniques, allows detection of low-abundance miRNA species but is restricted by the short length of miRNAs [6]. Moreover, these methods often require either fluorescent or radioactive dyes, which bring about limitations on reliability in that fluorescent dyes photobleach and radioactive probes suffer from difficulties in handling and disposal. Thus, there is a need to develop alternative diagnostic strategies that could offer more advantages over these conventional methods. Recently,

SERS for Point-of-care and Clinical Applications. https://doi.org/10.1016/B978-0-12-820548-8.00005-9

much effort has been devoted to developing practical detection strategies for nucleic acid biomarkers to overcome challenging analytical aspects in currently adopted laboratory-based methods.

Among recent developments in sensing, those based on surface-enhanced Raman scattering (SERS) are promising due their superior diagnostic accuracy and capability for multiplexed sensing, given by the sharp spectral features observed with Raman [7–9]. Raman spectroscopy provides unique chemical fingerprints produced from the inelastic scattering of light upon interaction with specific molecules. SERS, which increases the number Raman scattering photons providing amplification on the order of millions, enables the application of this process for extremely sensitive analyte detection [10]. Recently, Liu et al. have demonstrated endoscopic imaging of tumors via cell surface receptor–targeted SERS nanoparticles applied topically [11,12]. Other SERS-based diagnostic developments have focused on acquiring intrinsic SERS spectra (i.e., SERS signal from endogenous species) from extracted nucleic acids or tumor and liquid biopsies [13–16]. However, prior to this study, SERS analysis of validated nucleic acid biomarkers was limited to the detection of amplified targets, due to their low abundance of these biomarkers and the inadequate sensitivities of previous assays [17–20]. The use of amplified targets within a detection scheme brings about the same issues as PCR-based methods in that it requires a complex array of steps, ranging from target labeling or amplification to post-incubation reactions or washing steps, thus increasing the assay complexity and chances of inaccuracy [20,21]. More recently, a plasmonic platform based on a nanozyme and utilizing a SERS-signal amplification mechanism demonstrated promising results for the detection of a miRNA biomarker associated with prostate cancer following RNA extraction from patient urine samples [22].

Even with such advancements, there remains a technological gap between advancements in molecular diagnostics and their implementation in clinical or point-of-care (POC) settings. Although recent developments have attempted to close this gap, direct detection of cancer biomarkers has remained a significant challenge [23–28]. To overcome limitations, the Vo-Dinh group has developed the highly sensitive, specific, and multiplexed detection of nucleic acids in the development of unique one-step plasmonic nanobiosensor assays referred to as "molecular sentinel" (MS) [29–31] and "inverse molecular sentinel" (iMS) [32–37].

The MS probe consists of a DNA strand having a Raman label molecule at one end and a plasmonic metal nanoparticle at the other end (Fig. 5.1). The MS nanoprobe uses a hairpin-like stem-loop structure to recognize target DNA sequences. Hairpin DNA structures, first developed by Tyagi and Kramer in 1996 [38], have been used in molecular beacon systems that are based on optical and electrochemical detection [39–44]. The sensing principle of MS nanoprobe, however, is quite different from that of molecular beacons. In the normal configuration of MS nanoprobe (i.e., in the absence of target DNA), the DNA sequence forms a hairpin loop, which maintains the Raman label in close proximity of the metallic nanoparticle thereby inducing an intense SERS signal of the Raman label upon laser excitation due to the strong plasmonic effect, which is a short-range and distance-dependent

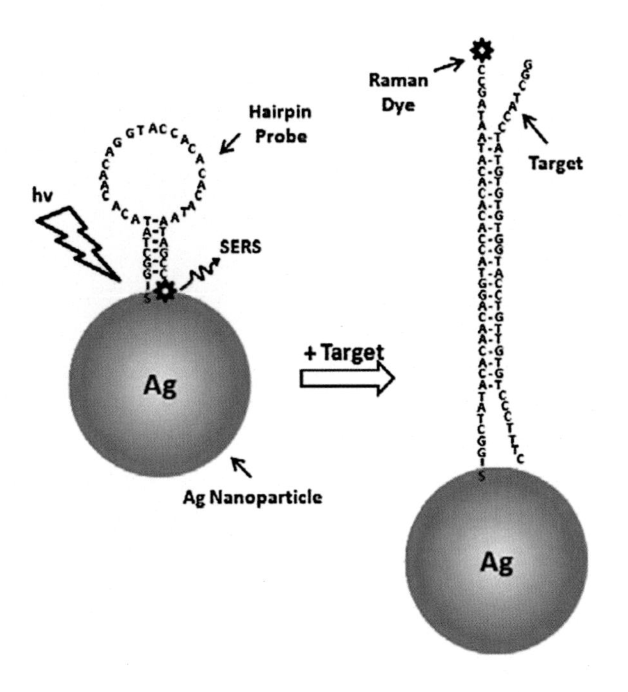

FIGURE 5.1

Detection scheme of the molecular sentinel (MS) nanoprobes.

Adapted with permission from H.N. Wang, et al., Surface-enhanced Raman scattering molecular sentinel nanoprobes for viral infection diagnostics, Anal. Chim. Acta 786 (2013) 153–158.

process. With introduction and hybridization of the complementary target DNA sequence to the nanoprobe hairpin loop, the Raman label molecule is physically separated from the plasmonic nanoparticle, thus leads to a decrease in the resulting SERS signal.

While the MS nanoprobe allows for homogenous biosensing, this technique uses a signal switch of ON-to-OFF which limits the assay's sensitivity. To overcome this limitation, we have developed the "inverse Molecular Sentinel (iMS)" nanoprobe for the detection of nucleic acid targets with an OFF-to-ON detection scheme [32,33]. This detection scheme and its development toward a POC diagnostic technique will be discussed in the following sections.

Development of the iMS nanoprobe for label-free homogenous biosensing

Silver-coated gold nanostars for SERS detection

The iMS nanobiosensor involves the use of a unique type of SERS-active nanoprobe, i.e., silver-coated gold nanostars (AuNS@Ag), as the sensing platform. Nanostars

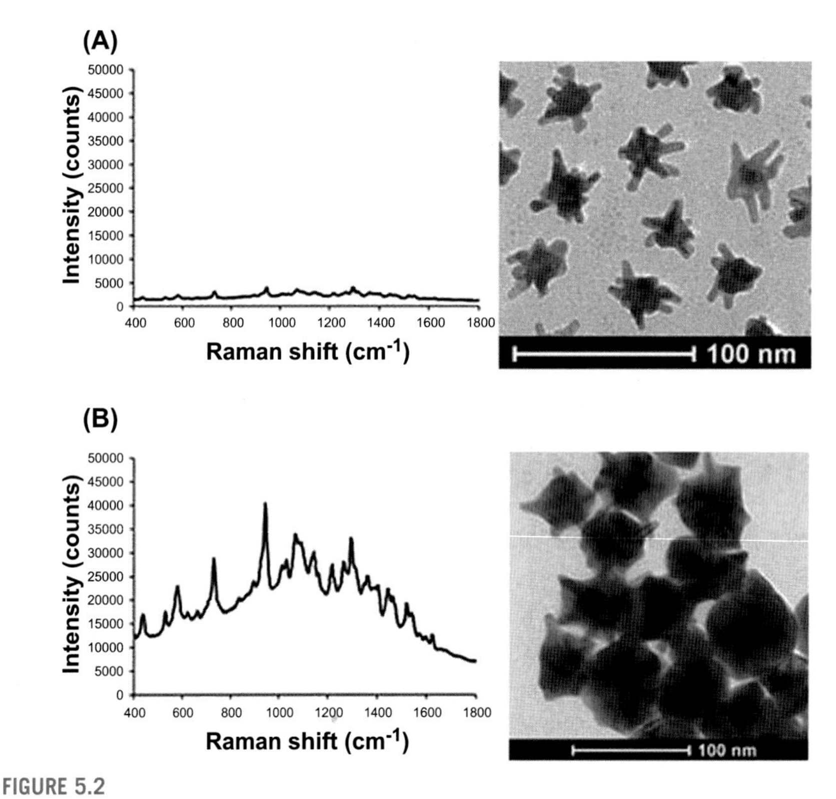

FIGURE 5.2

SERS Spectra and TEM images of (A) gold nanostars (AuNS) and (B) silver-coated gold nanostars (AuNS@Ag).

Adapted with permission from A.M. Fales, et al., J. Phys. Chem. 118 (7) (2014) 3708–15.

have emerged as one of the best geometries for producing strong SERS in a nonaggregated state due to their multiple sharp branches, each with a strongly enhanced electromagnetic field localized at its tip [45]. We have developed AuNS@Ag as a new hybrid bimetallic nanostar platform that exhibits superior resonant SERS (SERRS) properties [46]. The AuNS@Ag provided over an order of magnitude of signal enhancement compared to uncoated AuNS, rendering it an excellent SERS substrate (Fig. 5.2).

Detection scheme of the SERS iMS nanoprobe

As shown in Fig. 5.3, the iMS detection scheme utilizes a DNA probe, which has a Raman label at one end and is immobilized onto a nanostar via a metal-thiol bond. The probe is designed with a stem-loop, or hairpin, structure in order to produce a strong SERS signal when the loop is closed by bringing the Raman label close to the

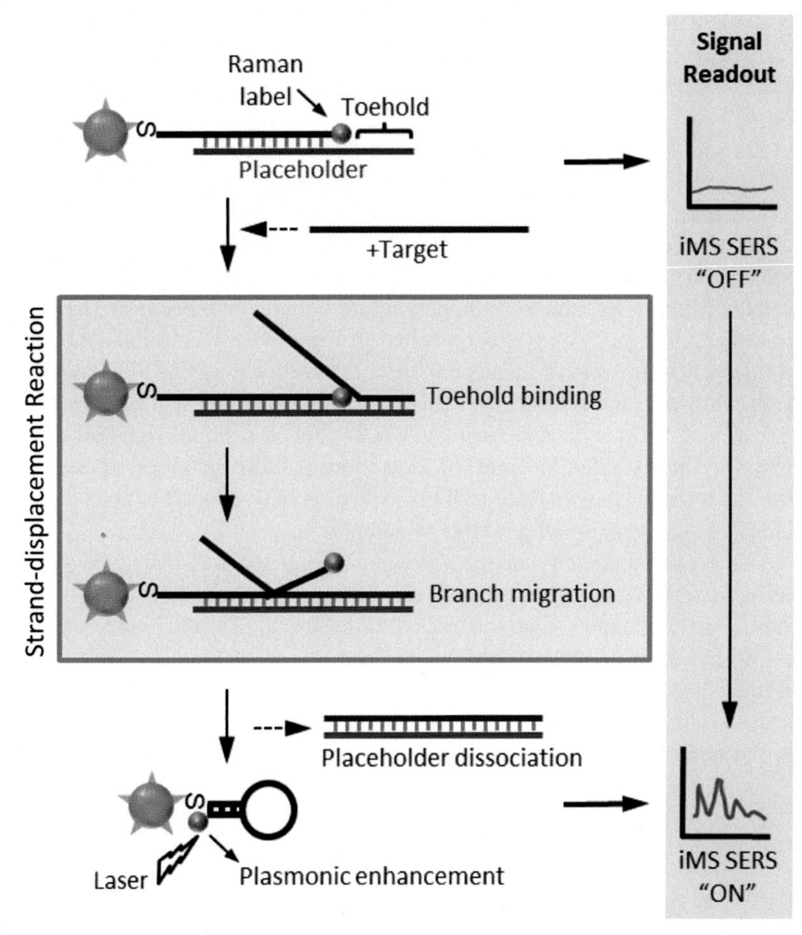

FIGURE 5.3

Detection scheme of the iMS OFF-to-ON SERS nanobiosensor.

Adapted with permission from H.N. Wang, et al., Nanomedicine 11 (4) (2015) 511–520.

surface of plasmonic gold nanostars. A single-stranded DNA, which serves both as a capture probe and a placeholder strand, is partially hybridized to the stem-loop probe via a placeholder-binding region. This probe–placeholder duplex disrupts the stem-loop structure and keeps the label away from the nanostar surface. In this linear configuration, the nanobiosensor exhibits low SERS intensity (i.e., OFF state) as the SERS enhancement decreases exponentially with increasing the separation between the label and the metallic surface. When target nucleic acid is present, the target first binds to the overhang region of the probe–placeholder duplex (i.e., the toehold) and begins displacing the stem-loop probe from the placeholder through a branch migration process and leads to the release of the placeholder from the iMS system. This nonenzymatic strand-displacement process [47] allows the stem-loop structure to close, thus moving the Raman label in close proximity to the plasmonics-active nanostar surface and yielding a strong SERS signal (i.e., ON state).

As a proof of concept, the human radical S-adenosyl methionine domain containing 2 (RSAD2) gene, known to be a host–response biomarker for infectious diseases, was used to demonstrate feasibility of this SERS nanobiosensor. As seen in Fig. 5.4A, in the presence of target (curve c), the SERS intensity is significantly increased, indicating that the hybridization between target and placeholder strands enabled the formation of the stem-loop structure of the nanobiosensor. The possibility for quantitative analysis is demonstrated in Fig. 5.4B, which illustrates that the SERS intensity of the main Raman peak increases with increasing amounts of target DNA. As shown in the inset of Fig. 5.4B, a linear trend line was fitted to the data between 0 and 5 nM with a slope ($s = 376.47$) and a residual standard deviation ($\sigma = 28.87$) that yielded a limit of detection (LOD; $3\sigma/s$) of approximately 0.1 nM. As a result, the absolute LOD is as low as 200 amol. It is noteworthy that this analysis used very small amounts of samples.

The detection specificity of the iMS nanobiosensor was evaluated using the perfectly matched target and sequences having one-, two-, or three-base mismatches. Fig. 5.5 shows that the SERS intensity at 557 cm^{-1} decreases with increasing numbers of mismatched bases. In the case of one-base mismatch, the SERS intensity is about 30% less than the perfectly matched target case. To further demonstrate the capability of our system to discriminate single nucleotide differences, the nanobiosensor was tested in solutions of different ionic strength and it

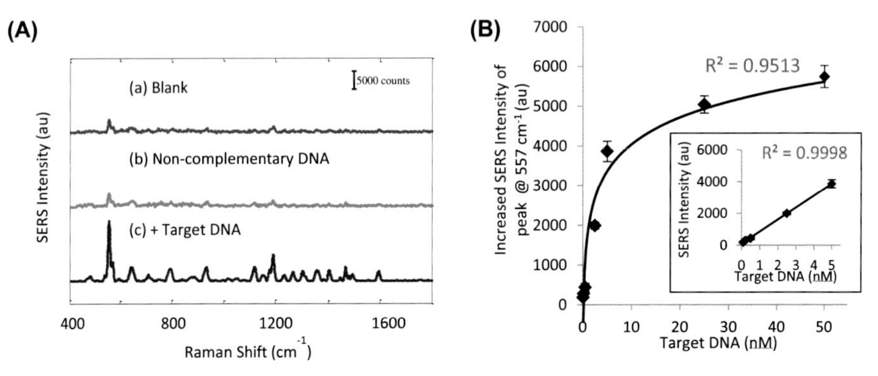

FIGURE 5.4

(A) SERS spectra of the iMS nanobiosensor in the presence or absence of target DNA. (B) Evaluation of the detection sensitivity of the nanobiosensor. A linear trend line was fitted to the data between 0 and 5 nM with a slope ($s = 376.47$) and a residual standard deviation ($\sigma = 28.87$). The limit of detection (LOD, $3\sigma/s$) for target DNA was determined to be approximately 0.1 nM.

Adapted with permission from H.N. Wang, et al., Nanomedicine 11 (4) (2015) 511–520.

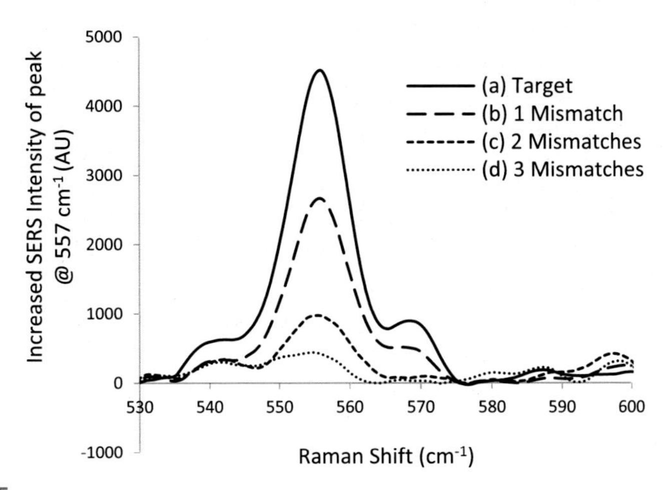

FIGURE 5.5

Evaluation of the detection specificity in the presence of perfectly matched or mismatched sequences.

Adapted with permission from H.N. Wang, et al., Nanomedicine 11 (4) (2015) 511–520.

was determined that the SERS signal of the nanobiosensor treated with one-base mismatched sequences decreases with decreasing NaCl concentration in the reaction buffer. The specificity of detection is critical for many clinical applications, such as single-nucleotide polymorphism (SNP) identification or microRNA detection. As such, reaction buffers can be optimized for SNP identification by reducing the ionic strength to generate the largest signal change between perfectly matched targets and those with mismatches, allowing for improved detection sensitivity. This is due to the fact that the mismatched target–placeholder duplex is thermodynamically unstable in low ionic strength solutions compared to the perfectly matched duplex. Additionally, according to a previous study [32], the kinetics of strand displacement can be modulated by changing the length or sequence composition (i.e., G/C content) of the toehold region which first binds the target. Thus, the effective reaction time for turning on the SERS signal will depend on the toeholds used, e.g., using a strong toehold with a higher G/C content would result in a faster reaction rate. These factors go into the design of the iMS nanoprobe for each individual application, whether that is for detecting biomarkers of infectious disease in humans or miRNA that regulate gene expression in the plant developmental processes [36].

Development of iMS for detection of microRNA biomarkers

To expand the utility of the iMS detection scheme, we further developed this nanostar-based SERS sensing technology for the detection of miRNAs [33]. MiRNAs are small, noncoding RNAs of approximately 20–25 nucleotides in length that bind to perfect or near-perfect complementary sequences in the untranslated regions of mRNA targets, thereby regulating gene expression at the posttranscriptional level [48]. Dysregulation of miRNAs is often observed in a wide range of cancers, including breast cancer. Recent studies also show that miRNAs regulate not just single oncogenes but entire signaling networks [49–51]. For instance, while miRNA-21 (miR-21) is overexpressed and plays an important oncogenic role, miR-34a is downregulated in triple-negative breast cancers, which has been recognized as the most aggressive subtype of breast cancer [52,53]. Therefore, miRNAs hold great potential to serve as an important class of biomarkers not only for early diagnosis of cancer but also for investigation of cancer initiation and progression [54–56]. However, these small molecules have not been adopted into clinical practice for early diagnostics because of the technical difficulties arising from the intrinsic characteristics of miRNAs previously mentioned [3].

The development of iMS for miRNA detection is aimed to overcome the detection challenge posed by the short length, low abundance, and high sequence similarity of miRNA sequences. An iMS nanoprobe was first designed labeled with Cy5 to detect the mature human miR-21. Previously, the physical length of the Raman-labeled stem-loop probes was designed to be around 10 nm for a low background SERS signal in their OFF state but the short sequence lengths of miRNAs would not afford as long of a stem-loop sequence. To overcome this challenge, a spacer sequence is added between the loop and the 5′-end stem (Fig. 5.6A). This spacer

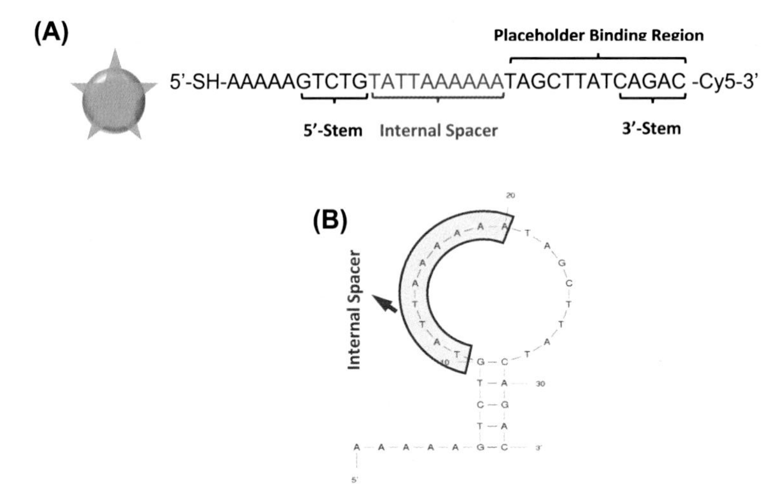

FIGURE 5.6

The design of the iMS nanoprobes for miRNA detection. (A) The sequence structure of the iMS stem-loop probe with an internal spacer for miR-21detection. (B) The stem-loop configuration of the probe showing the internal spacer is located within the loop region. The stem-loop structures are predicted using the two-state folding tool on the DINAMelt server [57].

Adapted with permission from H.N. Wang, et al., J. Phys. Chem. C 120 (2016).

is used to increase the physical length of the probe to keep the Raman label at an appropriate distance away from the nanoparticle surface while hybridizing to a placeholder strand. This internal spacer, however, does not affect the generation of a strong SERS signal when the probe is in a closed stem-loop configuration (Fig. 5.6B).

As for the OFF-to-ON sensing technique, the iMS nanoprobes should have a low SERS background signal in the OFF state through the formation of stable probe–placeholder duplexes to allow for the highest possible sensitivity. In the development of iMS nanoprobes for the detection of miRNA, different placeholder strands were investigated for their capability to turn off the SERS signal of the iMS nanoprobes. A short poly(T) tail (four to six thymine bases) was added to the 3′-end of the placeholder strand; as such, it can hybridize to a portion of the internal spacer in the probe. The red bars in Fig. 5.7A show the SERS background signal of the iMS nanoprobes in their OFF state with different placeholder strands. Without the poly(T) tail, a strong SERS background signal was observed indicating that the placeholder-0T strand did not effectively turn off the iMS SERS signal. This is due to the fact that the duplex with melting temperature (Tm) of 38 °C is thermodynamically unstable at the reaction temperature of 37 °C. In this case, the probe (with the hairpin stem Tm of 44.1 °C) has a high tendency to fold into a stem-loop structure, yielding a strong SERS background signal. In contrast, with the short poly(T) tails, low background signals were observed indicating that stable probe–placeholder duplexes were formed. The melting temperatures of these probe–placeholder duplexes are estimated to be 44.7 °C, 46.1 °C and 47.4 °C for Placeholder-4T, -5T, and -6T, respectively, which are higher than the reaction temperature, leading to a higher thermal stability.

To turn ON the nanobiosensor, the placeholder strands must be released from the iMS system with target binding. We investigated whether the additional poly(T) tail affects the dissociation of the placeholders from the iMS nanoprobes. The blue bars in Fig. 5.7A represent the SERS intensity of the main peak from the iMS nanoprobes after incubation of DNA targets. Fig. 5.7B shows an example of the SERS spectra for the detection of miR-21 targets using the Placeholder-6T. In the presence of miR-21 targets, the SERS intensity was significantly increased (signal ON) indicating that the hybridization between targets and placeholders enabled the formation of the stem-loop structure of the probes by triggering the strand displacement reaction and releasing the placeholder strands. Thus, because of the thermal instability of the short poly(A)–poly(T) duplexes (Tm < 0°C), the additional short poly(T) tail can spontaneously dissociate from the probe following branch migration during the strand displacement process. It was also observed that in the presence of target strands, the SERS intensity decreases slightly with increasing length of the poly(T) tail. These results demonstrate that a proper poly(T) tail should be added to the placeholder in order to minimize the background signal without affecting the sensor functionality. Thus, care needs to be taken in designing the placeholder so that it can provide the lowest baseline signal while still being able to be released following branch migration process.

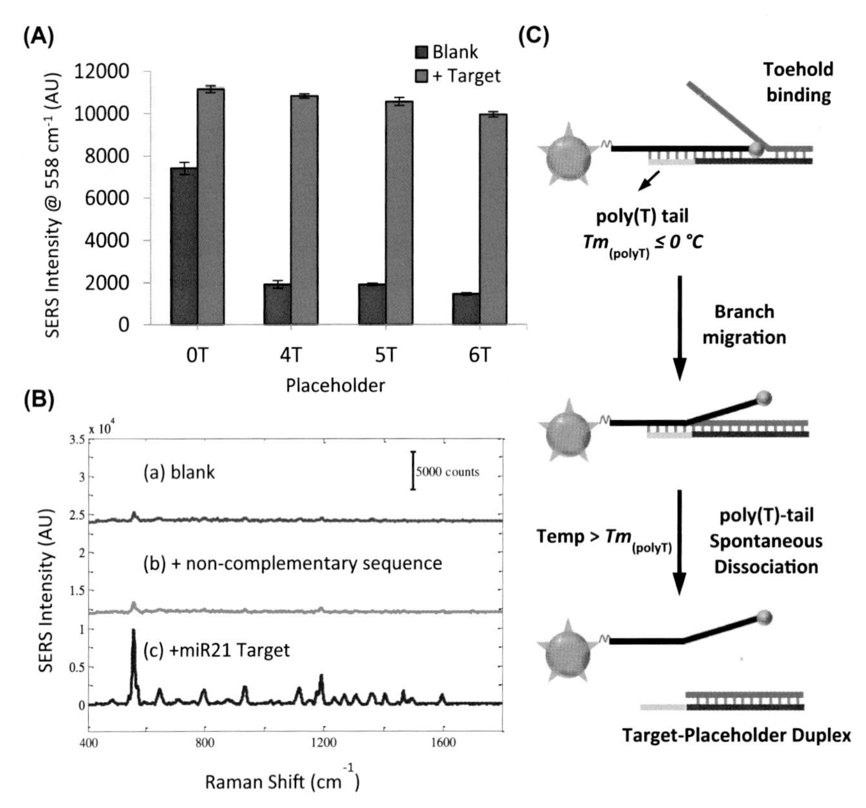

FIGURE 5.7

(A) Investigation of the use of different placeholders with a short poly(T) tail (four to six thymine bases) for turning off the iMS SERS signal. Placeholder-0T, -4T, -5T, and -6T contain 0, 4, 5, and 6 thymine base, respectively. (B) SERS spectra of the Cy5-labeled iMS nanoprobes with placeholder-6T in the presence or absence of miR-21 synthetic target. (C) Schematic representation of target–placeholder dissociation.

Adapted with permission from H.N. Wang, et al., J. Phys. Chem. C 120 (2016).

Detection of miRNA biomarkers within biological samples

RNA extracted from cancer cell lines

The proof-of-concept for using iMS nanoprobes to detect miRNAs in biological samples was performed with total small RNA extracted from breast cancer cell lines. To demonstrate the possibility of detecting miR-21 in real samples, preliminary experiments were performed using small RNA extracted from two breast cancer cell lines (AU565 and SUM149) that have been shown to exhibit different expression levels of miR-21 in our qRT-PCR experiments. Each sample was incubated with iMS nanoprobes followed directly by SERS measurements. Fig. 5.8A shows the

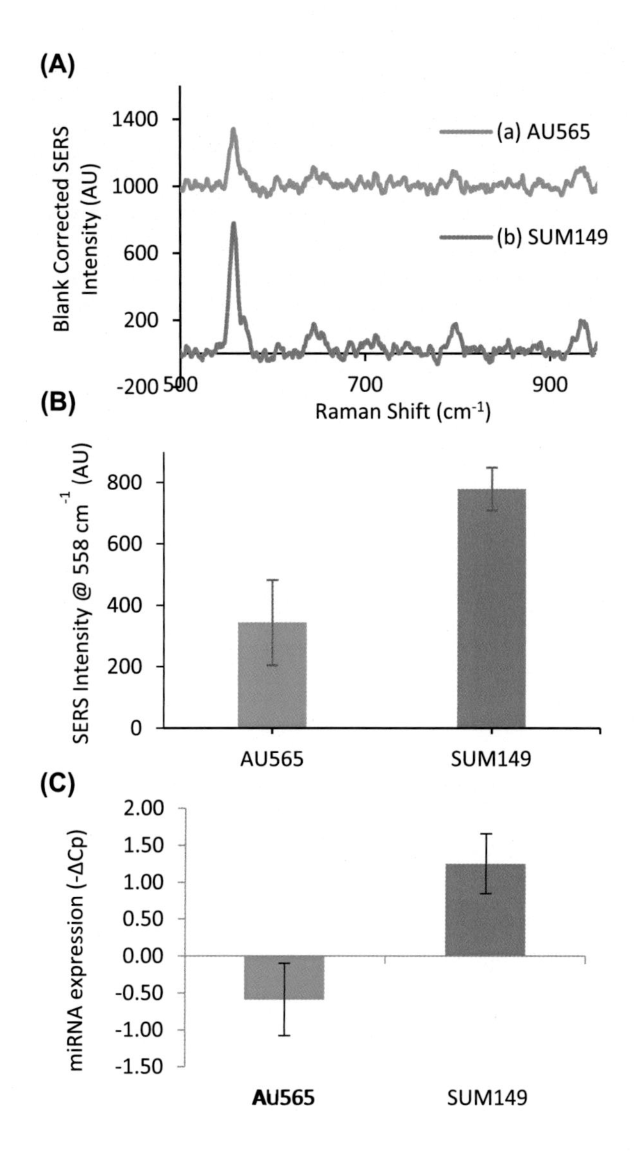

FIGURE 5.8

SERS detection of miR-21 from 200 ng total small RNA extracted from AU565 and SUM129 breast cancer cell lines. (A) Blank-corrected SERS spectra of the iMS nanoprobes in the presence of the RNA samples. (B) A bar graph showing the SERS signal of the major Raman peak at 558 cm^{-1} in the presence of the RNA samples. (C) QRT-PCR showing miR-21 expression in Au565 and SUM149 cell lines relative to MCF-10a. Values represent average ± s.d.

Adapted with permission from H.N. Wang, et al., J. Phys. Chem. C 120 (2016).

blank-corrected SERS spectra of the iMS nanoprobes in the presence of the total small RNA samples. The blue spectrum is the SERS signal measured when the iMS nanoprobes were mixed with the RNA sample extracted from AU565 cells, while the red spectrum shows the SERS signal detected from the RNA sample extracted from the SUM149 cells. The bar graph (Fig. 5.8B) shows that the SERS signal of the major SERS peak measured with RNA extracted from SUM149 cells was significantly higher than that extracted from AU565 cells. The SERS data were consistent with the qRT-PCR data showing that the miR-21 expression in SUM149 cells was higher than that in AU565 cells (Fig. 5.8C). These results demonstrate the possibility to detect and evaluate the expression levels of miRNAs from cell lysates that contain different amounts of miRNAs and provide the foundations for translating the detection of miRNAs into clinical practice.

The dose response of the iMS technique was assessed by using serial dilutions of total small RNA enriched from human breast adenocarcinoma MCF-7 total RNA. The MCF-7 RNA was used because it has been found that miR-21 is highly expressed in this cell line [58]. As shown in Fig. 5.9, the assay was linear in the range of 10–500 ng total small RNA sample ($R^2 = 0.9949$), and as little as 10 ng small RNA can be used if the target of interest is of sufficient abundance.

Clinical evaluation of miRNA cancer biomarker detection using iMS nanoprobes

We used esophageal adenocarcinoma (EAC) and Barrett's esophagus (BE), the asymptomatic premalignant metaplasia associated with EAC, as a model clinical

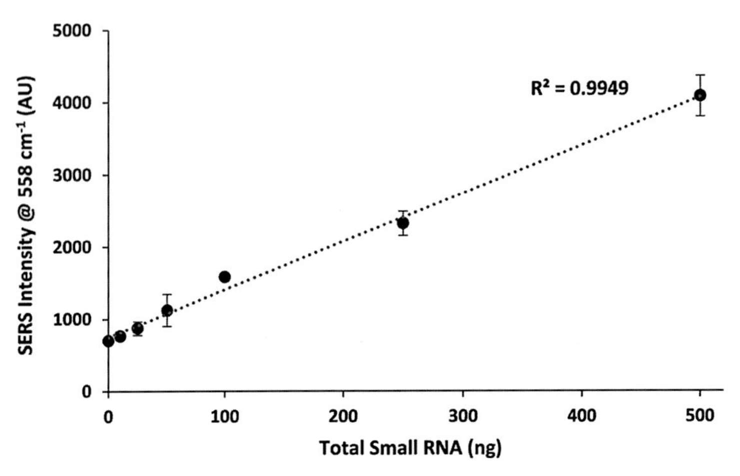

FIGURE 5.9

Evaluation of the dose response of the iMS technique using serial dilutions of total small RNA enriched from breast adenocarcinoma MCF-7 total RNA.

Adapted with permission from H.N. Wang, et al., J. Phys. Chem. C 120 (2016).

system to demonstrate the feasibility of iMS in cancer detection and diagnostics [59]. Similar to other cancers, esophageal cancers (ECs) have unique biomarker profiles that can be found in the tissue, as well as in peripheral blood [60]. The dysregulation of several miRNAs have been identified to be associated with EC [61]. The ubiquitously expressed miR-21 has been well established as an oncogenic miRNA due to its aberrant overexpression in various cancers, including EC [61]. Furthermore, miR-21 shows significant differences in expression between squamous and BE/EAC samples. In addition to being identified as an EC biomarker in tissue and blood, its association with other cancers makes the proposed technology relevant to other clinical areas as well [62]. As discussed in this section, we demonstrated the feasibility of the iMS nanoprobes for direct detection of the early cancer biomarker miR-21 in clinical tissue samples without the need for target amplification. To confirm the capability of iMS miR-21 detection in clinical samples, samples were tested in parallel using qRT-PCR. The implications of our miRNA detection technique are vast and significant as our method can provide rapid and simple direct diagnosis, which holds transformative potential for applications within cancer research, as well as in future clinical applications for the direct detection of miRNA in patient biopsies.

An evaluation of the detection sensitivity of the iMS nanobiosensors for miR-21 detection within clinical samples is shown in Fig. 5.10, which illustrates that the SERS intensity of the main Raman peak increases with increasing amounts of

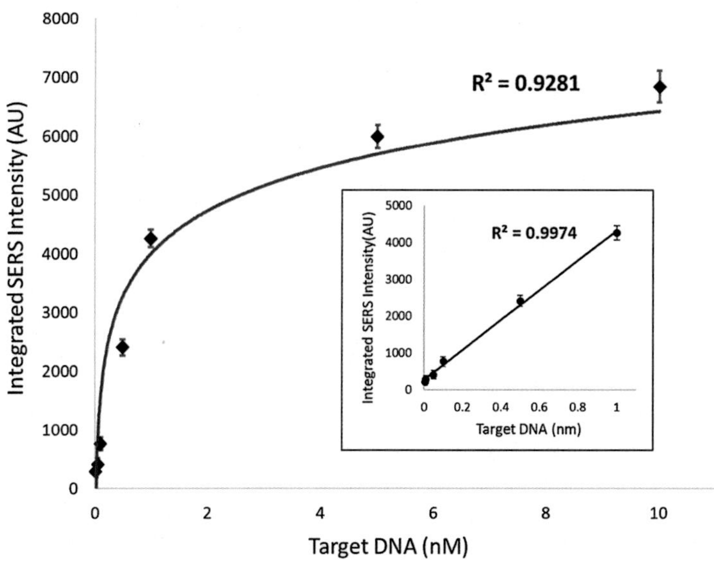

FIGURE 5.10

Evaluation of the detection sensitivity of the iMS nanobiosensors for miR-21 detection in clinical samples.

Adapted with permission from B.M. Crawford, et al., Analyst (2020).

miR-21 target RNA. A logarithmic fit was obtained and as shown in the inset, a linear trend line was fit to the data between 0 and 1 nM. Evaluation of the detection sensitivity provided a LOD of 4.6 pM (46 amol).

The iMS nanoprobe sensing mechanism was maintained within clinical samples, exhibited no particle dimerization, and allowed for the detection of miR-21 target. Such a response can be seen in Fig. 5.11, which shows SERS spectra exemplifying the iMS nanobiosensor response when testing unamplified enriched small RNA from a normal and tumor biopsy pair from a single patient. The SERS intensity of the prominent peak of the Raman label (Cy5, 557 cm^{-1}) is significantly increased when testing extracted RNA from tumor tissue as compared to normal tissue, indicating the greater abundance of miR-21 in the tumor tissue sample.

A blinded study was performed to determine diagnostic capabilities of the iMS nanoprobes via detection of miRNA in clinical samples. Using our miR-21 iMS nanoprobes, the iMS assay was applied in a blinded manner to 17 tissue samples in triplicate. The area under the curve (AUC) of the 557 cm^{-1} peak (as identified in Fig. 5.11) was used in analysis for each sample, as shown in Fig. 5.12. After unblinding, a threshold was determined for this pilot cohort. The iMS diagnostic results using this threshold were then compared to the independently conducted histopathological diagnosis of each tissue sample. Provided by this set threshold, the true-positive and true-negative rates were calculated (Tables 5.1 and 5.2). In this pilot cohort, the miR-21 iMS assay demonstrated notable diagnostic accuracy with 100% true positive rate and 100% true negative rate when discriminating tumor from normal tissue. When including BE tissue samples as positive (i.e., normal vs. BE and EC), the miR-21 iMS assay demonstrated 90% true-positive and 100% true-negative rates.

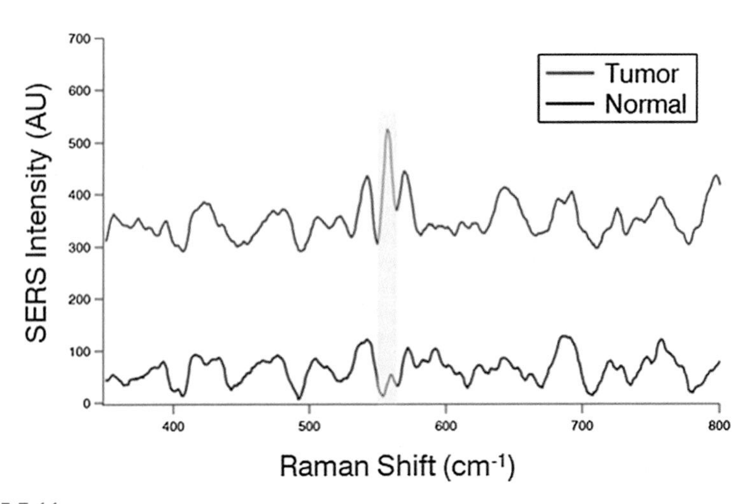

FIGURE 5.11

Representative SERS spectra of the iMS response to RNA extracted from a paired biopsy from a single patient; normal (black), tumor (red).

Adapted with permission from B.M. Crawford, et al., Analyst (2020).

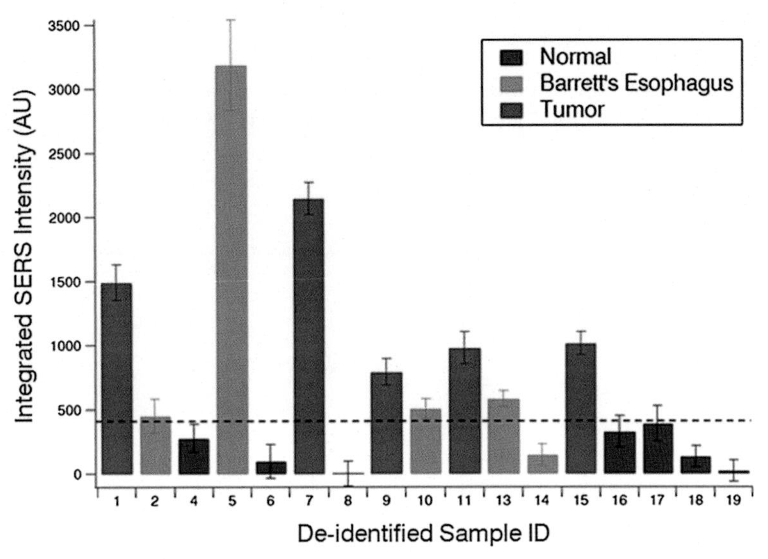

FIGURE 5.12

SERS intensities (AUC of the 557 cm^{-1} peak; arbitrary units) against the threshold for normal (black), Barrett's esophagus (BE, gray), and tumor (red) samples.

Adapted with permission from B.M. Crawford, et al., Analyst (2020).

Table 5.1 True positive (sensitivity) and true negative (specificity) of discriminating esophageal cancer (EC) from normal tissue.

	EC	Normal	Total
Total	5	7	12
Test positive	5	0	5
Test negative	0	7	7
	True positive 5/5 = 100%	**True negative** 7/7 = 100%	

Adapted with permission from B.M. Crawford, et al., Analyst (2020).

Table 5.2 True positive (sensitivity) and true negative (specificity) of discriminating EC and Barrett's esophagus (BE) from normal tissue.

	EC and BE	Normal	Total
Total	10	7	17
Test positive	9	0	9
Test negative	1	7	8
	True positive 9/10 = 90%	**True negative** 7/7 = 100%	

Adapted with permission from B.M. Crawford, et al., Analyst (2020).

To verify the iMS assay for miR-21 detection, Taqman qRT-PCR was performed on all samples. As seen in Fig. 5.13, SERS and qRT-PCR show agreeing trends when comparing the average response to miR-21 for the two detection methods, with an observed increase in miR-21 between normal, BE, and EC samples.

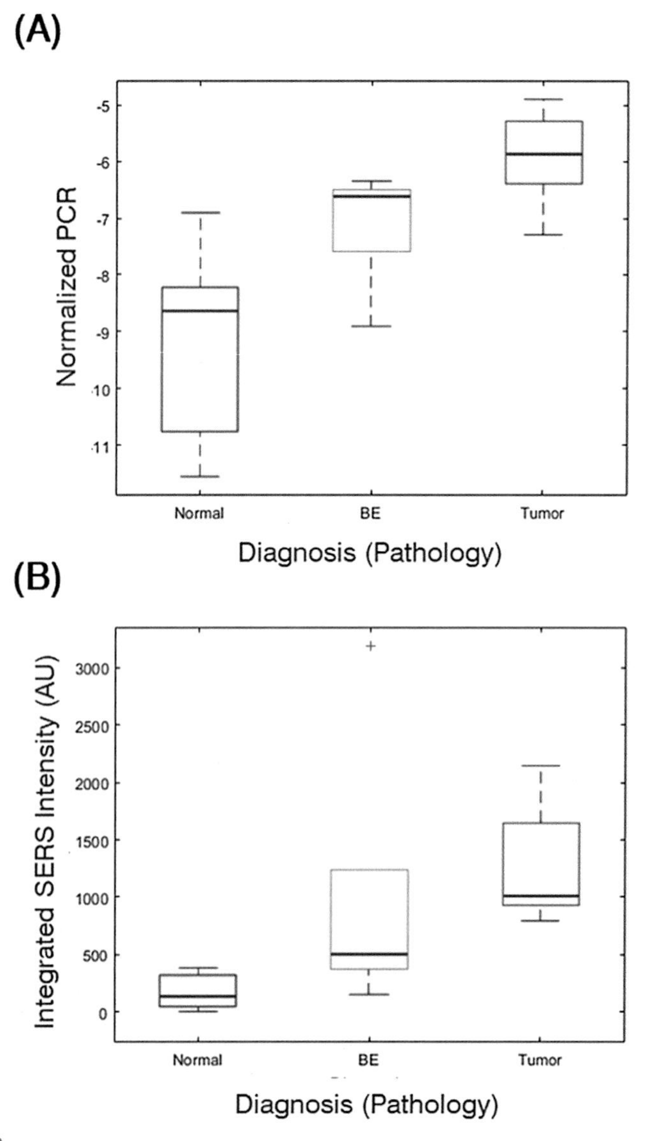

FIGURE 5.13

(A) Correlation of PCR delta CT values with histopathological diagnosis and (B) correlation of iMS detection of miR-21 (AUC) with pathology.

Adapted with permission from Crawford BM et al. Analyst (2020).

To further evaluate the relationship between RT-PCR and iMS, a linear regression was performed (Fig. 5.14). The concordance and correlation between the two measurements were evaluated using Kendall Tau [63] and Pearson correlation, respectively. Concordance was observed between SERS and qRT-PCR data although the strength was varied among normal, BE and EC. Table 5.3 provides concordance (Kendall Tau) and correlation (Pearson correlation) between miR-21 PCR results (normalized by cel-39) and SERS intensity (AUC of the 557 cm^{-1} peak) within

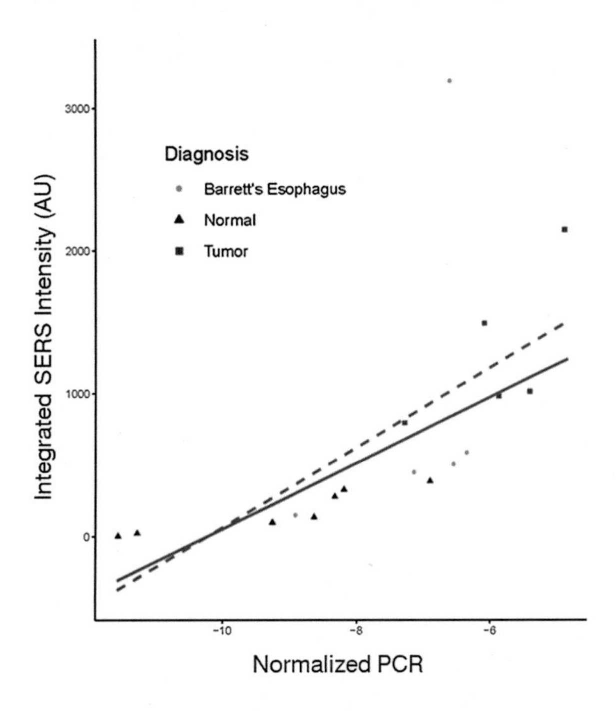

FIGURE 5.14

Linear regression between miR-21 PCR and iMS results. The solid line indicates a robust linear fit, while the *dashed line* indicates a normal linear fit.

Adapted with permission from B.M. Crawford, et al., Analyst (2020).

Table 5.3 Concordance and correlation between miR-21 PCR results and iMS response intensity within each tissue type.

	n	τ	*P*-value	ρ	*P*-value
All types combined	17	0.8	<0.001	0.63	0.007
Normal	7	1.0	<0.001	0.93	0.003
Barrett's esophagus	5	0.6	0.23	0.40	0.51
Tumor	5	0.6	0.23	0.70	0.19

Adapted with permission from B.M. Crawford, et al., Analyst (2020).

each tissue type as determined by histopathological diagnosis. It is noteworthy that one group BE data point lies outside the expected range. There is potential for this point to be an outlier; however, with a low sample number, we have hesitated to claim it as such.

When considering the combined data, the relatively small sample size may cause the relationship to appear nonlinear. An alternate reasoning could be that there is a nonlinear relationship between the two methods, which could potentially be due to the different response of the two assays as a function of concentration of miRNA. SERS-based sensors function by having the analyte directly interact with surface sites on the nanoparticle and therefore, the sensor response follows a surface saturation curve, similar to a Langmuir isotherm. Due to this saturation response, the target detection is achieved in a specific range of concentrations, the ends of which exhibit a dramatically spiked analytical sensitivity. This specific range depends on the parameters of the assay, including the number of particles and the probe strand loading. Such behavior produces a nonlinear response to the presence of target in certain concentration ranges, as previously reported [32]. On the other hand, the mechanism of qRT-PCR detection provides a large linear range, in which the number of cycles constantly increases depending on concentration. This difference in operating behaviors between the two methods can be partially observed in Fig. 5.14, where the SERS response does not follow linearly to the qRT-PCR response, due to the analytical sensitivity of SERS varying in the range of concentrations under analysis.

We report the promising results of this pilot cohort for the potential of the iMS SERS-based assay for detection of cancer biomarkers without the need for target amplification. It is worth noting that this pilot study was performed using a small sample size in order to establish future development of this technology. Further studies will expand the sample number to create a large training and validation cohort [59].

In the first clinical demonstration of direct detection of miRNA using SERS biosensors, our homogenous assay method allows for the detection of miRNA in unamplified small RNA extracts from human tissue biopsies with excellent diagnostic accuracy when discriminating between EC and normal tissue as well as notable diagnostic accuracy when including BE in the unhealthy sample set. The qRT-PCR data indicate the overlap of miR-21 present in BE tissue samples with that of normal and EC tissue samples. This suggests the inability for the single biomarker miR-21 to perfectly differentiate BE from either normal or EC tissue, as is consistent with previous research [61]. The great benefit of the iMS nanobiosensor is the potential for multiplexed detection. Future studies will incorporate multiple EC biomarkers for more robust diagnostic capability.

In addition to the simplicity of this homogenous assay technique, our method yields an output of a binary diagnosis based on SERS signal intensity following training and validation, rather than a qRT−PCR curve. We envision that future studies, along with advancements in PCR, will allow for determination of absolute miRNA copy number (rather than relative fold change) associated with disease state, which could then be used to develop normative values. While the current progress of our method requires RNA isolation, there are several methods to extract RNA within POC systems [64,65]. Compared to diagnosis by histopathology, this method can be completed in a matter of minutes following RNA isolation. We anticipate that ultimately this method will be incorporated into an integrated device for use in POC testing as a diagnostic alternative to histopathology, thereby simplifying the testing process and allowing clinicians to focus on providing quality patient care in a timely manner. Our method provides a user-friendly, resource-efficient alternative to gastrointestinal cancer diagnostics and warrants additional refinement as well as additional clinical testing and validation [59].

Multiplexed detection of miRNA biomarkers

Development of multiplexing technique

The multiplexing capability of SERS is an important feature due to the narrow Raman bandwidths. To demonstrate the multiplexing capability of the iMS technique, a second iMS nanoprobe labeled with a different Raman dye, Cy5.5, was designed to target miR-34a. MiR-21 and miR-34a miRNAs have both been recognized as critical biomarkers for various cancers [66–69]. Fig. 5.15 shows the blank corrected SERS signal of the miR-34a iMS nanoprobes in the presence or absence of target molecules. In the presence of miR-34a targets (spectrum b), the SERS

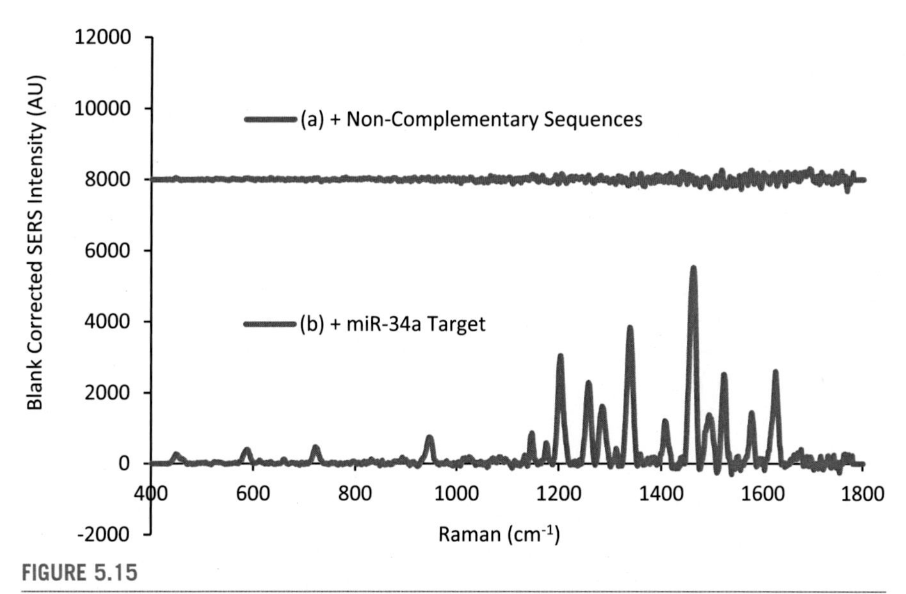

FIGURE 5.15

SERS spectra of the Cy5.5-labeled iMS nanoprobes in the presence or absence of miR-34a synthetic target.

Adapted with permission from H.N. Wang, et al., J. Phys. Chem. C 120 (2016).

intensity was significantly increased, comparing with the negative control in the presence of noncomplementary sequences (spectrum a), indicating that the miR-34a iMS nanoprobes were turned on upon target binding. It is noticed that the SERS spectrum of the Cy5.5-labeled miR-34a nanoprobes exhibits multiple characteristic Raman peaks, which are significantly different than those from the Cy5-labeled miR-21 nanoprobes. Thus, the SERS sensing modality provides specific spectral fingerprints with very sharp peaks allowing the sensing of multiple targets simultaneously in a single assay platform.

Demonstration of the multiplex capability of the iMS technique was performed with a mixture of the two iMS nanoprobes for simultaneous detection of miR-21 and miR-34a miRNAs. Fig. 5.16 shows the SERS spectra from the mixture in the presence of synthetic miR-21 (spectrum b), miR-34a (spectrum c), or both (spectrum d) targets. As can be seen, only the SERS peaks associated with the designated iMS nanoprobes were significantly increased only in the presence of its specific targets (spectra b and c). In the presence of both targets (spectra d), all major SERS peaks from both iMS nanoprobes (indicated by arrows) were greatly increased. The results of this study demonstrate the feasibility of using the iMS technique for multiplex

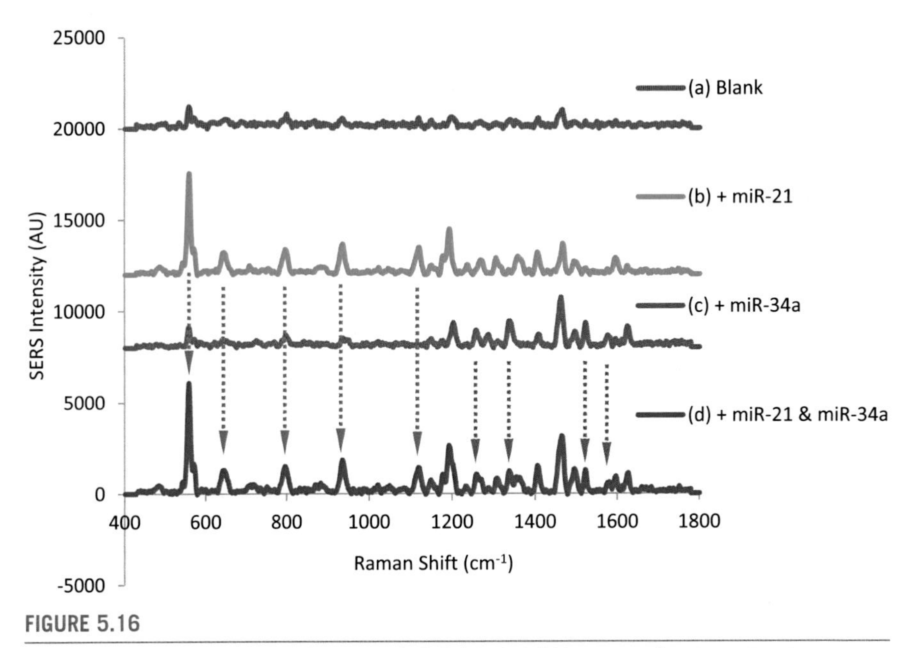

FIGURE 5.16

SERS spectra of a mixture of Cy5- and Cy5.5-labeled iMS nanoprobes targeting to synthetic miR-21 and miR-34a miRNAs. Spectrum (A): blank (in the absence of any targets). Spectrum (B): in the presence of miR-21 synthetic target. Spectrum (C): in the presence of miR-34a synthetic target. Spectrum (D): in the presence of both miR-21 and miR-34a synthetic targets. The arrows illustrate the increased intensity of the major SERS peaks in the presence of corresponding miRNA targets.

detection of miRNAs. Note that the SERS measurements were performed immediately following the incubation of target molecules without any washing steps, which greatly simplifies and accelerates the assay procedure.

Further experiments mixed different ratios of miR-21 and miR-34a nanoprobes (1:1, 1:2, and 1:4) while keeping the concentration of miR34a nanoprobes constant at 10 pM in a 10-μL assay volume. Fig. 5.17 shows the blank-corrected SERS spectra from the mixtures in the presence of both miR-21 and miR-34a synthetic targets.

To identify the SERS signal of each nanoprobe from the mixture, the spectra were analyzed using a spectral decomposition method. As previously described, the spectral decomposition procedure, which was adapted from Lutz et al. [70], was processed using Matlab (R2015b, MathWorks, MA). The decomposition is based on the assumption that the multiplex spectrum is comprised of the reference spectra and an unknown polynomial. SERS spectra were collected for each probe alone (reference spectra) and for the mixture. The acquired SERS spectra were background subtracted and smoothed using a Savitsky–Golay filter (five-point window and first-order polynomial). The entire spectra ranging from 400 to 1800 cm^{-1}, which contains the distinctive Raman signature of each dye, were loaded into Matlab's workspace. The decomposition processes utilized Matlab functions lsqnonneg and fmincon to determine the best fit of the reference spectra to the mixture spectrum. A free-fitting polynomial was introduced to reduce the fitting error [70]. Matlab analysis subsequently generated the minimally constrained coefficients (i.e., the fractions) for each reference spectrum, which were normalized to 1. The best fit

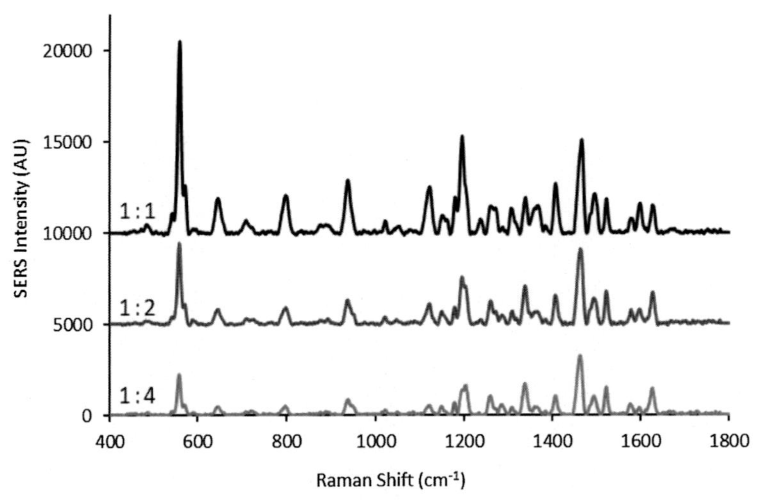

FIGURE 5.17

SERS spectra from mixtures of Cy5-labeled miR-21 and Cy5.5-labeled miR-34a nanoprobes (1:1, 1:2, and 1:4) in the presence of 1 μM of both targets.

Adapted with permission from H.N. Wang, et al., J. Phys. Chem. C 120 (2016).

curve for the multiplex probe mixture is shown in Fig. 5.18. As can be seen in Fig. 5.19, the SERS spectra of the mixture can be decomposed into the contributions

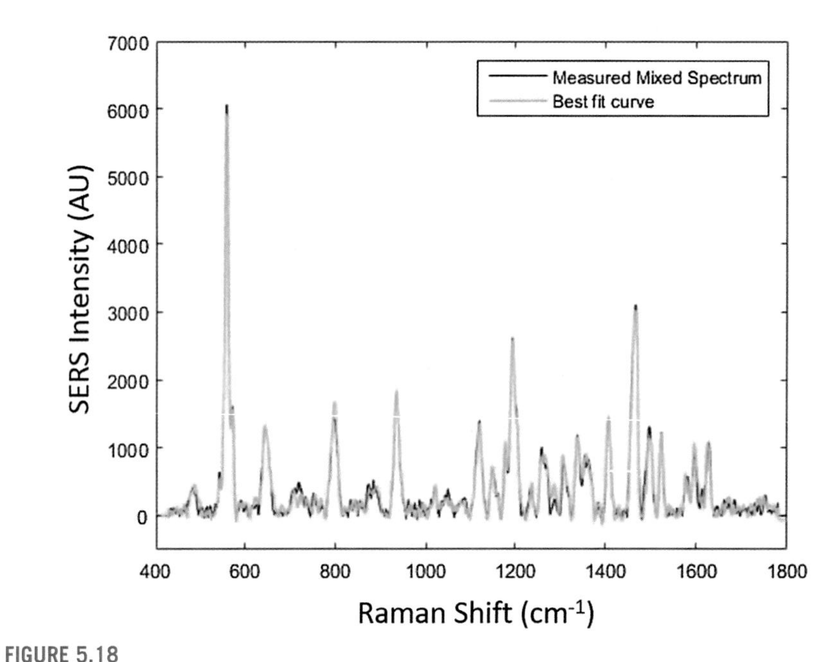

FIGURE 5.18

SERS signal fractions obtained from the spectral decomposition procedure.

Adapted with permission from H.N. Wang, et al., J. Phys. Chem. C 120 (2016).

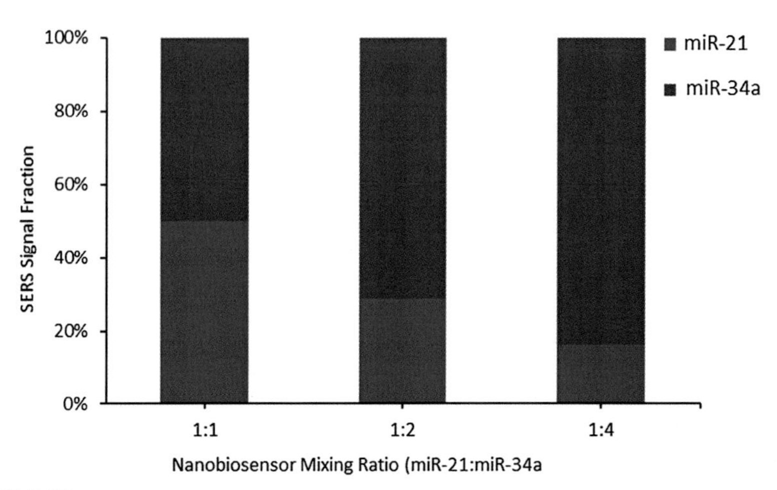

FIGURE 5.19

Best fit curve for multiplex mixture of nanoprobes in the presence of both targets using spectral decomposition method.

Adapted with permission from H.N. Wang, et al., J. Phys. Chem. C 120 (2016).

of the two distinct reporters, Cy5 (miR-21 nanoprobes) and Cy5.5 (miR-34a nanoprobes). The signal fractions were in good agreement with the predetermined ratios of the two nanoprobes.

Multiplex detection of endogenous targets extracted from breast cancer cell lines

The demonstration of using iMS nanoprobes to detect miRNAs in real biological samples was performed with total small RNA extracted from breast cancer cell lines. The multiplex capability of the iMS technique was demonstrated using a mixture of the two differently labeled nanoprobes to detect miR-21 and miR-34a miRNA, which have been recognized as critical biomarkers for breast cancer [66–69]. To demonstrate the detection of the two real miRNA targets in RNA extracts, different amounts (250 ng, 500 ng and 1 μg) of total small RNA extracted from the MCF-7 breast cancer cell line were added to a nanoprobe mixture containing 5 pM of miR-21 nanoprobes and 10 pM of miR-34a nanoprobes. Fig. 5.20A shows a typical SERS spectrum of the mixture in the presence of 1 μg of total small RNA sample. The characteristic SERS peaks of the miR-21 nanoprobes at 558 cm^{-1} and 797 cm^{-1}, and of the miR-34a nanoprobes at 1523 cm^{-1} and 1626 cm^{-1} were indicated by the arrows. The integrated SERS intensity of these peaks shown in Fig. 5.20B and C were given by the AUC over the spectral range of 540–569 cm^{-1}, 780–810 cm^{-1}, 1512–1530 cm^{-1}, and 1615–1640 cm^{-1} for peaks at 558 cm^{-1}, 797 cm^{-1},

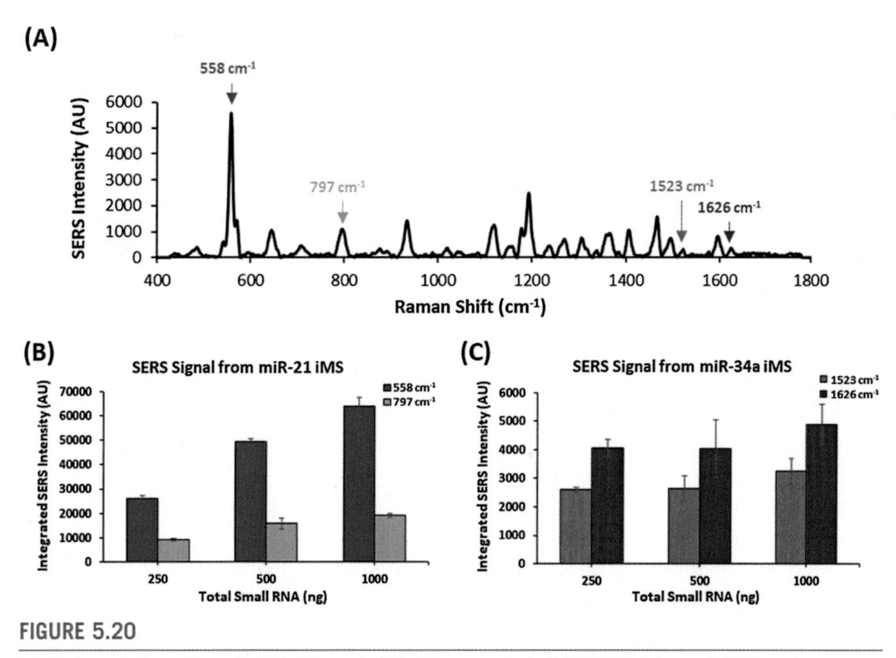

FIGURE 5.20

Multiplexed detection of miR-21 and miR-34a targets in MCF-7 total small RNA extracts.

1523 cm^{-1}, and 1626 cm^{-1}, respectively. The SERS intensity of the miR-21 nanoprobes at 558 cm^{-1} and 797 cm^{-1} (Fig. 5.20B) was found to increase significantly with increasing amounts of small RNA input. For miR-34a SERS signal, a slight increase in the intensity at both 1523 cm^{-1} and 1626 cm^{-1} was observed only when 1 μg of the RNA sample was added (Fig. 5.20C). Due to the significant difference in the amounts of RNA samples required for a detectable signal change, the abundance of miR-21 in MCF-7 cells was found to be higher than that of miR-34a. This result was consistent with that reported by Fix et al. [58] using microarray. The results of this study demonstrate the feasibility of using the iMS technique for multiplexed detection of miRNAs in real biological samples. Note that the SERS measurements were performed immediately following the incubation of target molecules without any washing steps, which greatly simplifies and accelerates the assay procedure.

iMS bioassay-on-chip

A unique SERS platform consisting of a metal film over close-packed nanosphere substrate was first developed in our laboratory in 1984 [71]. With this substrate, we first demonstrated the analytical potential of the SERS effect and its practical use for trace analysis. This type of substrate, referred to Nanowave due its periodic wave-like structure, was later investigated and used by Van Duyne and coworkers, who employed the term metal film over nanosphere (MFON) to describe it [72]. The Nanowave or MFON substrate has been shown to be particularly effective for SERS applications [73]. Due to its relatively low fabrication cost and high enhancement factor, it has been used in a wide variety of chemical and biological sensing applications [74–76].

Recently, we showed that the SERS-based MS DNA detection strategy can be applied to SERS-active chips, referred to as molecular sentinel-on-chip (MSC) technique [77–79]. The SERS chips used in these studies were the triangular-shaped nanowire and more recently the Nanowave, also known as MFON. The wafer scale MFON concept has been used by other groups due to its particularly high SERS enhancement factor, 10^6–10^8, and facile fabrication [72,80–84]. Using bimetallic film (gold and silver), Nanowave's SERS enhancement can be further improved compared to single gold film while still maintaining the stability [85]. A large area of bimetallic Nanowave chip can be fabricated with high reproducibility (Fig. 5.21A). The higher magnification SEM image shows the periodic hexagonal pattern of the chip and crevices between metal-coated PS (Fig. 5.21B). The image also indicates considerable surface roughness on the chip. To confirm this, we conducted AFM measurements (Fig. 5.21C and D). Deep nanosize crevices between metal-coated polystyrene beads and substantial surface roughness can be clearly observed. All of these characteristics are believed to contribute to the high SERS enhancement of the bimetallic SERS Nanowave chip. Compared to single-metal (gold) Nanowave chip, the bimetallic Nanowave chip has a 3.6 times higher SERS intensity. Meanwhile, compared to a commercially available SERS substrate

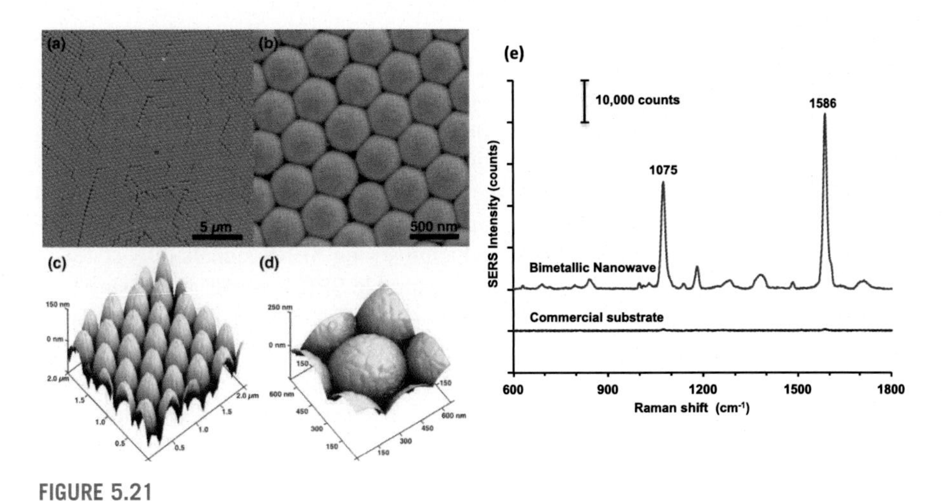

FIGURE 5.21

SEM images (A and B) and AFM images (C and D) of bimetallic Nanowave chip. (E) SERS spectra of p-mercaptobenzoic acid on bimetallic Nanowave chip and Klarite commercial substrate.

Adapted from Ngo, et al., Analyst 139 (22) (2014) 5656.

(Klarite), the bimetallic Nanowave chip has approximately 100 times higher SERS intensity (Fig. 5.21E).

By functionalizing the bimetallic Nanowave with iMS biosensor duplexes, we were able to develop a unique homogeneous bioassay on-chip using SERS detection. The detection strategy (Fig. 5.22) is similar to that of the colloidal iMS sensing mechanism in that in the absence of complementary target, the reporter probes

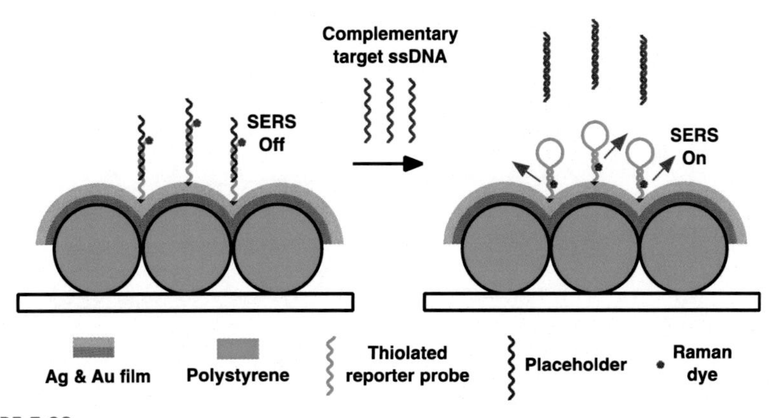

FIGURE 5.22

iMS bioassay-on-chip detection scheme.

Adapted from Ngo, et al., Analyst 139 (22) (2014) 5656.

and placeholders maintain partial duplex structures, keeping Raman dyes tagged at 3′ ends of the reporter probes away from the Nanowave chip metal surface. At such distance, the SERS signal is low due to the fact that SERS enhancement exponentially decreases with increase in Raman dye-metal surface distance (OFF state). When complementary targets are introduced, they hybridize with the placeholder strands and the reporter probes are free to form hairpin structures. With hairpin structures, the Raman label is brought into close proximity of the plasmonic Nanowave metallic nanostructured surface, inducing strong SERS signals (ON state).

This homogenous bioassay allows for measurable SERS signals after a single incubation step without any washing to remove unreacted components, making it simple-to-use and reducing reagent cost. Compared to the MSC technique where SERS intensity decreases in the presence of complementary target ssDNA (signal-off), the iMS bioassay has signal-on with SERS intensity increasing in the presence of target. Hence, the iMS on chip method is less susceptible to false-positive responses.

The capability of detecting Dengue virus oligonucleotides is demonstrated in Fig. 5.23. To demonstrate the specificity of the method, the functionalized Nanowave chips were tested against three different samples including: synthetic DENV 4 ssDNA (complementary target), 1 noncomplementary ssDNA, and buffer only (blank). As shown in Fig. 5.23, for blank and noncomplementary ssDNA samples, the SERS intensities were low and similar to SERS intensity before these samples

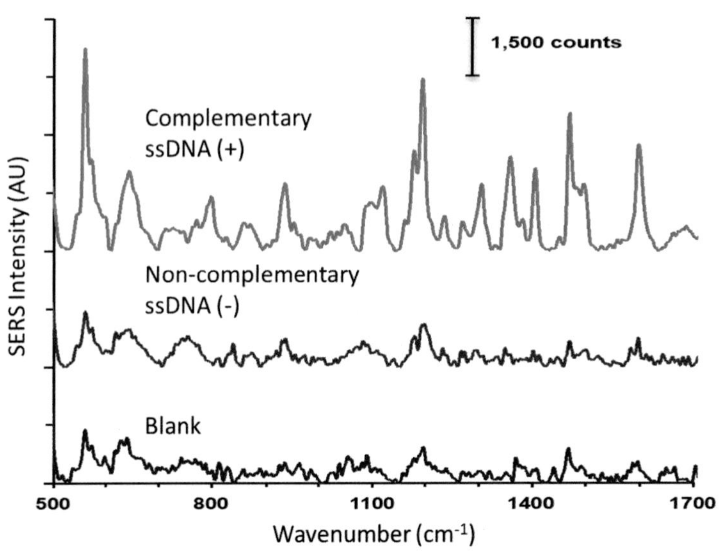

FIGURE 5.23

SERS spectra after incubation of functionalized Nanowave chip in blank, noncomplementary ssDNA, and complementary DENV 4 target ssDNA samples.

Adapted from Ngo, et al., Analyst 139 (22) (2014) 5656.

had been applied. This indicates that the reporter probe—placeholder partial duplexes were not disturbed by the noncomplementary ssDNA. The design of the iMS nanoprobe for detection of Dengue virus allows for a separation of Raman label from metal surface of 40 nucleotides (approximately 13.5 nm) when in the linear conformation. At such a separation, SERS enhancement is weak, resulting in low SERS background (OFF state). With the addition of complementary target ssDNA, the SERS intensity increased (Fig. 5.23). This increase is explained by the hybridization between the complementary target ssDNA and the placeholders, leaving the reporter probes free to form hairpin structures. As a result, the Cy5 Raman labels were brought close to the Nanowave chip's surface, resulting in increased SERS intensity (ON state).

The possibility for quantitative analysis is shown in Fig. 5.24. Sample solutions of complementary target ssDNA in buffer at different known concentrations were tested. The calibration curve was plotted based on SERS intensity of the main Raman peak. A linear trend line was fitted to the data points using linear regression. Based on slope ($s = 128.56$) and residual standard deviation ($\sigma = 255.02$) of the regression line, as low as 6 amol of the target ssDNA (DENV 4) inside the probed area (as defined by the laser spot size, approximately 3 μm diameter) could be detected. The presented method of iMS-on-chip is simple, sensitive, and suitable for low resource settings with use of a handheld Raman reader, making it an attractive candidate for integration into portable platforms for POC molecular diagnostics.

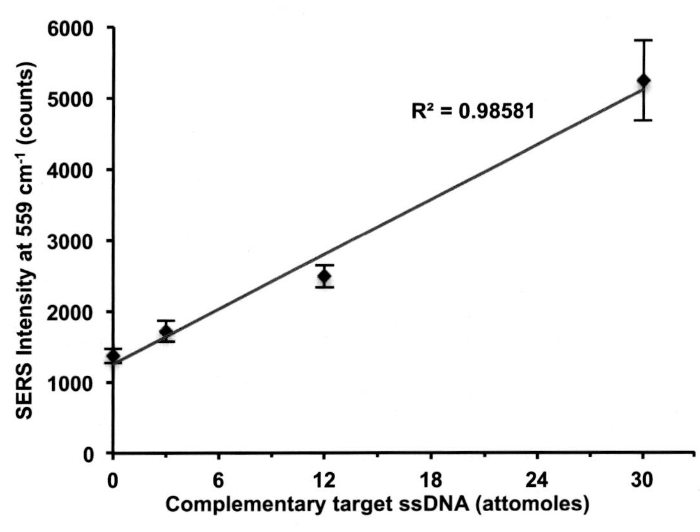

FIGURE 5.24

SERS intensity in the existence of different amounts of complementary target ssDNA inside probed area.

Adapted from Ngo, et al., Analyst 139 (22) (2014) 5656.

Conclusion

As technology advancements continue to progress diagnostic capabilities in clinical care, SERS-based methods have become of great interest in nucleic acid detection due to their high sensitivity [20,86,87]. However, most of the proposed methods still require amplification of extracted RNA for the potential of clinical application. We have demonstrated the development of both colloidal and on-chip forms of an OFF-to-ON SERS iMS nanobiosensor as a homogeneous assay for multiplex detection of miRNAs in a single sensing platform. As it does not require target labeling and any subsequent washing steps, the positive-readout iMS technique will offer a simple, rapid, monitor for a wide ranging number of sensing applications. As illustrated in this chapter, the iMS nanobiosensor can be used for diagnostics of a diverse set of diseases, including various cancers, with high potential for use in POC settings. The great benefit of the iMS nanobiosensor is the potential for multiplexed detection and the homogeneous bioassay format that does not require sample separation of hybridized and unhybridized probes using microfluidics. The iMS technology can also be applied to nonmedical applications such as monitoring plant biomarkers for renewable energy research. The future demonstration of clinical applicability using multiple biomarkers is of great interest in the development of a more robust diagnostic capability. The iMS biosensing platform provides a versatile and powerful tool for diagnostics and sensing for a wide variety of applications [88–90].

Acknowledgments

This material is based upon work supported by the National Institute of Health, under award numbers 1R01GM135486-01, 1R21CA196426, and 1R01EB028078-01A1, and the US Department of Energy Office of Science, under award number DE-SC0019393.

References

[1] F. Degliangeli, P.P. Pompa, R. Fiammengo, Nanotechnology-based strategies for the detection and quantification of MicroRNA, Chem.– Eur. J. 20 (31) (2014) 9476–9492.

[2] B.N. Johnson, R. Mutharasan, Biosensor-based microRNA detection: techniques, design, performance, and challenges, Analyst 139 (7) (2014) 1576–1588.

[3] H. Wang, R.A. Ach, B. Curry, Direct and sensitive miRNA profiling from low-input total RNA, RNA 13 (1) (2007) 151–159.

[4] E. Koscianska, J. Starega-Roslan, K. Czubala, W.J. Krzyzosiak, High-resolution northern blot for a reliable analysis of microRNAs and their precursors, Sci. World J. 11 (2011) 102–117.

[5] C.-G. Liu, G.A. Calin, B. Meloon, N. Gamliel, C. Sevignani, M. Ferracin, C.D. Dumitru, M. Shimizu, S. Zupo, M. Dono, An oligonucleotide microchip for genome-wide microRNA profiling in human and mouse tissues, Proc. Natl. Acad. Sci. 101 (26) (2004) 9740–9744.

[6] K. Lao, N.L. Xu, V. Yeung, C. Chen, K.J. Livak, N.A. Straus, Multiplexing RT-PCR for the detection of multiple miRNA species in small samples, Biochem. Biophys. Res. Commun. 343 (1) (2006) 85–89.

[7] R.A. Alvarez-Puebla, L.M. Liz-Marzán, SERS-based diagnosis and biodetection, Small 6 (5) (2010) 604–610.

[8] A.-I. Henry, B. Sharma, M.F. Cardinal, D. Kurouski, R.P. Van Duyne, Surface-enhanced Raman spectroscopy biosensing: in vivo diagnostics and multimodal imaging, Anal. Chem. 88 (13) (2016) 6638–6647.

[9] S. Laing, K. Gracie, K. Faulds, Multiplex in vitro detection using SERS, Chem. Soc. Rev. 45 (7) (2016) 1901–1918.

[10] D.L. Jeanmaire, R.P. Van Duyne, Surface Raman spectroelectrochemistry: Part I. Heterocyclic, aromatic, and aliphatic amines adsorbed on the anodized silver electrode, J. Electroanal. Chem. Interf. Electrochem. 84 (1) (1977) 1–20.

[11] Y.W. Wang, S. Kang, A. Khan, P.Q. Bao, J.T.C. Liu, In vivo multiplexed molecular imaging of esophageal cancer via spectral endoscopy of topically applied SERS nanoparticles, Biomed. Opt. Express 6 (10) (2015) 3714–3723.

[12] S. Kang, Y. Wang, N.P. Reder, J.T. Liu, Multiplexed molecular imaging of biomarker-targeted SERS nanoparticles on fresh tissue specimens with channel-compressed spectrometry, PLoS One 11 (9) (2016) e0163473.

[13] S. Feng, J. Lin, M. Cheng, Y.-Z. Li, G. Chen, Z. Huang, Y. Yu, R. Chen, H. Zeng, Gold nanoparticle based surface-enhanced Raman scattering spectroscopy of cancerous and normal nasopharyngeal tissues under near-infrared laser excitation, Appl. Spectrosc. 63 (10) (2009) 1089–1094.

[14] S. Feng, R. Chen, J. Lin, J. Pan, G. Chen, Y. Li, M. Cheng, Z. Huang, J. Chen, H. Zeng, Nasopharyngeal cancer detection based on blood plasma surface-enhanced Raman spectroscopy and multivariate analysis, Biosens. Bioelectron. 25 (11) (2010) 2414–2419.

[15] Y. Chen, G. Chen, S. Feng, J. Pan, X. Zheng, Y. Su, Y. Chen, Z. Huang, X. Lin, F. Lan, R. Chen, H. Zeng, Label-free serum ribonucleic acid analysis for colorectal cancer detection by surface-enhanced Raman spectroscopy and multivariate analysis, J. Biomed. Opt. 17 (6) (2012) 1–7.

[16] C.A. Jenkins, R.A. Jenkins, M.M. Pryse, K.A. Welsby, M. Jitsumura, C.A. Thornton, P.R. Dunstan, D.A. Harris, A high-throughput serum Raman spectroscopy platform and methodology for colorectal cancer diagnostics, Analyst 143 (24) (2018) 6014–6024.

[17] K.M. Koo, J. Wang, R.S. Richards, A. Farrell, J.W. Yaxley, H. Samaratunga, P.E. Teloken, M.J. Roberts, G.D. Coughlin, M.F. Lavin, P.N. Mainwaring, Y. Wang, R.A. Gardiner, M. Trau, Design and clinical verification of surface-enhanced Raman spectroscopy diagnostic technology for individual cancer risk prediction, ACS Nano 12 (8) (2018) 8362–8371.

[18] E.J.H. Wee, Y. Wang, S.C.-H. Tsao, M. Trau, Simple, sensitive and accurate multiplex detection of clinically important melanoma DNA mutations in circulating tumour DNA with SERS nanotags, Theranostics 6 (10) (2016) 1506–1513.

[19] K.M. Koo, E.J.H. Wee, P.N. Mainwaring, Y. Wang, M. Trau, Toward precision medicine: a cancer molecular subtyping nano-strategy for RNA biomarkers in tumor and urine, Small 12 (45) (2016) 6233–6242.

[20] P. Vohra, P. Strobbia, H. Ngo, W. Lee, T.V. Dinh, Rapid nanophotonics assay for head and neck cancer diagnosis, Sci. Rep. 8 (1) (2018) 11410.

[21] H.T. Ngo, E. Freedman, R.A. Odion, P. Strobbia, A.S. De Silva Indrasekara, P. Vohra, S.M. Taylor, T. Vo-Dinh, Direct detection of unamplified pathogen RNA in blood lysate using an integrated lab-in-a-stick device and ultrabright SERS nanorattles, Sci. Rep. 8 (1) (2018) 4075.

[22] J. Li, K.M. Koo, Y. Wang, M. Trau, Native MicroRNA targets trigger self-assembly of nanozyme-patterned hollowed nanocuboids with optimal interparticle gaps for plasmonic-activated cancer detection, Small 15 (50) (2019) 1904689.

[23] H.B. Grossman, E. Messing, M. Soloway, K. Tomera, G. Katz, Y. Berger, Y. Shen, Detection of bladder cancer using a point-of-care proteomic assay, JAMA 293 (7) (2005) 810–816.

[24] M. Tsujimoto, K. Nakabayashi, K. Yoshidome, T. Kaneko, T. Iwase, F. Akiyama, Y. Kato, H. Tsuda, S. Ueda, K. Sato, Y. Tamaki, S. Noguchi, T.R. Kataoka, H. Nakajima, Y. Komoike, H. Inaji, K. Tsugawa, K. Suzuki, S. Nakamura, M. Daitoh, Y. Otomo, N. Matsuura, One-step nucleic acid amplification for intraoperative detection of lymph node metastasis in breast cancer patients, Clin. Cancer Res. 13 (16) (2007) 4807–4816.

[25] K. Hsieh, A.S. Patterson, B.S. Ferguson, K.W. Plaxco, H.T. Soh, Rapid, sensitive, and quantitative detection of pathogenic DNA at the point of care through microfluidic electrochemical quantitative loop-mediated isothermal amplification, Angew. Chem. Int. Ed. 51 (20) (2012) 4896–4900.

[26] K. Zagorovsky, W.C.W. Chan, A plasmonic DNAzyme strategy for point-of-care genetic detection of infectious pathogens, Angew. Chem. Int. Ed. 52 (11) (2013) 3168–3171.

[27] M. Soler, M.C. Estevez, R. Villar-Vazquez, J.I. Casal, L.M. Lechuga, Label-free nanoplasmonic sensing of tumor-associate autoantibodies for early diagnosis of colorectal cancer, Anal. Chim. Acta 930 (2016) 31–38.

[28] S. Mabbott, S.C. Fernandes, M. Schechinger, G.L. Cote, K. Faulds, C.R. Mace, D. Graham, Detection of cardiovascular disease associated miR-29a using paper-based microfluidics and surface enhanced Raman scattering, Analyst 145 (3) (2020) 983–991, https://doi.org/10.1039/C9AN01748H.

[29] M.B. Wabuyele, T. Vo-Dinh, Detection of human immunodeficiency virus type 1 DNA sequence using plasmonics nanoprobes, Anal. Chem. 77 (23) (2005) 7810–7815.

[30] H.N. Wang, A.M. Fales, A.K. Zaas, C.W. Woods, T. Burke, G.S. Ginsburg, T. Vo-Dinh, Surface-enhanced Raman scattering molecular sentinel nanoprobes for viral infection diagnostics, Anal. Chim. Acta 786 (2013) 153–158.

[31] H.-N. Wang, T. Vo-Dinh, Multiplex detection of breast cancer biomarkers using plasmonic molecular sentinel nanoprobes, Nanotechnology 20 (6) (2009) 065101.

[32] H. Wang, A. Fales, T. Vo-Dinh, Plasmonics-based SERS nanobiosensor for homogeneous nucleic acid detection, Nanomed. Nanotechnol. Biol. Med. 11 (4) (2015) 811.

[33] H.-N. Wang, B.M. Crawford, A.M. Fales, M.L. Bowie, V.L. Seewaldt, T. Vo-Dinh, Multiplexed detection of MicroRNA biomarkers using SERS-based inverse molecular sentinel (iMS) nanoprobes, J. Phys. Chem. C 120 (37) (2016) 21047–21055.

[34] H.-N. Wang, B.M. Crawford, T. Vo-Dinh, Molecular SERS nanoprobes for medical diagnostics, in: Nanotechnology in Biology and Medicine: Methods, Devices, and Applications, 2017, p. 289.

[35] H.-N. Wang, J.K. Register, A.M. Fales, N. Gandra, E.H. Cho, A. Boico, G.M. Palmer, B. Klitzman, T. Vo-Dinh, Surface-enhanced Raman scattering nanosensors for in vivo detection of nucleic acid targets in a large animal model, Nano Res. (2018) 1–12.

[36] B.M. Crawford, P. Strobbia, H.-N. Wang, R. Zentella, M.I. Boyanov, Z.-M. Pei, T.-P. Sun, K.M. Kemner, T. Vo-Dinh, Plasmonic nanoprobes for in vivo multimodal sensing and bioimaging of microRNA within plants, ACS Appl. Mater. Interfaces 11 (8) (2019) 7743–7754.
[37] P. Strobbia, Y. Ran, B.M. Crawford, V. Cupil-Garcia, R. Zentella, H.-N. Wang, T.-P. Sun, T. Vo-Dinh, Inverse molecular sentinel-integrated fiberoptics sensor for direct and in situ detection of miRNA targets, Anal. Chem. 91 (9) (2019) 6345–6352.
[38] S. Tyagi, F.R. Kramer, Molecular beacons: probes that fluoresce upon hybridization, Nat. Biotechnol. 14 (3) (1996) 303–308.
[39] Y. Zhang, Z. Tang, J. Wang, H. Wu, A. Maham, Y. Lin, Hairpin DNA switch for ultrasensitive spectrophotometric detection of DNA hybridization based on gold nanoparticles and enzyme signal amplification, Anal. Chem. 82 (15) (2010) 6440–6446.
[40] H.-I. Peng, C.M. Strohsahl, K.E. Leach, T.D. Krauss, B.L. Miller, Label-free DNA detection on nanostructured Ag surfaces, ACS Nano 3 (8) (2009) 2265–2273.
[41] J.E.N. Dolatabadi, O. Mashinchian, B. Ayoubi, A.A. Jamali, A. Mobed, D. Losic, Y. Omidi, M. de la Guardia, Optical and electrochemical DNA nanobiosensors, Trends Anal. Chem. 30 (3) (2011) 459–472.
[42] M. Bercovici, G. Kaigala, K. Mach, C. Han, J. Liao, J. Santiago, Rapid detection of urinary tract infections using isotachophoresis and molecular beacons, Anal. Chem. 83 (11) (2011) 4110–4117.
[43] W.R. Algar, M. Massey, U.J. Krull, The application of quantum dots, gold nanoparticles and molecular switches to optical nucleic-acid diagnostics, Trends Anal. Chem. 28 (3) (2009) 292–306.
[44] A. Bonanni, M. Pumera, Graphene platform for hairpin-DNA-based impedimetric genosensing, ACS Nano 5 (3) (2011) 2356–2361.
[45] H. Yuan, C.G. Khoury, H. Hwang, C.M. Wilson, G.A. Grant, T. Vo-Dinh, Gold nanostars: surfactant-free synthesis, 3D modelling, and two-photon photoluminescence imaging, Nanotechnology 23 (7) (2012) 075102.
[46] A.M. Fales, H. Yuan, T. Vo-Dinh, Development of hybrid silver-coated gold nanostars for nonaggregated surface-enhanced Raman scattering, J. Phys. Chem. C 118 (7) (2014) 3708–3715.
[47] D.Y. Zhang, G. Seelig, Dynamic DNA nanotechnology using strand-displacement reactions, Nat. Chem. 3 (2) (2011) 103.
[48] D.P. Bartel, MicroRNAs: genomics, biogenesis, mechanism, and function, Cell 116 (2) (2004) 281–297.
[49] X. Ma, L.E. Becker Buscaglia, J.R. Barker, Y. Li, MicroRNAs in NF-κB signaling, J. Mol. Cell Biol. 3 (3) (2011) 159–166.
[50] T. Schepeler, Emerging roles of microRNAs in the Wnt signaling network, Crit. Rev. Oncog. 18 (4) (2013).
[51] A. Bischoff, M. Bayerlová, M. Strotbek, S. Schmid, T. Beissbarth, M.A. Olayioye, A global microRNA screen identifies regulators of the ErbB receptor signaling network, Cell Commun. Signal. 13 (1) (2015) 5.
[52] E. D'Ippolito, M.V. Iorio, MicroRNAs and triple negative breast cancer, Int. J. Mol. Sci. 14 (11) (2013) 22202–22220.
[53] G. Dong, X. Liang, D. Wang, H. Gao, L. Wang, L. Wang, J. Liu, Z. Du, High expression of miR-21 in triple-negative breast cancers was correlated with a poor prognosis and promoted tumor cell in vitro proliferation, Med. Oncol. 31 (7) (2014) 57.
[54] G. Di Leva, M. Garofalo, C.M. Croce, MicroRNAs in cancer, Annu. Rev. Pathol. Mech. Dis. 9 (2014) 287–314.

[55] J. Hayes, P.P. Peruzzi, S. Lawler, MicroRNAs in cancer: biomarkers, functions and therapy, Trends Mol. Med. 20 (8) (2014) 460–469.
[56] J. Lu, G. Getz, E.A. Miska, E. Alvarez-Saavedra, J. Lamb, D. Peck, A. Sweet-Cordero, B.L. Ebert, R.H. Mak, A.A. Ferrando, MicroRNA expression profiles classify human cancers, Nature 435 (7043) (2005) 834–838.
[57] N.R. Markham, M. Zuker, UNAFold, in Bioinformatics, Springer, 2008, pp. 3–31.
[58] L.N. Fix, M. Shah, T. Efferth, M.A. Farwell, B. Zhang, MicroRNA expression profile of MCF-7 human breast cancer cells and the effect of green tea polyphenon-60, Cancer Genom. Proteom. 7 (5) (2010) 261–277.
[59] B.M. Crawford, H.-N. Wang, C. Stolarchuk, R.J. von Furstenberg, P. Strobbia, D. Zhang, X. Qin, K. Owzar, K.S. Garman, T. Vo-Dinh, Plasmonic nanobiosensors for detection of microRNA cancer biomarkers in clinical samples, Analyst 145 (13) (2020) 4587–4594.
[60] K. Zhang, X. Wu, J. Wang, J. Lopez, W. Zhou, L. Yang, S.E. Wang, D.J. Raz, J.Y. Kim, Circulating miRNA profile in esophageal adenocarcinoma, Am. J. Cancer Res. 6 (11) (2016) 2713–2721.
[61] K.S. Garman, K. Owzar, E.R. Hauser, K. Westfall, B.R. Anderson, R.F. Souza, A.M. Diehl, D. Provenzale, N.J. Shaheen, MicroRNA expression differentiates squamous epithelium from Barrett's esophagus and esophageal cancer, Dig. Dis. Sci. 58 (11) (2013) 3178–3188.
[62] V.S. Nair, L.S. Maeda, J.P. Ioannidis, Clinical outcome prediction by microRNAs in human cancer: a systematic review, J. Natl. Cancer Inst. 104 (7) (2012) 528–540.
[63] M.G. Kendall, A new measure of rank correlation, Biometrika 30 (1/2) (1938) 81–93.
[64] R. Boom, C. Sol, M. Salimans, C. Jansen, P. Wertheim-van Dillen, J. Van der Noordaa, Rapid and simple method for purification of nucleic acids, J Clin Microbiol 28 (3) (1990) 495–503.
[65] H. Bordelon, N.M. Adams, A.S. Klemm, P.K. Russ, J.V. Williams, H.K. Talbot, D.W. Wright, F.R. Haselton, Development of a low-resource RNA extraction cassette based on surface tension valves, ACS Appl. Mater. Interfaces 3 (6) (2011) 2161–2168.
[66] A.M. Krichevsky, G. Gabriely, miR-21: a small multi-faceted RNA, J. Cell. Mol. Med. 13 (1) (2009) 39–53.
[67] M. Mackiewicz, K. Huppi, J.J. Pitt, T.H. Dorsey, S. Ambs, N.J. Caplen, Identification of the receptor tyrosine kinase AXL in breast cancer as a target for the human miR-34a microRNA, Breast Cancer Res. Treat. 130 (2) (2011) 663.
[68] S.D. Selcuklu, M.T. Donoghue, C. Spillane, miR-21 as a Key Regulator of Oncogenic Processes, Portland Press Limited, 2009.
[69] M. Weiland, X.-H. Gao, L. Zhou, Q.-S. Mi, Small RNAs have a large impact: circulating microRNAs as biomarkers for human diseases, RNA Biol. 9 (6) (2012) 850–859.
[70] B.R. Lutz, C.E. Dentinger, L.N. Nguyen, L. Sun, J. Zhang, A.N. Allen, S. Chan, B.S. Knudsen, Spectral analysis of multiplex Raman probe signatures, ACS Nano 2 (11) (2008) 2306–2314.
[71] T. Vo-Dinh, M. Hiromoto, G. Begun, R. Moody, Surface-enhanced Raman spectrometry for trace organic analysis, Anal. Chem. 56 (9) (1984) 1667–1670.
[72] L.A. Dick, A.D. McFarland, C.L. Haynes, R.P. Van Duyne, Metal film over nanosphere (MFON) electrodes for surface-enhanced Raman spectroscopy (SERS): improvements in surface nanostructure stability and suppression of irreversible loss, J. Phys. Chem. B 106 (4) (2002) 853–860.
[73] K.A. Willets, R.P. Van Duyne, Localized surface plasmon resonance spectroscopy and sensing, Annu. Rev. Phys. Chem. 58 (2007) 267–297.

[74] T. Vo-Dinh, Surface-enhanced Raman spectroscopy using metallic nanostructures, Trends Anal. Chem. 17 (8–9) (1998) 557–582.

[75] P.L. Stiles, J.A. Dieringer, N.C. Shah, R.P. Van Duyne, Surface-enhanced Raman spectroscopy, Annu. Rev. Anal. Chem. 1 (2008) 601–626.

[76] X. Zhang, M.A. Young, O. Lyandres, R.P. Van Duyne, Rapid detection of an anthrax biomarker by surface-enhanced Raman spectroscopy, J. Am. Chem. Soc. 127 (12) (2005) 4484–4489.

[77] H.N. Wang, A. Dhawan, Y. Du, D. Batchelor, D.N. Leonard, V. Misra, T. Vo-Dinh, Molecular sentinel-on-chip for SERS-based biosensing, Phys. Chem. Chem. Phys. 15 (16) (2013) 6008–6015.

[78] H.T. Ngo, H.-N. Wang, A.M. Fales, T. Vo-Dinh, Label-free DNA biosensor based on SERS molecular sentinel on nanowave chip, Anal. Chem. 85 (13) (2013) 6378–6383.

[79] H.T. Ngo, H.-N. Wang, T. Burke, G.S. Ginsburg, T. Vo-Dinh, Multiplex detection of disease biomarkers using SERS molecular sentinel-on-chip, Anal. Bioanal. Chem. 406 (14) (2014) 3335–3344.

[80] J.-F. Masson, K.F. Gibson, A. Provencher-Girard, Surface-enhanced Raman spectroscopy amplification with film over etched nanospheres, J. Phys. Chem. C 114 (51) (2010) 22406–22412.

[81] H. Im, K.C. Bantz, N.C. Lindquist, C.L. Haynes, S.-H. Oh, Vertically oriented sub-10-nm plasmonic nanogap arrays, Nano Lett. 10 (6) (2010) 2231–2236.

[82] H. Im, K.C. Bantz, S.H. Lee, T.W. Johnson, C.L. Haynes, S.H. Oh, Self-assembled plasmonic nanoring cavity arrays for SERS and LSPR biosensing, Adv. Mater. 25 (19) (2013) 2678–2685.

[83] J.-A. Huang, Y.-Q. Zhao, X.-J. Zhang, L.-F. He, T.-L. Wong, Y.-S. Chui, W.-J. Zhang, S.-T. Lee, Ordered Ag/Si nanowires array: wide-range surface-enhanced Raman spectroscopy for reproducible biomolecule detection, Nano Lett. 13 (11) (2013) 5039–5045.

[84] D. Kim, A.R. Campos, A. Datt, Z. Gao, M. Rycenga, N.D. Burrows, N.G. Greeneltch, C.A. Mirkin, C.J. Murphy, R.P. Van Duyne, Microfluidic-SERS devices for one shot limit-of-detection, Analyst 139 (13) (2014) 3227–3234.

[85] H.T. Ngo, H.-N. Wang, A.M. Fales, B.P. Nicholson, C.W. Woods, T. Vo-Dinh, DNA bioassay-on-chip using SERS detection for dengue diagnosis, Analyst 139 (22) (2014) 5655–5659.

[86] H.T. Ngo, N. Gandra, A.M. Fales, S.M. Taylor, T. Vo-Dinh, Sensitive DNA detection and SNP discrimination using ultrabright SERS nanorattles and magnetic beads for malaria diagnostics, Biosens. Bioelectron. 81 (2016) 8–14.

[87] T. Donnelly, W.E. Smith, K. Faulds, D. Graham, Silver and magnetic nanoparticles for sensitive DNA detection by SERS, Chem. Commun. 50 (85) (2014) 12907–12910.

[88] B.M. Crawford, H.-N. Wang, P. Strobbia, R. Zentella, Z.-M. Pei, T.-p. Sun, T. Vo-Dinh, Plasmonic Nanobiosensing: from in situ plant monitoring to cancer diagnostics at the point of care, J. Phys. Photon. 2 (3) (2020) 034012.

[89] P. Strobbia, R.A. Odion, M. Maiwald, B. Sumpf, T. Vo-Dinh, Direct SERDS sensing of molecular biomarkers in plants under field conditions, Analy. Bioanal. Chem. 412 (14) (2020) 3457–3466.

[90] V. Cupil-Garcia, P. Strobbia, B.M. Crawford, H.-N. Wang, H. Ngo, Y. Liu, T. Vo-Dinh, Plasmonic nanoplatforms: from surface-enhanced Raman scattering sensing to biomedical applications, J. Raman Spectrosc. 52 (2) (2021) 541–553.

CHAPTER

SERS detection of oral and gastrointestinal cancers

6

Alexander Czaja, M.S. [1,2], **Cristina Zavaleta, PhD** [1,2]

[1]*Department of Biomedical Engineering, University of Southern California, Los Angeles, CA, United States;* [2]*USC Michelson Center for Convergent Bioscience, Los Angeles, CA, United States*

Introduction

Raman spectroscopy has become an indispensable analytical tool for identification and quantification of chemicals and materials. This "new radiation" heralded by Chandrasekhara Venkata Raman in 1928 is a process that inelastically scatters photons in a pattern that is characteristic of the material's bonds and permitted vibrational states [1]. A Raman spectral measurement is made up of these inelastically scattered photons, allowing Raman measurements to be rich in detail, nondestructive to the sample, and relatively simple to prepare. Imaging using fluorescent labels is plagued by broad absorption and emission bands that occupy similar wavelength bands, imposing the use of multiple excitation wavelengths and limitations on the set of labels one can apply simultaneously. By contrast, Raman spectral peaks are highly specific and unique, invulnerable to the excitation wavelength used, and thus are eminently separable and ideal for multiplexed imaging, especially in biological applications where many chemical species are present. With the advent of the laser, it is now feasible to obtain enough of these inelastically scattered photons to form a Raman spectrum, the so-called molecular fingerprint, without prohibitively long exposures. Coupled with the subsequent discovery of the localized plasmon resonance effect on metallic nanotextured surfaces and its strengthening of the local electric field beginning with Fleischman et al. [2], Raman spectra can now be greatly amplified and measured with high degrees of confidence, offering highly sensitive and specific detection (Fig. 6.1).

For several years, Raman spectroscopy had previously been confined to analytical chemistry laboratories due to these technological roadblocks and the lack of knowledge to interpret the spectral information. Engineering disciplines are now clamoring to research and utilize surface-enhanced Raman spectroscopy (SERS) for a wide range of applications. The uniqueness and intensity of SERS spectra are vital to the platform's robustness and precision to quantify femtomolar levels of analytes. In biological and diagnostic applications, surface-enhanced Raman analysis is being employed to detect cellular events and specific biomarkers of disease. As we will explore, the burgeoning space of Raman-based diagnostics is driving the development of new device architectures and nanotextured materials

SERS for Point-of-care and Clinical Applications. https://doi.org/10.1016/B978-0-12-820548-8.00005-9

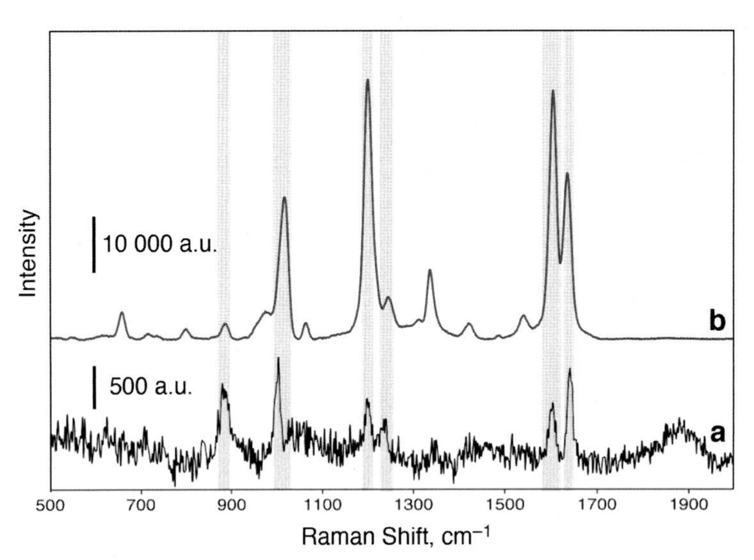

FIGURE 6.1

Comparison of the nonenhanced and enhanced Raman spectrum of a SERS reporter molecule. These are Raman spectral readings of trans-1,2-bis(4-pyridyl)-ethylene (BPE) (A) without enhancement and (B) with enhancement by being adsorbed onto 60 nm gold nanoparticles and encapsulated inside a shell of silica. The enhanced spectrum of the reporter exhibits greater intensity and signal to noise ratio, both of which are important for elucidating more of the molecule's excited vibrational modes and for enabling multiplexed measurement of simultaneous Raman signals.

Figure provided by Olga Eremina.

to achieve amplified Raman scattering. Noninvasive and nontoxic applications of SERS for diagnosis in live organisms are a complementary avenue of design that opens the door to a new suite of exploratory and diagnostic clinical procedures.

As we will explore, SERS techniques can measure the presence of a known Raman reporter molecule or measure the fingerprints of the endogenous biomolecules in a fluid or tissue specimen (Fig. 6.2). A procedure that uses a Raman reporter label involves targeted binding of the reporter to a cancer cell or biomarker substance one wishes to quantify. The intensity of the known Raman label's spectrum, when enhanced by a metallic nanostructure, correlates to the presence of the desired target, just as fluorescence indicates biomarker concentration in an enzyme-linked immunosorbent assay. Alternatively, one may wish to detect cancer by virtue of the altered microenvironment and balance of biomolecules caused by the disease, which can be detected using nanotextured metal to enhance the intensity of the biological specimen's intrinsic Raman scattering. Here, the indicator of cancer is the signature of how the Raman peaks have diverged from those exhibited by normal tissue. Because these spectral changes are subtle and hidden among the Raman contributions of many biomolecules, classifier models can be trained to determine the

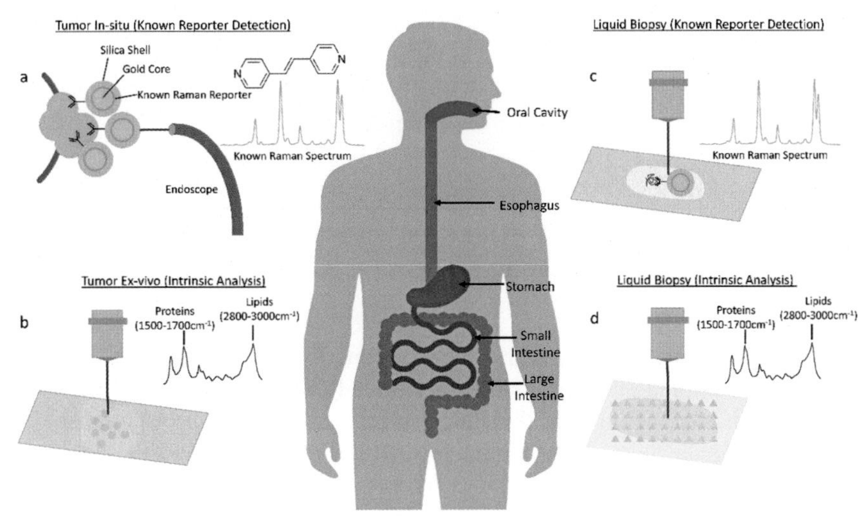

FIGURE 6.2

Primary classes of SERS diagnostic methods that have been developed to investigate and characterize gastrointestinal cancer. (A) Tumors in the lumen may be coated with SERS nanoparticles carrying a Raman reporter and be formulated to preferentially bind to cancer cells. Loci of particle binding can be discovered using a Raman attachment to a conventional endoscope. (B) A tumor may remain in situ or be excised from the gastrointestinal tract and placed onto a Raman-enhancing nanotextured metallic substrate or coated with a colloid of metallic nanoparticles to readily measure the Raman bands of the tissue itself. These enhanced intrinsic Raman measurements are typically subjected to further computational analysis to reach a diagnostic decision. (C) Some fluid sample, like saliva or blood, may contain a biomarker of interest one wishes to assay, e.g., an important enzyme or RNA sequence. The biomarker can be isolated and bound by labeled SERS nanoparticles to indicate the substance's presence. (D) Fluid samples can also be interrogated for their endogenous biomolecule balances by enhancing the fluid's intrinsic Raman signal using metallic nanoparticles or a nanotextured metallic substrate.

diagnostically relevant changes in Raman peak intensities that signify the presence of cancer. Models that process complex data and provide rapid feedback are referred to as computer-aided diagnostic (CAD) systems. Labeled and unlabeled SERS strategies for acquiring clinically relevant information can each be deployed in in vivo and ex vivo fashions.

Detection and diagnosis of cancers is a high impact application space for applying these enhanced Raman measurement techniques. Gastrointestinal cancers are particularly serendipitous targets because 1) collection of samples for interrogation is noninvasive and relatively simple via saliva, stool, or blood, 2) most of the GI tract is easily accessible via endoscopy for examination and imaging, 3) oral delivery of tumor-targeting SERS nanoparticles to the GI tract is easy and avoids systemic toxicity concerns. The routes of detection discussed here can be in vivo or in vitro

Raman diagnosis techniques according to whether they operate within or outside of a living subject, respectively (Fig. 6.2). In this chapter, we will discuss each gastrointestinal organ's set of design challenges and the ways the field is leveraging SERS to engineer improved cancer detection and characterization.

Oral cancer

Introduction

We begin our discussion of SERS applications for diagnosing gastrointestinal cancers with a look at the work on oral cancer. Primarily seeking to detect oral squamous cell carcinoma and parotid (salivary) gland cancer, groups have developed in vitro methods to acquire sensitive Raman measurements on tissue and saliva samples to investigate the spectral biomarkers of the disease and identify malignant cases. Targeted binding of nanoparticles involves a targeting molecule with high affinity for some excreted or overexpressed cell surface biomarker. The use of a strong Raman reporter allows spectral measurements to easily identify the presence of a malignant cell by virtue of the reporter's known signal. Typically, a Raman reporter's spectrum will exhibit peaks not characteristically observed in biological substances, and will be of a much greater magnitude, both of which aid in confidently declaring the presence of a Raman reporter nanoparticle on the surface of a malignant cell. Metallic nanoparticles without a Raman reporter dye adsorbed onto their surface have also been used to enhance the intrinsic Raman signature of the biological substances to interrogate the changes occurring in cancer cells that set them apart from normal cells.

Optimization and design considerations

It is recognized that the global prevalence of oral cancer is on the rise, thought to be the result of environmental factors and exposure to unhealthy foods and substances [3]. Combined with the clinical admission that oral lesions are difficult to observe and diagnose upon visual examination, the need has arisen for dependable and inexpensive methods to diagnose oral cancer [3]. Because the oral cavity is so readily accessible, it is entirely noninvasive to collect saliva and exfoliated tissue samples. However, since these samples are very complex biological mixtures, and the biomarkers tend to be present at low concentrations, assay methods need to be very sensitive and well characterized to confidently diagnose a patient with cancer. Due to the degree of Raman signal amplification provided by metallic surface enhancement, SERS-based diagnostic devices achieve very low limits of detection.

Working on a sensitive assay to analyze oral mucosal lesions to aide in pathology assessment, Han et al. developed SERS nanoparticles conjugated to two species of oligonucleotides complementary to the RNA sequence for S100 calcium binding protein P (S100P), which is a biomarker for oral squamous cell carcinoma [3].

The two oligonucleotides conjugated to the gold nanoparticles' surfaces were designed to be complementary to the two ends of the known RNA sequence of S100P, and one of these oligonucleotides was additionally conjugated to malachite green isothiocyanate to serve as the Raman reporter. In a sandwich format assay, particles were designed to hybridize to excreted S100P RNA and, under excitement by a laser beam, emit the Raman signal of the reporter, indicating the presence of the excreted RNA. In testing, Han's group reports a limit of detection of 3 nM of S100P RNA.

This same group employs these DNA-conjugated gold nanoparticles in a vertical flow device intended for point-of-care use, where a top layer preincubated with DNA-conjugated gold nanoparticles with malachite green isothiocyanate was situated atop a wicking membrane of filter paper and waste pad [4]. Whole saliva samples were collected from normal subjects and patients who had recently received a diagnosis of oral squamous cell carcinoma and were deposited onto their device. After allowing time for hybridization to occur between the gold nanoparticles and the S100P RNA, the top layer was washed with buffer and drawn away by the lower layers, causing the unbound nanoparticles to be washed away from the top layer of the device. The device's middle layer of filter paper prevented any large complexes of S100P RNA bound to pairs of DNA-conjugated gold nanoparticles. Raman measurements of the device's top layer achieved a limit of detection of 10 nM and managed to provide semiquantitative results despite most paper-based diagnostic devices only being capable of providing binary readings.

In the presence of cells overexpressing human epidermal growth factor receptor (EGFR), Huang et al. observed that gold nanorods conjugated to an anti-EGFR monoclonal antibody would assemble and align along their long edges to a greater degree than when in the presence of normal cells [5]. Rod-shaped gold nanoparticles allow for the tuning of their aspect ratio during synthesis to elicit more surface plasmon resonance to better suit one's optical Raman system. Due to the high degree of alignment along the long edge of the nanorods, more surface plasmon resonance occurs longitudinally across the nanorods, vastly increasing the opportunities for Raman enhancement. Their gold nanorods were capped with hexadecyltrimethylammonium bromide (CTAB), a common capping agent to prevent gold nanoparticle aggregation. When introduced to cultures of HSC3 cells, a line of human oral squamous cell carcinoma, CTAB served as a Raman reporter, and its Raman signature was greatly enhanced, but when introduced to cultures of normal human keratinocytes, far less nanorod binding and alignment occurred, and the CTAB signature was not observed to such a degree. Targeted binding at sites of EGFR expression created a domain at the cancer cells' surface where a high degree of gold nanorod packing could be achieved to facilitate enhanced Raman scattering. The nonuniform, or anisotropic, packing of aggregated gold nanorods, while eliciting a nonlinear degree of Raman enhancement, does also provide the benefit of eliciting minimal enhancement unless aggregation to a suitably overexpressing cell has occurred. Gold nanospheres are capable of packing isotropically, and do not have to be arranged in a particular fashion to serve as Raman enhancement surfaces.

Unlabeled metallic nanoparticles have been introduced to patient samples to enhance the intrinsic Raman scattering of the cells' endogenous substances. Yan et al. mixed gold nanoparticles with serum derived from peripheral blood of patients diagnosed with varying types of parotid gland cancer, including pleomorphic adenoma, Warthin's tumor, and mucoepidermoid carcinoma, as well as from normal subjects [6]. The spectral differences that occurred in samples from nonhealthy patients were attributed to marked increase in protein and nucleotide content, consistent with the idea that cancer cells never cease to proliferate, thus needing to form genetic copies and produce cellular machinery. They trained a support vector machine (SVM) model to output probabilities that a serum sample's Raman spectrum indicated one of the three investigated forms of parotid gland cancer. Shortly thereafter, Tan et al. performed a similar study using serum of patients diagnosed with oral squamous cell carcinoma and mucoepidermoid carcinoma using unlabeled gold nanoparticles, again noting spectral changes associated with increased presence of proteins, nucleic acids, and lipids [7]. A principal component analysis and linear discriminant analysis–based (PCA-LDA) model was trained to automatically discriminate between normal and healthy patients. Kah et al. instead opted to collect saliva samples from patients diagnosed with oral squamous cell carcinoma for microscopic analysis as well as SERS analysis by mixing samples with anti-EGFR conjugated gold nanoparticles [8]. Not only did the gold nanoparticles provide optical contrast during microscopy, but it was observed that the Raman spectra exhibited peaks not present when the procedure was performed using healthy saliva.

Surface-enhanced intrinsic Raman measurements have also been performed by assaying excised tissue sections using unlabeled silver nanoparticles. Kartha et al. sectioned biopsies from patients diagnosed with oral cancer and performed Raman spectroscopy on tissue sections treated with silver nanoparticles and noted that the Raman spectra were markedly different in regions of malignancy compared to normal tissue at the periphery of the specimen [9]. Namely, that Raman spectra of normal tissue exhibit increased peaks associated with lipids, while malignant tissue exhibit increased Amide I, Amide II, and phenylalanine bands. A PCA-based classifier model was also employed to demonstrate that such a computer-aided workflow could be achieved and had the potential to reduce the effort and training needed to deploy Raman systems in clinical settings.

In efforts to develop substrates as devices, solutions include creating thin films of metallic nanoparticles on glass surfaces, depositing them onto filter paper, and depositing them onto preformed nanotextured surfaces to further encourage surface plasmon resonance. Olivo and Kho et al. showed that saliva samples can be placed onto dried films of gold nanoparticles on glass slides for SERS analysis [10,11]. Both groups observed increased Raman peaks at wavenumber shifts of 670, 1097, and 1627 cm^{-1} in spectra measured from saliva samples deposited onto the gold nanoparticle film from patients diagnosed with oral cancer that were not present when measuring saliva of healthy patients. Liu et al. deposited gold nanorods onto a filter paper substrate and performed Raman analysis after depositing cells exfoliated from oral tissue and suspicious oral lesions [12]. They subsequently developed diagnostic criteria for using ratios of Raman peak intensities as biomarkers of disease.

Dai et al. produced a gold nanoparticle–coated silicon substrate for SERS analysis of cultured cells [13]. Using normal fibroblast cells as a control, they further measured the SERS Raman spectra of SAS and SCC4 oral cancer cell lines, observing that the cancer cells exhibited Raman peaks known to be caused by adenine's ring-breathing vibrational mode. This peak's intensity was normalized against the Raman peak at 1460 cm^{-1} associated with the C–H bending mode, which resulted in a more quantitative assay readout. Additionally, the substrate onto which the noble metal particles are deposited can be nanotextured to provide more surface for plasmon resonance. Connoly et al. produced a gold nanoparticle–coated nanopillar-textured silicon substrate for SERS analysis of saliva from oral squamous cell carcinoma patients. Using the Raman data, they further trained PCA-LDA models and performed logistic regression, resulting in diagnostic accuracy of 73% and sensitivity of 89% for models trained on Raman spectra of cells extracted from saliva [14]. Girish et al. further investigated the efficacy of different substrate nanotextures, including needular, bipyramidal, and leaf-like structures [15]. After coating their TiO_2 substrate with silver nanoparticles, the degree of SERS enhancement was compared by depositing crystal violet as a Raman reporter. Leaf-like nanotextured TiO_2 provided the most enhancement and was used for further Raman analysis of tissue sections taken from lesions of patients diagnosed with oral squamous cell carcinoma. A PCA-LDA model was also used to demonstrate the ability to computationally differentiate between normal and malignant tissue samples (Fig. 6.3).

Discussion and future directions

Many of these works on applying SERS to detect and characterize oral cancer recognize there are avenues that require further investigation. Huang's work on anti-EGFR conjugated gold nanorods notes that gold nanorods produce an anisotropic alignment, as opposed to the isotropic aggregation that occurs with gold nanospheres, which introduces the ability to acquire polarized Raman spectral measurements. In tests where samples of cells were allowed to bind to anti-EGFR conjugated gold nanorods and rotated in 20° increments with respect to the laser field, some orientations produced Raman peaks with greater intensities than other orientations. Polarized Raman analysis may allow for selection of desired Raman contributions while filtering others or provide the opportunity to choose the optimal orientation in which Raman enhancement produced the clearest signal on a per-experiment basis to optimize data quality.

Kang et al. utilized gold nanoparticles to enable SERS analysis of molecular and morphological changes of oral squamous cell carcinoma cells during cell death [16]. They produced gold nanoparticles conjugated to the nuclear localization signal peptide and to an arginine–glycine–aspartic acid (RGD) peptide to localize the particles to the cell nuclei. Raman measurements followed treatment with hydrogen peroxide of cells previously given these nucleus-binding gold nanoparticles, resulting in reliable Raman enhancement of the biomolecular events occurring within the

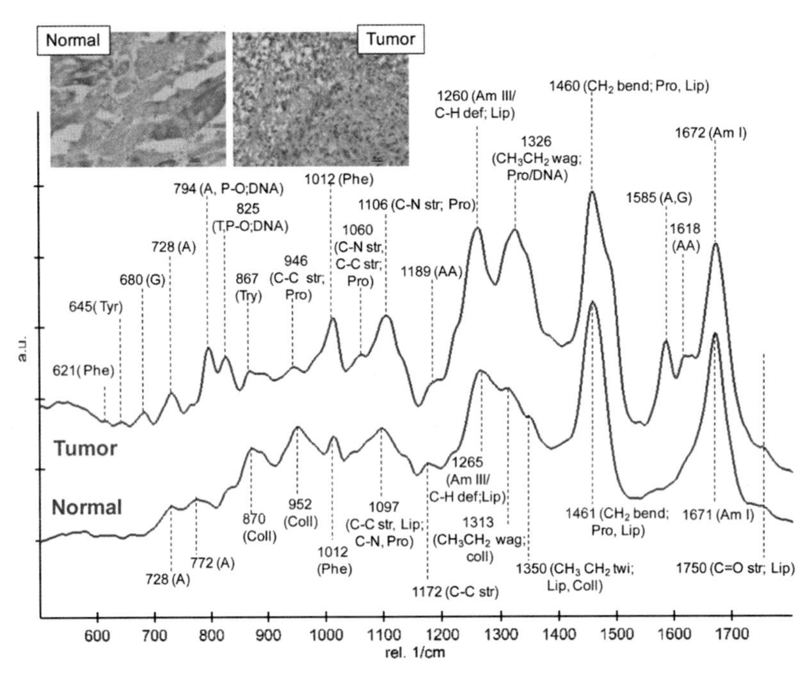

FIGURE 6.3

Comparison of averaged SERS spectra of normal oral tissue and oral cancer. Hematoxylin and eosin (H&E) stained inset images of each tissue type confirm the pathology of the tissue specimens. Enhancement was achieved by depositing the tissue sections onto an engineered substrate fabricated with a silver nano-leaf surface to achieve Raman enhancement. The Raman bands of molecular structures known to occur in biomolecules are annotated at the respective wavenumber coordinate. Shifts and altered intensities of these Raman peaks can be highly characteristic of cancer tissue compared to normal tissue and can be statistically characterized and modeled to build automated classifier tools.

Reproduced from C.M. Girish, S. Iyer, K. Thankappan, V.V.D. Rani, G.S. Gowd, D. Menon, S. Nair, M. Koyakutty, Rapid detection of oral cancer using Ag–TiO_2 nanostructured surface-enhanced Raman spectroscopic substrates, J. Mater. Chem. B 2 (8) (2014) 989–998 with permission from the Royal Society of Chemistry.

cells, enabling real-time Raman analysis of the protein denaturation, proteolysis, and DNA fragmentation stages of apoptosis. This time-resolved Raman characterization of cellular response to treatment could prove to be an incredibly useful basis for subtyping patients' tumors as well as automated assessment of treatment response of patient samples to batteries of potential chemotherapies. RGD-conjugated gold nanoparticles have also been shown to specifically bind to the $\alpha5\beta1$ and $\alpha v\beta3$ integrins on the nuclear membrane of cultured cancer cells by Xiao et al., enabling tip-enhanced Raman spectroscopy (TERS) of the nucleus [17]. TERS, which uses an electrified needle tip in addition an incident laser, can produce very high spatial resolution Raman maps due to the added control over the site of Raman enhancement. Though the technique requires additional hardware, the oral cavity may permit the space required to employ some TERS system for future in vivo clinical assessment of lesions.

In vivo applications of SERS nanoparticles remain to be investigated in the context of oral cancer. In mice, Campbell et al. administered Raman-labeled SERS gold nanoparticles also conjugated to ^{64}Cu, serving as a PET tracer while also being detectable by Raman spectroscopy [18]. The group used gamma emission counting and PET imaging to measure overall radionuclide emission from mice over time as well as inductively coupled plasma mass spectrometry (ICP-MS) and Raman analysis to assess nanoparticle distribution in different organs and tissues. The study revealed complete clearance of the SERS nanoparticles within 24 h and no systemic toxicity post oral administration demonstrating great potential for their clinical translation. Kartha noted in their discussion that fiberoptic Raman devices could be employed to perform spectroscopy of lesions directly in vivo. Campbell also discusses that since total clearance may be achievable in humans post oral administration, in vivo tumor targeting SERS analysis and screening of oral cancer could greatly impact their early detection rates and treatment efficacy. Further work must be done to assess the safety of gold nanoparticles administered topically to the gastrointestinal system to ensure no adverse effects or long-term toxicity ensues. Such a SERS system would still be noninvasive and would provide an incredibly quick clinical guide as to the malignancy of any oral lesion, especially when paired with CAD models, to be followed with further testing and biopsy in the case of positive test results.

Esophageal cancer

Introduction

Proceeding through the gastrointestinal system, we discuss the work applying SERS analysis to esophageal cancer. Unlike the oral cavity where it is accessible and noninvasive to perform visual examinations, acquire exfoliated samples and biopsies, and perform topical fiber optic measurements, esophageal examination techniques are more restricted. Assessment of esophageal disease currently entails white light endoscopy, which is performed by an experienced operator looking for suspicious lesions. However, dysplastic lesions are not always large or raised enough to be seen, and flat lesions can evade notice altogether. Groups have developed a variety of fiber optic Raman systems that are compatible with an endoscope's accessory channel to enable Raman measurements during white light endoscopy. The goal of SERS analysis during endoscopy is to interrogate the cellular identity of the tissue along the esophagus to more accurately and reliably identify diseased regions. Additionally, work has been done to leverage the SERS effect in devices to measure the presence of secreted cancer biomarkers as well as to characterize the local Raman spectral changes exhibited by diseased esophageal tissue. The merits of different classes of mathematical models for stratifying patients based on Raman spectral measurements have also been performed, providing a basis for designing and comparing computer-aided diagnosis methods.

Optimization and design considerations

Endoscopy is routinely used in the diagnosis of esophageal cancer, and endoscopes tend to include accessory channels for additional equipment during the procedure. For these reasons, research has been carried out to develop SERS methodology and Raman optics compatible with current endoscopic products. Minimizing the complexity of including SERS tools into existing clinical workflows will be crucial to the technique's adoption.

Harmsen et al. utilized gold nanoparticles labeled with the Raman reporter IR780 perchlorate injected into mice intravenously and allowed time for them to passively accumulate in malignant and premalignant lesions of mice [19]. Their nanoparticles and methods were tested on bile acid–treated transgenic human interleukin-1β mice that develop esophagitis, Barrett-like metaplasia, and esophageal carcinoma to emulate human esophageal cancer. Passive accumulation in the gastrointestinal system was not found to occur in mice who had not developed cancer but was observed in mice that developed disease. Circumferential scanning of the animals' gastrointestinal organs using their Raman endoscope system was successful and was validated by histology to confirm the malignancy of the lesions identified by the presence of labeled SERS particles. The group opted for intravenous SERS particle administration due to concerns that topical administration would have led to excessive nonspecific mucosal adhesion of the particles and would have necessitated prior knowledge of the targeting moieties with which to functionalize the SERS particles' surface. It should be noted that the FDA has yet to approve intravenous administration of metallic-based nanoparticles in humans over concerns of prolonged retention and long-term organ damage, so this approach of intravenously administering SERS nanoparticles for identifying metaplastic and dysplastic gastrointestinal tissue is limited to preclinical investigation and is not clinically viable at the moment.

Investigators are developing new topical Raman imaging strategies that circumvent the toxicity concerns associated with IV administration of SERS nanoparticles. The evaluation of SERS nanoparticle biodistribution by Zavaleta et al. demonstrated that local routes of administration via oral or rectal delivery can offer significant advantages. Studies revealed that SERS nanoparticles were confined to the GI tract with no evidence of systemic toxicity and were cleared from the body within 24 h in mice [20]. Garai et al. and Wang et al. achieved SERS Raman imaging by topical administration of labeled SERS particles as well as assessed the multiplexing ability of the systems and the interference of endogenous contributions to the Raman signal [21,22]. The studies validated their methods by subcutaneously injecting excised porcine colon with labeled SERS nanoparticles and by measuring Matrigel injected with SERS nanoparticle mixtures, respectively. The endoscope systems consisted of a fiber optic component, whose central fiber conducted the excitation laser and surrounding fibers served to collect inelastically scattered photons that represent the Raman spectrum, which was paired with a rotating mirror or glass prism, respectively, to enable circumferential spectral scanning of the gastrointestinal organ. The ability to detect multiple molecular targets in Raman endoscopic screening is noted by both groups to significantly contribute to the value and specificity of lesion assessment. Both works successfully demonstrated that multiplexed Raman localization of SERS particle presence was possible in porcine and rat models, and both demonstrated that scanning of the intrinsic Raman signal prior to SERS particle administration improved demultiplexing. Garai et al. also showed that the first principal component could be used to emulate topographic rendering of uneven surfaces from their Raman dataset. Jokerst et al. proposed and tested EGFR-specific Raman analysis of tissues by testing anti-EGFR affibody-functionalized SERS particles ability to A431 xenograft tumors, using EGFR nonoverexpressing MDA-MB-435S tumors as a control [23]. Multimodal gold particles coated with Alexa Fluor 647 along with Raman reporter BPE enable reliable cell-binding affinity by flow cytometry as well as distinct Raman signal localization by micro-Raman scanning. Wang et al. went on to demonstrate their Raman endoscopic system's ability to demultiplex the presence of labeled SERS particles conjugated to anti-HER2 and anti-EGFR antibodies in rats presenting esophageal cancer [24].

SERS Raman analysis has also been applied to patients' fluid samples to identify the presence of esophageal cancer. Studies by Li et al. and Huang et al. mixed silver nanoparticles with serum and urine samples, respectively [25,26]. Both studies collected samples from patients who had been diagnosed with esophageal cancer as well as from healthy subjects, and both trained mathematical models to differentiate the enhanced Raman spectra from the patient groups. Li utilized PCA-aided hierarchical clustering to develop an automated diagnosis system, while Huang trained a PCA-LDA model, revealing that urine of esophageal cancer patients exhibited decreased Raman bands of urea and increased bands of uric acid compared to healthy patients. Increased nucleic acid and phenylalanine SERS bands with decreased saccharide and protein bands were discovered by Lin et al. in the plasma of esophageal cancer patients. In the same group, Feng et al. found that the increased presence of modified nucleosides in urine, acknowledged as a general cancer biomarker and known to be a normal nucleic acid degradation product, could be measured by gold particle—enhanced Raman spectroscopy [27]. Purification of this class of analytes would traditionally require labor-intensive HPLC or capillary electrophoresis and are normally difficult to measure due to low abundance. The spectral specificity and amplified signal in SERS spectroscopy address both concerns, enabling noninvasive urinalysis screening. A partial least squares-linear discriminant analysis (PLS-LDA) model, retaining the first six principal components, yielded an overall diagnostic accuracy of 99.2%. Serum had been acquired from both esophageal cancer and nasopharyngeal cancer patients, and the model was additionally able to discriminate between these patient groups. The normalized Raman band at 725 cm^{-1} was identified as the primary differentiator between nasopharyngeal and esophageal cancer SERS spectra, indicating that RNA base metabolism could vary significantly between these conditions.

Li et al. evaluated the utility of enhanced intrinsic SERS analysis of serum samples across patient groups diagnosed with a wider array of cancers, including liver, colon, esophageal, nasopharyngeal, and gastric cancer, and found that an SVM model could achieve 95.4% diagnostic accuracy [28]. Additionally, they directly compared the performance of PCA and SVM models using serum samples from normal patients and patients diagnosed with varying stages of esophageal cancer [29]. Performance of classifier models is expected to depend on the nature of the problem space and dataset, with PCA being well suited for linearly separable data and SVM models capable of learning nonlinear boundaries by inclusion of an appropriate basis function. Serum was separated from whole blood, mixed with silver nanoparticles, dried onto microscope slides, and the enhanced Raman spectra were preprocessed by normalizing by the area under the curve and baseline subtraction according to a fifth-order polynomial using the Vancouver algorithm prior to modeling. Models trained using the same dataset included PCA-LDA, conventional SVM using a linear or a Gaussian radial basis function, and PCA-aided SVM using a linear or Gaussian radial basis function. Of all the models, the PCA-aided SVM using a Gaussian radial basis function achieved the best performance of 85.2% diagnostic accuracy, 86.7% specificity, and 83.3% sensitivity.

Micro-Raman SERS mapping of silver colloid—coated tissue excised from surgery to treat esophageal cancer was conducted by Feng et al. [30]. As a result of the silver nanoparticle coating, SERS enhancement of the intrinsic tissue signature was significant (Fig. 6.4) and maps revealed that lesions exhibited elevated Raman peaks at 478 cm^{-1} associated with exocyclic deformation of thymine compared to healthy tissue. The normalized and baseline subtracted dataset was used to train a PCA-LDA model that achieved 90.9% sensitivity and 97.8% specificity. Additional investigation by the group found increased Raman signal due to tryptophan, collagen, and phenylalanine with decreased signal of lactose in micro-Raman analysis of esophageal cancer specimens [31].

Assay techniques that utilize the SERS effect to quantify intrinsic changes in cell populations are vital to our understanding of the phenotypic changes in cancer cells. SERS measurements of CE48T and SCC-4 malignant esophageal cancer cell lines, using HFF fibroblast cells as a negative control, exhibited increased SERS peaks at 735 cm^{-1} associated with adenine's ring breathing mode in the same study that examined intrinsic SERS biomarkers of oral cancer [13]. Enzyme biomarkers for esophageal cancer have also been quantified using a SERS device. Matrix metalloproteinase (MMP) enzymes are biomarkers known to be associated with the breakdown of the extracellular matrix to promote tumor invasion and metastasis. An assay scheme was devised by Gong et al. to assay for the MMP-2 and MMP-7 to indicate the degradation of the enzymes' target peptides [32]. A silver- and gold-coated substrate with PEGylated biotin and neutravidin along with MMP-2 and MMP-7 peptides was fabricated, and gold nanoparticles coated with PEGylated biotin and MMP2 and MMP-7 peptides were also coated separately with ATP and NT as

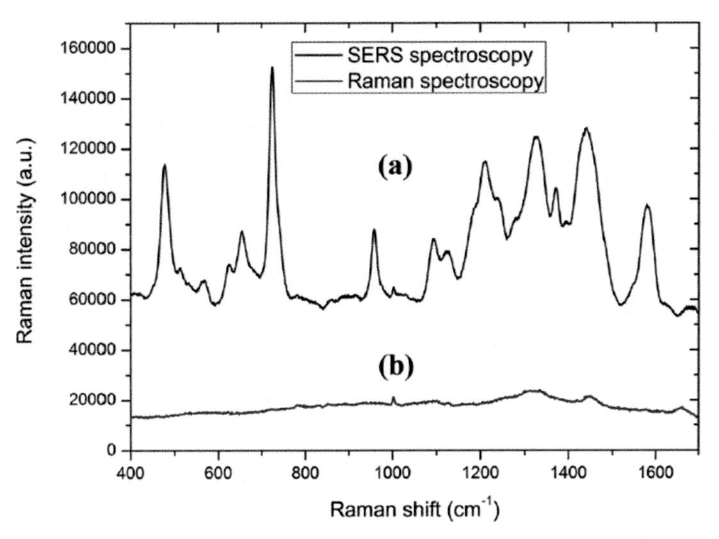

FIGURE 6.4

Raman signatures of sectioned esophageal tissue A) without enhancement and B) with enhancement after coating with colloid of silver nanoparticles. The Raman bands of important biomolecules are made apparent once enhancement is utilized.

Reproduced from Appl. Phys. Lett. 102 (2013) 043702; https://doi.org/10.1063/1.4789996 the permission of AIP Publishing.

Raman reporters. Presence of the MMP enzymes would degrade the MMP peptides hindering the particles from reaching the substrate's surface, resulting in strong enhancement of the reporters' Raman signals.

Discussion and future directions

SERS endoscopy shows a clear path forward for Raman-augmented screening of esophageal disease. Intraprocedural identification of raised and flat lesions by cell surface phenotype using labeled metallic nanoparticles can enable sensitive screening of precursor conditions such as Barrett's esophagus and more precise ablation and biopsy collection for pathologic assessment. However, it must also be noted that there is not a single cellular makeup and presentation of a premalignant esophagus. Kendall et al. collected micro-Raman datasets of jumbo biopsies from Barrett's esophagus patients that were snap frozen and stained with hematoxylin and eosin (H&E) for pathology assessment [33]. The samples were homogenously classified as consisting of healthy tissue or one type of dysplasia: either normal squamous cell epithelium, cardiac-type mucosa, fundic-type mucosa, intestinal-type Barrett's mucosa, low-grade dysplasia, high-grade dysplasia, squamous dysplasia, adenocarcinoma, or squamous cell carcinoma. A PCA-LDA model was trained on the Raman spectra collected from the mucosal surface of the biopsy. Two-way kappa analysis was used to judge the agreement of Raman-based classification of biopsies to the expert opinions of two pathologists blinded to patient identity and diagnosis, finding very high degrees of agreement across all patient subgroups. Clearly, the group's PCA-LDA model was capable of sufficiently learning the boundaries between the groups, demonstrating that there is more to learn about the cellular and tissue level changes that occur in premalignant and malignant disease. Medically relevant distinctions could be made during patient examination if the instrumentation and software are engineered with the knowledge of the possible pathophysiological presentations. Similar to how Li et al. demonstrated that mathematical models trained on identical Raman datasets can vary widely in performance, we learn that exhaustive comparison of models and adequately pairing the model to suit the disease's complexity is crucial to achieve acceptable clinical outcomes. Raman assessment of tumors stands to gain much from the integration of our clinical knowledge of disease presentation with the engineering approach to training computer-aided diagnosis systems.

Stomach cancer

Introduction

The stomach poses some unique design challenges to employing SERS-based cancer diagnostic methods. Lack of ready access to the stomach's interior demands that any in vivo diagnostic procedure involve gastroscopy to administer the metallic Raman-enhancing agent. However, due to the body's consistent accumulation of acids and enzymes in the stomach, the nanoparticle's stability and accumulation at a cancerous site would be difficult to control and measure. Other gastrointestinal organs such as the esophagus and intestines are, however, candidates for endoscopic SERS procedures because there are already clinically established ways to cleanse and prepare these regions for administration of SERS substrates (i.e., acetic acid wash). Therefore, much of the work on SERS assessment of stomach cancer involves analysis of patient serum, excised lesions, or secreted substances such as cancer biomarkers present in exhaled breath.

Optimization and design considerations

SERS analysis of serum, as we have seen, presents a noninvasive avenue for cancer detection that relies on cancer patients' measurable changes in the concentration of natural blood constituents. Because of the multitude of classes of molecules present in blood, the spectra measured are quite complicated due to the simultaneous contributions of all these constituents, so the trend continues in the diagnosis of stomach cancer of pairing some classification model with the SERS procedure to detect the spectral changes associated with the disease state. Li et al. and Guo et al. have collected whole blood from patients with and without stomach cancer, mixed the separated serum with metallic nanoparticles and deposited them onto glass slides for SERS analysis, and successfully trained models for future identification of stomach cancer patients [34,35]. Li's group additionally demonstrated that an iterative process can be used to determine which Raman shifts exhibited the statistically significant changes between patient groups. They also compared the performance of PCA-LDA, SVM, and classification and regression tree algorithms, concluding that PCA-SVM models yielded the best results with an overall diagnostic accuracy of 96.5%.

Classifier model performance is known to be a function of the training set's specificity and diversity. Feng et al. explored the differences in performance of a PCA-LDA model when trained with SERS spectra of patient serum measured with a laser that was either nonpolarized, linearly polarized, right-handed circularly polarized, and left-handed circularly polarized [36]. In their testing, a PCA-LDA model achieved the best diagnostic accuracy using SERS spectra collected with left-handed circularly polarized light. It remains to be explained, however, why altering the polarization has any effect on the Raman scattering process, but certain spectral differences were observed, such as a band associated with the C=C stretching of tryptophan and tyrosine only being present in the difference spectra of serum from stomach cancer—positive patients when measured with circularly polarized laser light. It is speculated that molecules' chirality plays a role in the selection rules for available vibrational modes, and the phase content, i.e., polarization, of light that can promote the molecules to certain classes of those vibrational states.

Serum constituents can also be separated chemically prior to SERS analysis to remove unwanted contributions and enable more direct quantification of a known Raman peak for diagnostic correlation. Chen et al. separated the RNA from serum of patients using the Trizol reagent for SERS quantification, as to cite a correlation between stomach cancer occurrence with *H. pylori* infection which may cause upregulation of miR-222, an RNA sequence that may be associated with increased cell proliferation [37]. Their PCA-LDA model achieved 100% sensitivity and 94.1% specificity using the patient serum samples they collected.

Engineering of devices or substrates to facilitate SERS enhancement of biological specimens has been applied to the diagnosis of stomach cancer, as seen earlier applied to the diagnosis of previous gastrointestinal organs. Serum of stomach cancer patients has been analyzed on nanotextured substrates fabricated by Wei et al. and Ito et al. [38,39]. Wei developed a gold nanoparticle—coated array of silicon nanowires and verified its efficacy by measuring the enhanced spectrum of rhodamine 6G prior to Raman assessment of stomach cancer patients. Consistent with other groups' findings, the increased Raman bands associated with nucleic acids and decreased bands associated with proteins were observed in the serum of stomach cancer patients relative to healthy subjects. Ito's substrate consisted of a phosphor bronze chip onto which hexagonal silver nanocolumns were synthesized, facilitating SERS enhancement of substances deposited onto the device. The group proved a statistically significant difference in the SERS peaks observed in serum of stomach cancer patients relative to the control group. It was also discussed that their device was made to more consistently facilitate surface enhancement of Raman scattering if the substrate was treated with sodium hypochlorite prior to applying a test sample.

Work is ongoing in the development of SERS sensors to detect particular substances that are biomarkers of stomach cancer. Chen et al. identified, using gas chromatography, 14 volatile organic compounds present in elevated levels in the exhaled breath of stomach cancer patients relative to healthy subjects [40]. They then developed a substrate by reducing graphene oxide onto a surface using hydrazine vapor and forming gold nanoparticles in situ onto the graphene oxide layer. Simulated breaths and the breaths of 200 patients were used to validate and identify Raman bands associated with those 14 volatile biomarkers to distinguish the patients' disease state. Despite limited access to the stomach itself, the small quantities of naturally exuded substances could be sensitively detected by exploiting the significant signal amplification of a metallic SERS substrate.

Lee et al. developed a substrate that can be reliably and consistently fabricated consisting of an array of gold nanobowls containing a nanotextured silver surface, and used it to quantify a micro-RNA biomarker for stomach cancer [41]. It has been reported that elevation of miR-34a has been observed in cases of stomach cancer but has been difficult to detect and quantify due to its low abundance and chemical similarity to other naturally occurring miRNAs. They optimized and validated their substrate using scanning and transmission electron microscopy and X-ray diffraction to confirm crystal structure and properly sized nanogaps between the silver formations and ensured uniform SERS performance by applying and repeatedly measuring the signal of a reporter dye across the substrate surface. They reported a SERS enhancement factor of 4.80×10^9 using the SERS spectrum of rhodamine 6G on the substrate. Finally, to assay the presence of miR-34a, an RNA sequence complementary to miR-34a was attached to the silver surface hybridized with a dummy RNA sequence conjugated to Cy3 dye as a Raman reporter. Introduction of actual miR-34a in a sample resulted in more favorable hybridization of the silver-bound target, resulting in release of the Cy3-bound RNA sequence from the silver surface and a proportional decrease in the Cy3 SERS signal. The experiments were repeated with RNA extracted from MKN-45 cells, which are a line of miR-34a-positive human gastric carcinoma cells, and miR-34a-negative SNU-1 cells, and assay signal for miR-34a presence corresponded to the amount measured using traditional quantitative RT-PCR.

Targeting another well-known cancer biomarker, Eom et al. developed an assay for telomerase utilizing their SERS substrate and tested it on a set of cell lines and stomach cancer xenografts [42]. Telomerase works to extend the telomeres on the ends of the cell's genetic material, protecting the genome from the aging effect of DNA shortening after cell division. In normal somatic cells, however, telomerase's activity is repressed, resulting in the eventual DNA damage and apoptosis. In many cancers, telomerase is overexpressed, allowing continual extension of the telomeres and prevention of apoptosis. Traditional estimation of telomerase activity requires PCR amplification of the telomere primer, but this method is time consuming and prone to false results due to amplifying undesired DNA or possible mutations by the polymerase enzyme. The group developed a substrate consisting of gold nanowires synthesized on a sapphire surface, onto which telomere primers were conjugated. Introduction of a sample containing telomerase would result in elongation of the telomeres extending from the substrate's surface. To measure the amount of telomere elongation, the substrate could be treated with a potassium solution to form G-quadruplexes, followed by introduction of methylene blue as a Raman reporter which can intercalate into these structures. This strategy makes use of the known $(TTAGGG)_n$ sequence of telomeres and the fact that G-rich DNA regions are capable of winding into quadruplex formations. SERS signal of methylene blue was demonstrated to be proportional to telomere elongation and amount of telomerase deposited onto the assay substrate. The substrate was validated by measuring methylene blue's SERS spectrum across the substrate and was found to uniformly facilitate enhancement. Telomerase activity was assessed using enzyme of equal quantities extracted from 293T, HepG2, MCF-7, SNU-484, MDA-MB-231, and NIH3T6.7 cells and compared with the signal of negative control assays run using telomerase-negative HeCaT cells and heat-treated HeLa cells. Tests were repeated using cells extracted from mouse xenografts of NCI-N87, MDA-MB-231, and stomach-implanted SNU-484 stomach cancer cells followed by comparison of assay results using specimens of normal tissue and heat-treated tumor tissue, demonstrating that telomerase activity can reliably be detected from excised tumor tissue.

Without needing to assay for a particular biomarker such as telomerase, excised tissue from a stomach cancer lesion can also be analyzed using SERS techniques to confirm a cancer diagnosis. Ma et al. prepared homogenized tissue samples from healthy stomachs and from cancerous stomach lesions [43]. The resulting SERS spectra were shown to exhibit different peaks that enabled reliable classification by a PCA-LDA model. If routine care identifies that a patient is a candidate for examination and biopsy for stomach cancer, the SERS assays can identify the subtle abnormalities in the tissue to aid in diagnosis, avoiding the challenges of in vivo SERS assessment in the stomach.

Discussion and future directions

Going forward, as many proposed diagnostic methods rely on a CAD model trained on datasets of SERS spectra, it will be important to the clinical adoption of SERS-based diagnostics that these models are adequately and rigorously trained. The work by Feng et al. highlights a phenomenon that has largely been unacknowledged among projects that depend on CAD models to stratify patients based on SERS measurements, namely that there may be diagnostically crucial analytes present in a patient's sample that may go unnoticed without proper substrate and excitation conditions. Substrates consisting of nanotextured surfaces with a variety of spatial features and textures in addition to measurement protocols that use multiple polarizations to excite more classes of biomolecules should be investigated to ensure thorough probing of cancerous biomarkers.

SERS also has the potential to answer more fundamental questions about how cancer cells operate and respond to chemotherapy treatment. Liang et al. reported work on gold nanorods coated with nuclear localizing signal peptide used to enhance the intrinsic Raman signal of cultured stomach cancer cells treated with DNA-binding chemotherapies Hoechst33342 and doxorubicin [44]. Both drugs elicit fluorescence when bound, enabling an imaging scheme where dark-field microscopy could be used to guide Raman measurements directly to the nucleus to begin interrogating the reactions and changes brought on by these chemotherapeutic drugs. Such a workflow has the potential to influence diagnosis as well as personalized medicine by characterization of subcellular changes to cancer cells under treatment pressure.

Intestinal cancer

Introduction

Finally, in our tour of the SERS applications through the gastrointestinal system, we arrive at the small intestine, large intestine, and rectum. Being similar in structure and access routes, many of the diagnostic mechanisms to detect cancers of these organs resemble each other. Endoscopy is the only clinical procedure we have for directly and noninvasively examining the intestinal lumen. The majority of the small intestinal lumen can currently only be visualized with swallowable imaging technology. Viewing the intestinal lumen without the long optical apparatus of an endoscope would require some abdominal or laparoscopic surgical approach to gain access. The classical white light endoscopy has undergone much engineering in recent years to explore augmentation of the procedure with varying forms of Raman analysis, making use of SERS nanoparticles to locate cancerous lesions (Fig. 6.5). While these studies have shown promise, metallic nanoparticle–based SERS endoscopy has yet to be conducted in humans over concerns of prolonged accumulation and potential toxicity of the particles. Investigation in animal models has accompanied the work on SERS endoscopy, including monitoring of organ accumulation, evidence of toxicity, and outward signs of well-being, in order to begin answering questions of the extent and type of physiological response that can result from topical administration of metallic nanoparticles to the intestinal lumen. Here, we will discuss the innovations in SERS applications to endoscopy, SERS assessment of excised colon tumor tissue, and diagnostic CAD models based on enhanced intrinsic Raman spectral measurements of serum content, as well as studies on cultured colon cancer cell lines that attempt to investigate their intracellular proteome involved in rapid tumor growth.

Optimization and design considerations

Periodic endoscopic examination has long been a crucial component of intestinal cancer screening. White light endoscopy enables a trained endoscopist to visualize and cauterize polypoid growths in lumen of the rectum and large intestine. Such raised and discolored growths can likely be precancerous or cancerous, and so are

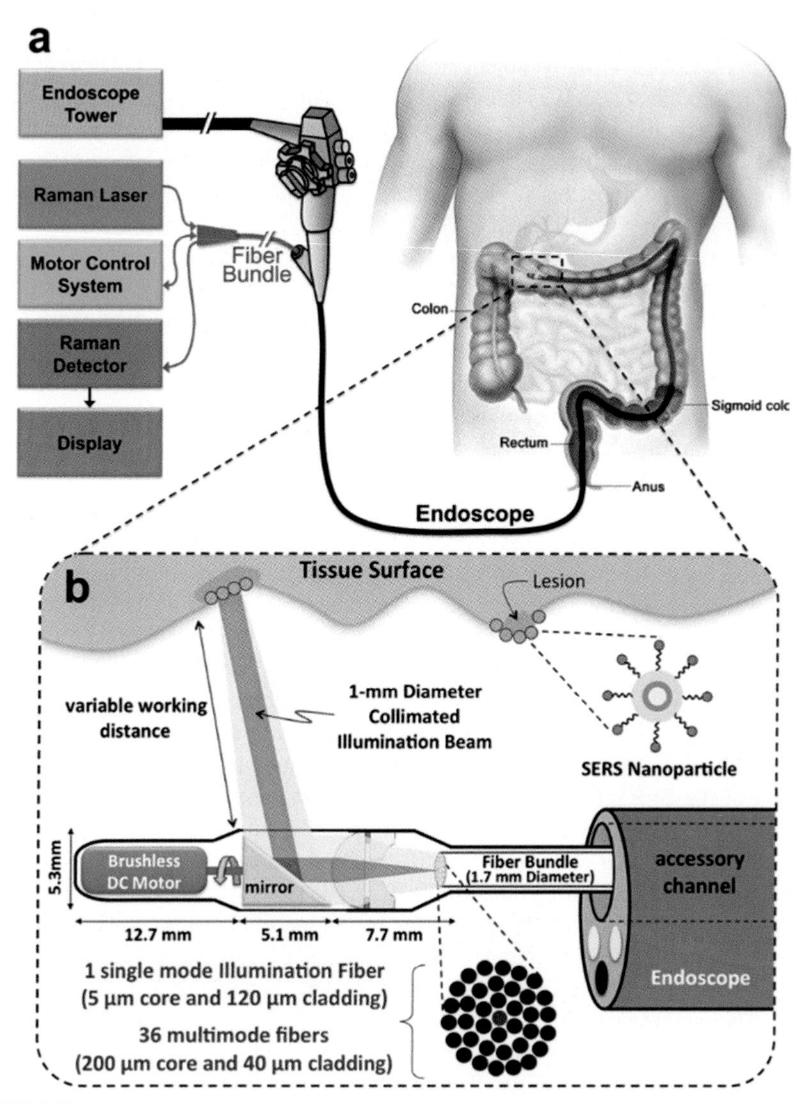

FIGURE 6.5

Conceptual design of a fiber optic-based Raman imaging tool engineered to fit within the accessory channel of a clinical endoscope. It consists of a rotating reflective element to illuminate the colon wall with a laser and optical fibers to collect Raman-scattered photons. Imaging modalities such as this can be built to noninvasively perform circumferential scanning of an organ's lumen such as the colon.

Attribution E. Garai, S. Sensarn, C.L. Zavaleta, N.O. Loewke, S. Rogalla, M.J. Mandella, S.A. Felt, S. Friedland, J.T. Liu, S.S. Gambhir, A real-time clinical endoscopic system for intraluminal, multiplexed imaging of surface-enhanced Raman scattering nanoparticles, PloS One 10 (4) (2015).

treated directly if thought not to be a case of invasive disease. However, not all cancerous foci are visually apparent. Lesions may be flat and nondiscolored or may be raised but so small as to be easily missed upon examination. Recent innovations in endoscopic imaging are addressing the need to better visualize evasive intestinal cancers by utilizing SERS scanning of the intestinal surface to measure the presence and location of Raman-labeled metallic nanoparticles or cancer-associated Raman spectral changes.

A clinical Raman endoscope instrument was developed and formalized by Zavaleta et al. and Garai et al., designed to be compatible with the accessory channel of commercially available endoscopes [45,46]. Working toward a clinical workflow of topical administration of tumor-surface targeting SERS-labeled nanoparticles followed by Raman endoscopic detection and quantification, they characterized and studied the reliability of the endoscopic instrument as well as ensured the system's ability to demultiplex the Raman signatures of nanoparticles in real time. The device's laser power delivery stability, light collection efficiency over the range of working distances, and adherence to the ANSI safety standard for maximum permissible laser exposure were all examined prior to further use. It was able to measure SERS spectra as deep as 5 mm inside a depth phantom and demultiplex 10 SERS spectral signatures simultaneously on prepared quartz slides as well as in excised human colon tissue. Though the SERS particles are not currently approved for use in humans, a pilot study was approved to investigate the system's safety for human use as well as to collect preliminary intrinsic Raman spectra to gain knowledge of potential background signals and to test demultiplexing strategies. Demultiplexing was simulated by mixing the intrinsic measurements with combinations of known spectra of the SERS particles' spectra and applying spectral unmixing algorithms to ensure accurate estimation of the SERS labels' spectral contributions (Fig. 6.6).

Garai et al. further developed their Raman endoscope to demonstrate circumferential scanning of the entire colon lumen [22]. The study reaches closer to the goal of full functional imaging of the molecular landscape of the lumen. The instrument paired with data collection and demultiplexing software was capable of rapid scanning of the whole colon and presenting both 2D and 3D renderings of the SERS mapping. The endoscope device and software component were first tested on a cylindrical paper phantom painted with the labeled SERS nanoparticles to ensure correct localization and SERS flavor identification and unmixing (Fig. 6.6). A freshly excised porcine colon was also acquired and subcutaneously injected with the same set of SERS nanoparticles. Not only was the system capable of correct nanoparticle localization and demultiplexing in the phantom colon but also a preliminary topographical map that included the natural ridges and folds of the colon wall was able to be rendered by utilizing the weighting factor of the first principal component, demonstrating further capabilities of the system to provide important orientation and context to a clinician. Multiplexed Raman mapping in vivo holds the promise of strongly labeling foci of tissue that overexpress sets of cell membrane biomarkers using targeted binding of Raman-labeled particles. Targeting moieties traditionally include selections of antibodies to achieve preferential binding of

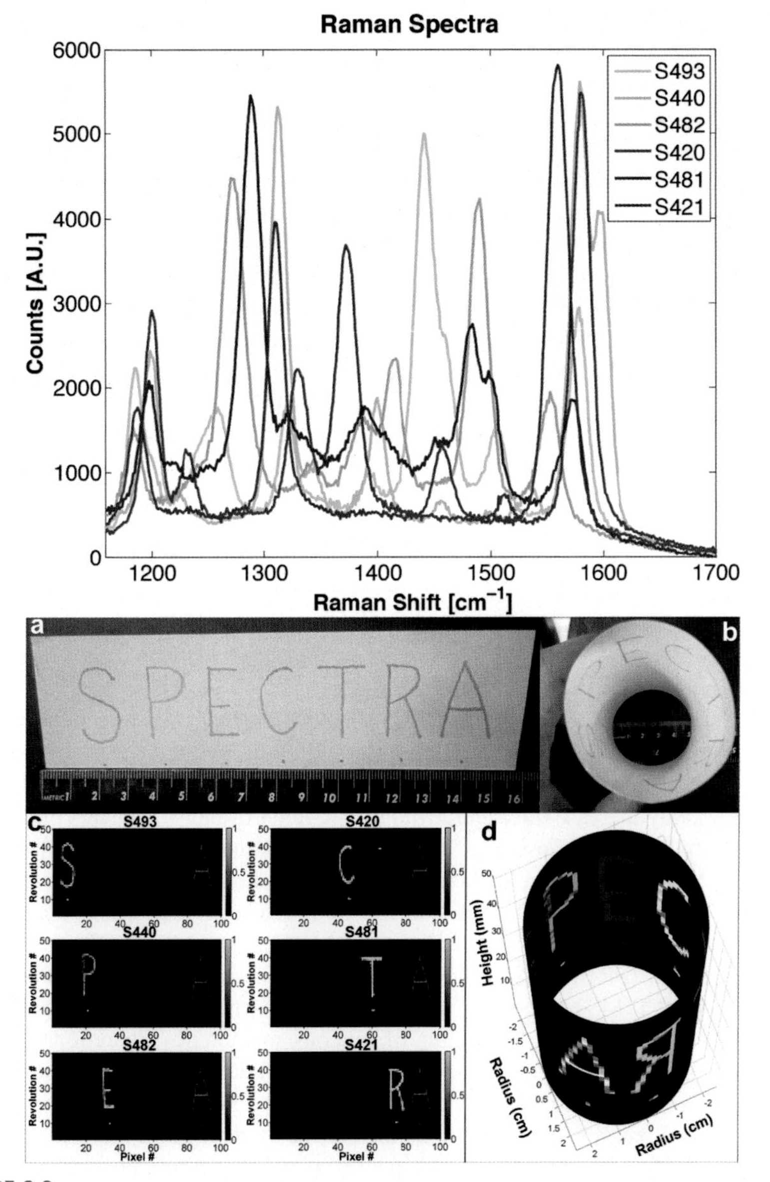

FIGURE 6.6

Library of spectra of Raman reporter molecules that can be adsorbed onto metallic nanoparticles and example of multiplexed luminal multiplexed mapping. Each spectrum crucially exhibits a unique set of peaks to distinguish them from one another. These "flavors" of SERS nanoparticles can bind to desired cell types based on cell surface phenotype, resulting in a Raman measurement containing a linear combination of these spectral fingerprints. A multiplexed Raman measurement can be unmixed using a library of particle flavor spectra to calculate the concentration of particle flavor present at a site.

Attribution E. Garai, S. Sensarn, C.L. Zavaleta, N.O. Loewke, S. Rogalla, M.J. Mandella, S.A. Felt, S. Friedland, J.T. Liu, S.S. Gambhir, A real-time clinical endoscopic system for intraluminal, multiplexed imaging of surface-enhanced Raman scattering nanoparticles, PloS One 10 (4) (2015).

particles to certain cells. Multiplexing results have been improved by the introduction of a nonspecific binding SERS flavor to the mixture as a negative control, allowing calculations to account for the extent of nonspecific binding of the other flavors to undesired targets. Targeted binding of nanoparticles can be improved through future work by the replacement of antibodies with small peptides, or affibodies [23], engineered or discovered by molecular display processes, to potentially achieve greater reproducibility and higher affinity and more specific binding to malignant surfaces than may consistently be achieved with antibodies.

The in vivo study of IV administered nanoparticles in mice to locate cancerous tissue via labeled SERS nanoparticles by Harmsen et al. also included the study of $Apc^{Min/+h}$ engineered to develop colon cancer [19]. As this study depended on the enhanced permeability and retention effect of tumor spaces to passively accumulate nanoparticles over time, the procedure requires the additional step of allowing circulation of the nanoparticles prior to circumferential endoscopic analysis. The IR780 perchlorate-labeled nanoparticles were injected IV into mice without disease to ensure no accumulation in the gastrointestinal organs would occur without the presence of disease. The lesions discovered by SERS scanning were confirmed to be malignancies by histology. IV administration of gold colloid solutions to humans remains unapproved by the FDA, however, so this approach would be unfeasible to attempt in a clinical study.

Functional mapping of the cellular landscape of the intestinal lumen by scanning for SERS nanoparticles engineered to preferentially accumulate or bind to the surface of malignant cells overexpressing some characteristic biomarker has the potential to markedly increase the accuracy and specificity of intestinal cancer diagnosis. All these studies into endoscopic SERS mapping presume that the use of the necessary Raman enhancing agent, i.e., gold nanoparticles, are safe to administer to humans. Thakor et al. conducted studies in cultured cells and mice to assess the changes and fate of PEGylated, silica-coated gold nanoparticles administered IV and per rectum (PR) [47,48]. The presence of gold was measured in the organs and blood by ICP-MS and the mice were monitored for physiological and behavioral signs of wellness, such as the status of mucous membranes, secretions, posture, breathing, gait, blood pressure, and heart rate. In mice given gold nanoparticles IV, the presence of gold in the liver decreases over the course of 2 weeks while reaching a maximum in the spleen at 24 h followed by steady decrease over 2 weeks. Gold could only be detected in the blood within the first 5 min postinjection. Fine black particulates were observed in H&E stained liver section in the sinusoids and resident macrophages, or Kupffer cells, within minutes after injection, and gradually could only be observed in the Kupffer cells over the course of 2 weeks. These black particulates are thought to be aggregations of the gold nanoparticles, supported by the ruling out of them being some iron-containing substance due to their appearance being different from hemosiderin, a naturally found iron-containing pigment, and not being stained by traditional dyes that would normally label iron-containing compounds. Despite no outward signs of negative reaction to the gold nanoparticle IV injection, it was discovered that a number of oxidative stress and apoptosis markers were overexpressed in the liver. Mice-administered gold nanoparticles PR did not show these signs of liver toxicity and gold was not detected in the blood or organs, suggesting that topical administration of gold SERS nanoparticles should be a much safer approach to enable SERS endoscopic diagnosis of intestinal cancer. Zavaleta et al. also found that gold SERS nanoparticles labeled with both Raman reporter dyes and radioactive tags administered to mice PR were cleared from the body within 24 h [20].

Raman enhancement of the intrinsic components of solid colon tumors has been documented by Pînzaru et al. [49]. Deposition of silver nanoparticles onto the surface of excised colon cancer tissue reveals elevated Raman peaks associated with DNA and RNA bases. Their early exploration of SERS measurements of excised tissue demonstrated both visual and spectral changes in the colon tissue at the boundary of the tumor as they assessed the specimens and sectioned them transverse to the colon wall orientation. Intraprocedural section-wise Raman analysis, in the fashion of real time histology, relates to the workflow already used in procedures such as Mohs surgery, which is a tissue-conserving surgery that precisely removes tissue and conducts histology until no positive margins remain detectable. As some colon cancer cases are assessed to be not locally resectable, it may be possible to develop a slice-wise intraoperative Raman analysis workflow during colon cancer surgery to conservatively and accurately remove the disease, allowing the surgeon the option to avoid removing and ligating sections of bowel.

Diagnosis of colon cancer ex vivo has the flexibility of utilizing a precisely fabricated, nanotextured SERS substrate device for performing cell population analysis. Liang et al. has shown that HCT116 human colon cancer cells, which natively overexpressed p53, can be distinguished from cells grown as a knockout variant that does not express p53 [50]. The function of p53 is currently thought to be to halt the cell cycle when irreparable DNA damage has occurred to prevent propagation of significant mutations, so mutation and altered expression of p53 is treated as a strong cancer biomarker. Using the group's SERS device consisting of gold nanopyramids synthesized on a graphene-coated silicon surface, living, dead, and burst cells of both the p53 overexpressing and knockout cell types could all be differentiated based on the acquired SERS measurements. However, they note that further work should be done to elucidate the cellular changes that resulted from the knockout procedure and the altered p53 expression. Despite the contribution of many biomolecules simultaneously to a SERS measurement of whole cell populations, the spectral changes corresponding to altered expression and proteome demonstrates the superior sensitivity and information content of these techniques. One can imagine, like we have accumulated libraries of Raman spectra of known compounds, perhaps a library of cellular spectral changes could be compiled to quickly probe the type and behavior of a cell population.

Testing genome-level alterations in cancerous cells has also been accomplished in addition to these innovations in detecting expression-level changes. An assay technique developed by Li et al. utilizes traditional PCR and a library of primer sequences targeting known mutation subsequences in the *BRAF*, *KRAS*, and *PIK3CA* genes [51]. The assay concludes with hybridization of any amplified PCR product with complementary strands preconjugated to a Raman reporter dye followed by SERS analysis with silver nanoparticles. Typically, when information is desired regarding a patient's particular constellation of gene mutations that could be driving their cancer's growth, real-time PCR, and DNA sequencing technology have been required to obtain enough genetic material and recover the patient's genomic data, not to mention access to reference of healthy genomes and software or skilled analysts to determine if the patient's gene is abnormal. The group's PCR-SERS method serves as a template for easily performing normal PCR to test for subsets of known mutations for more rapid feedback on a cancer's subtype. Such a development could be an inexpensive yet accurate way to enable personalized medicine in regions without access to comprehensive genomic profiling technology.

SERS analysis of serum has also been shown to be capable of discriminating between healthy patients and those who had developed colon cancer. Lin et al. and Feng et al. included serum from subjects who had been diagnosed with colon cancer, and found again that a particular polarization of laser light was superior for measurement of serum mixed with silver nanoparticles [52–54]. Hong et al. was able to mix patient serum with gold nanoparticles and allow drops of the mixture to dry on glass slides and allow them to naturally dry, forming a coffee ring pattern so serum constituents and particles could collect at the periphery [55]. The SERS measurements were found to be more repeatable using this method, and an SVM model was capable of adequately identifying colorectal cancer patients, especially when the dataset was augmented with the patients' blood level of carcinoembryonic antigen. SERS measurements of serum from healthy patients and patients with breast, colorectal, lung, ovarian, and lung cancer could all be accurately distinguished from each other in the study conducted by Moisoiu et al. [56].

Discussion and future directions

The small intestine currently poses the greatest design challenge to SERS endoscopic assessment due to its sheer length and unpredictable winding within the abdominal cavity. The beginning of the cecum and the end of the ilium can be visualized, respectively, during an upper endoscopy or colonoscopy, but the remaining length of the small intestine remains difficult to examine. For these hard to reach areas within the small intestine, capsule endoscopy is being further developed to collect important optical data and could potentially be coupled with SERS nanoparticles for Raman imaging [57]. Swallowable white light endoscopy devices, such as the Pill Cam SB system, and accompanying sensor arrays for wirelessly collecting measurement data through the skin are already clinically available.

Another potential source of insight into disease of the small intestine as well as the large intestine could be a SERS assay of stool samples. As the samples would have had contact with both normal and cancerous regions of lumen, they could accumulate cellular debris and exfoliated cells indicative of cancer somewhere along the intestinal tract. Appropriate assay design to prevent measurement of unwanted substances in the stool samples and to account for nonspecific binding or enhancement of constituent biomolecules could enable such an assay to be more targeted and specific than a serum assay where constituents from the whole body compete with the cancer biomarkers in the spectral measurements.

Concluding remarks

Detection and diagnosis of cancers occurring in the oral cavity and gastrointestinal tract have traditionally relied on visual inspection by a trained clinician and cytological analysis by a pathologist. Positive diagnosis depends on reliably noticing a cancerous lesion and accurately acquiring a solid biopsy for confirmation. It currently takes years of training and experience to reliably discover the many and subtle manifestations of lesions and polyps among the normal gastrointestinal anatomy. Difficulty in gaining adequate access to these organs to conduct a thorough examination exacerbates the need for advanced tools to identify regions of interest to a clinician for further inspection. SERS-based diagnostic assays and new imaging approaches are expanding the array of tools that may be available to clinicians to screen for and pinpoint gastrointestinal cancers while offering important molecular information. By taking advantage of the sensitivity of SERS measurements, these new methods are addressing the challenges posed by the gastrointestinal system's anatomy to assay for rare cancer biomarkers in body fluids and to target cell surface biomarkers expressed on cancerous cells to enable direct Raman surface scanning. The high statistical diagnostic sensitivity of SERS-based assay protocols and the specificity of designated SERS nanoparticle flavors are vital and unique features of this class of techniques. SERS clearly has great potential as a point-of-care diagnostic tool in the clinical setting, offering a multitude of detection and imaging strategies to better inform clinicians about the most effective treatment options for their cancer patients.

References

[1] C.V. Raman, A new radiation, Indian J. Phys. 2 (1928) 387−398.

[2] M. Fleischmann, P.J. Hendra, A.J. McQuillan, Raman spectra of pyridine adsorbed at a silver electrode, Chem. Phys. Lett. 26 (2) (1974) 163−166.

[3] S. Han, A. Locke, L. Oaks, Y.-S.L. Cheng, G. Coté, Development of a free-solution SERS-based assay for point-of-care oral cancer biomarker detection using DNA-conjugated gold nanoparticles, SPIE 10501 (2018).

[4] S. Han, A.K. Locke, L.A. Oaks, Y.L. Cheng, G.L. Cote, Nanoparticle-based assay for detection of S100P mRNA using surface-enhanced Raman spectroscopy, J. Biomed. Opt. 24 (5) (2019) 1−9.

[5] X. Huang, I.H. El-Sayed, W. Qian, M.A. El-Sayed, Cancer cells assemble and align gold nanorods conjugated to antibodies to produce highly enhanced, sharp, and

polarized surface Raman spectra: a potential cancer diagnostic marker, Nano Lett. 7 (6) (2007) 1591–1597.

[6] B. Yan, B. Li, Z. Wen, X. Luo, L. Xue, L. Li, Label-free blood serum detection by using surface-enhanced Raman spectroscopy and support vector machine for the preoperative diagnosis of parotid gland tumors, BMC Cancer 15 (1) (2015) 650.

[7] Y. Tan, B. Yan, L. Xue, Y. Li, X. Luo, P. Ji, Surface-enhanced Raman spectroscopy of blood serum based on gold nanoparticles for the diagnosis of the oral squamous cell carcinoma, Lipids Health. Dis. 16 (1) (2017) 73.

[8] J.C. Kah, K.W. Kho, C.G. Lee, C. James, R. Sheppard, Z.X. Shen, K.C. Soo, M.C. Olivo, Early diagnosis of oral cancer based on the surface plasmon resonance of gold nanoparticles, Int. J. Nanomed. 2 (4) (2007) 785–798.

[9] V. Kartha, R. Jyothi Lakshmi, K.K. Mahato, C. Murali Krishna, C. Santhosh, K. Venkata Krishna, Raman spectroscopy in clinical investigations, Indian J. Phys. 77 (2003) 113–123.

[10] M. Olivo, In Photodetection in the non-fluorescent regime with gold nanostructures, in: 2008 IEEE Photonics Global@Singapore, 8–11 Dec. 2008, 2008, pp. 1–3.

[11] K. Kho, O. Malini, Z.X. Shen, K.C. Soo, Surface enhanced Raman spectroscopic (SERS) study of saliva in the early detection of oral cancer, SPIE 5702 (2005).

[12] Q. Liu, J. Wang, B. Wang, Z. Li, H. Huang, C. Li, X. Yu, P.K. Chu, Paper-based plasmonic platform for sensitive, noninvasive, and rapid cancer screening, Biosens. Bioelectron. 54 (2014) 128–134.

[13] W.-Y. Dai, S. Lee, Y.-C. Hsu, Discrimination between oral cancer and healthy cells based on the adenine signature detected by using Raman spectroscopy, J. Raman Spectrosc. 49 (2) (2018) 336–342.

[14] J.M. Connolly, K. Davies, A. Kazakeviciute, A.M. Wheatley, P. Dockery, I. Keogh, M. Olivo, Non-invasive and label-free detection of oral squamous cell carcinoma using saliva surface-enhanced Raman spectroscopy and multivariate analysis, Nanomed. Nanotechnol. Biol. Med. 12 (6) (2016) 1593–1601.

[15] C.M. Girish, S. Iyer, K. Thankappan, V.V.D. Rani, G.S. Gowd, D. Menon, S. Nair, M. Koyakutty, Rapid detection of oral cancer using Ag–TiO_2 nanostructured surface-enhanced Raman spectroscopic substrates, J. Mater. Chem. B 2 (8) (2014) 989–998.

[16] B. Kang, L.A. Austin, M.A. El-Sayed, Observing real-time molecular event dynamics of apoptosis in living cancer cells using nuclear-targeted plasmonically enhanced Raman nanoprobes, ACS Nano 8 (5) (2014) 4883–4892.

[17] L. Xiao, H. Wang, Z.D. Schultz, Selective detection of RGD-integrin binding in cancer cells using tip enhanced Raman scattering microscopy, Anal. Chem. 88 (12) (2016) 6547–6553.

[18] J.L. Campbell, E.D. SoRelle, O. Ilovich, O. Liba, M.L. James, Z. Qiu, V. Perez, C.T. Chan, A. de la Zerda, C. Zavaleta, Multimodal assessment of SERS nanoparticle biodistribution post ingestion reveals new potential for clinical translation of Raman imaging, Biomaterials 135 (2017) 42–52.

[19] S. Harmsen, S. Rogalla, R. Huang, M. Spaliviero, V. Neuschmelting, Y. Hayakawa, Y. Lee, Y. Tailor, R. Toledo-Crow, J.W. Kang, J.M. Samii, H. Karabeber, R.M. Davis, J.R. White, M. van de Rijn, S.S. Gambhir, C.H. Contag, T.C. Wang, M.F. Kircher, Detection of premalignant gastrointestinal lesions using surface-enhanced resonance Raman scattering–nanoparticle endoscopy, ACS Nano 13 (2) (2019) 1354–1364.

[20] C.L. Zavaleta, K.B. Hartman, Z. Miao, M.L. James, P. Kempen, A.S. Thakor, C.H. Nielsen, R. Sinclair, Z. Cheng, S.S. Gambhir, Preclinical evaluation of Raman nanoparticle biodistribution for their potential use in clinical endoscopy imaging, Small 7 (15) (2011) 2232–2240.

[21] Y.W. Wang, A. Khan, S.Y. Leigh, D. Wang, Y. Chen, D. Meza, J.T.C. Liu, Comprehensive spectral endoscopy of topically applied SERS nanoparticles in the rat esophagus, Biomed. Opt. Express 5 (9) (2014) 2883–2895.

[22] E. Garai, S. Sensarn, C.L. Zavaleta, N.O. Loewke, S. Rogalla, M.J. Mandella, S.A. Felt, S. Friedland, J.T. Liu, S.S. Gambhir, A real-time clinical endoscopic system for intraluminal, multiplexed imaging of surface-enhanced Raman scattering nanoparticles, PloS One 10 (4) (2015).

[23] J.V. Jokerst, Z. Miao, C. Zavaleta, Z. Cheng, S.S. Gambhir, Affibody-functionalized gold–silica nanoparticles for Raman molecular imaging of the epidermal growth factor receptor, Small 7 (5) (2011) 625–633.

[24] Y.W. Wang, S. Kang, A. Khan, P.Q. Bao, J.T.C. Liu, In vivo multiplexed molecular imaging of esophageal cancer via spectral endoscopy of topically applied SERS nanoparticles, Biomed. Opt. Express 6 (10) (2015) 3714–3723.

[25] X. Li, T. Yang, S. Li, D. Wang, D. Guan, Detecting esophageal cancer using surface-enhanced Raman spectroscopy (SERS) of serum coupled with hierarchical cluster analysis and principal component analysis, Appl. Spectrosc. 69 (11) (2015) 1334–1341.

[26] S. Huang, L. Wang, W. Chen, S. Feng, J. Lin, Z. Huang, G. Chen, B. Li, R. Chen, Potential of non-invasive esophagus cancer detection based on urine surface-enhanced Raman spectroscopy, Laser Phys. Lett. 11 (11) (2014) 115604.

[27] S. Feng, Z. Zheng, Y. Xu, J. Lin, G. Chen, C. Weng, D. Lin, S. Qiu, M. Cheng, Z. Huang, L. Wang, R. Chen, S. Xie, H. Zeng, A noninvasive cancer detection strategy based on gold nanoparticle surface-enhanced Raman spectroscopy of urinary modified nucleosides isolated by affinity chromatography, Biosens. Bioelectron. 91 (2017) 616–622.

[28] S. Li, Y. Zhang, Q. Zeng, L. Li, Z. Guo, Z. Liu, H. Xiong, S. Liu, Potential of cancer screening with serum surface-enhanced Raman spectroscopy and a support vector machine, Laser Phys. Lett. 11 (6) (2014) 065603.

[29] S.-X. Li, Q.-Y. Zeng, L.-F. Li, Y.-J. Zhang, M.-M. Wan, Z.-M. Liu, H.-L. Xiong, Z.-Y. Guo, S.-H. Liu, Study of support vector machine and serum surface-enhanced Raman spectroscopy for noninvasive esophageal cancer detection, J. Biomed. Opt. 18 (2) (2013) 027008.

[30] S. Feng, J. Lin, Z. Huang, G. Chen, W. Chen, Y. Wang, R. Chen, H. Zeng, Esophageal cancer detection based on tissue surface-enhanced Raman spectroscopy and multivariate analysis, Appl. Phys. Lett. 102 (4) (2013) 043702.

[31] L. Chen, Y. Wang, N. Liu, D. Lin, C. Weng, J. Zhang, L. Zhu, W. Chen, R. Chen, S. Feng, Near-infrared confocal micro-Raman spectroscopy combined with PCA–LDA multivariate analysis for detection of esophageal cancer, Laser Phys. 23 (6) (2013) 065601.

[32] T. Gong, K.V. Kong, D. Goh, M. Olivo, K.-T. Yong, Sensitive surface enhanced Raman scattering multiplexed detection of matrix metalloproteinase 2 and 7 cancer markers, Biomed. Opt. Express 6 (6) (2015) 2076–2087.

[33] C. Kendall, N. Stone, N. Shepherd, K. Geboes, B. Warren, R. Bennett, H. Barr, Raman spectroscopy, a potential tool for the objective identification and classification of neoplasia in Barrett's oesophagus, J. Pathol. 200 (5) (2003) 602–609.

[34] X. Li, T. Yang, S. Li, D. Wang, Y. Song, K. Yu, Different classification algorithms and serum surface enhanced Raman spectroscopy for noninvasive discrimination of gastric diseases, J. Raman Spectrosc. 47 (8) (2016) 917–925.
[35] L. Guo, Y. Li, F. Huang, J. Dong, F. Li, X. Yang, S. Zhu, M. Yang, Identification and analysis of serum samples by surface-enhanced Raman spectroscopy combined with characteristic ratio method and PCA for gastric cancer detection, J. Innovative Opt. Health Sci. 12 (02) (2019) 1950003.
[36] S. Feng, R. Chen, J. Lin, J. Pan, Y. Wu, Y. Li, J. Chen, H. Zeng, Gastric cancer detection based on blood plasma surface-enhanced Raman spectroscopy excited by polarized laser light, Biosens. Bioelectron. 26 (7) (2011) 3167–3174.
[37] Y. Chen, G. Chen, X. Zheng, C. He, S. Feng, Y. Chen, X. Lin, R. Chen, H. Zeng, Discrimination of gastric cancer from normal by serum RNA based on surface-enhanced Raman spectroscopy (SERS) and multivariate analysis, Med. Phys. 39 (9) (2012) 5664–5668.
[38] Y. Wei, Y.-y. Zhu, M.-l. Wang, Surface-enhanced Raman spectroscopy of gastric cancer serum with gold nanoparticles/silicon nanowire arrays, Optik 127 (19) (2016) 7902–7907.
[39] H. Ito, H. Inoue, K. Hasegawa, Y. Hasegawa, T. Shimizu, S. Kimura, M. Onimaru, H. Ikeda, S.-e. Kudo, Use of surface-enhanced Raman scattering for detection of cancer-related serum-constituents in gastrointestinal cancer patients, Nanomed. Nanotechnol. Biol. Med. 10 (3) (2014) 599–608.
[40] Y. Chen, Y. Zhang, F. Pan, J. Liu, K. Wang, C. Zhang, S. Cheng, L. Lu, W. Zhang, Z. Zhang, X. Zhi, Q. Zhang, G. Alfranca, J.M. de la Fuente, D. Chen, D. Cui, Breath analysis based on surface-enhanced Raman scattering sensors distinguishes early and advanced gastric cancer patients from healthy persons, ACS Nano 10 (9) (2016) 8169–8179.
[41] T. Lee, J.-S. Wi, A. Oh, H.-K. Na, J. Lee, K. Lee, T.G. Lee, S. Haam, Highly robust, uniform and ultra-sensitive surface-enhanced Raman scattering substrates for microRNA detection fabricated by using silver nanostructures grown in gold nanobowls, Nanoscale 10 (8) (2018) 3680–3687.
[42] G. Eom, H. Kim, A. Hwang, H.-Y. Son, Y. Choi, J. Moon, D. Kim, M. Lee, E.-K. Lim, J. Jeong, Y.-M. Huh, M.-K. Seo, T. Kang, B. Kim, Nanogap-rich Au nanowire SERS sensor for ultrasensitive telomerase activity detection: application to gastric and breast cancer tissues diagnosis, Adv. Funct. Mater. 27 (37) (2017) 1701832.
[43] J. Ma, H. Zhou, L. Gong, S. Liu, Z. Zhou, W. Mao, R.-e. Zheng, Distinction of gastric cancer tissue based on surface-enhanced Raman spectroscopy, SPIE 8553 (2012).
[44] L. Liang, D. Huang, H. Wang, H. Li, S. Xu, Y. Chang, H. Li, Y.-W. Yang, C. Liang, W. Xu, In situ surface-enhanced Raman scattering spectroscopy exploring molecular changes of drug-treated cancer cell nucleus, Anal. Chem. 87 (4) (2015) 2504–2510.
[45] E. Garai, S. Sensarn, C.L. Zavaleta, D. Van de Sompel, N.O. Loewke, M.J. Mandella, S.S. Gambhir, C.H. Contag, High-sensitivity, real-time, ratiometric imaging of surface-enhanced Raman scattering nanoparticles with a clinically translatable Raman endoscope device, J. Biomed. Opt. 18 (9) (2013) 096008.
[46] C.L. Zavaleta, E. Garai, J.T.C. Liu, S. Sensarn, M.J. Mandella, D. Van de Sompel, S. Friedland, J. Van Dam, C.H. Contag, S.S. Gambhir, A Raman-based endoscopic strategy for multiplexed molecular imaging, Proc. Natl. Acad. Sci. 110 (25) (2013) E2288.

[47] A.S. Thakor, R. Luong, R. Paulmurugan, F.I. Lin, P. Kempen, C. Zavaleta, P. Chu, T.F. Massoud, R. Sinclair, S.S. Gambhir, The fate and toxicity of Raman-active silica-gold nanoparticles in mice, Sci. Transl. Med. 3 (79) (2011) 79ra33.
[48] A.S. Thakor, R. Paulmurugan, P. Kempen, C. Zavaleta, R. Sinclair, T.F. Massoud, S.S. Gambhir, Oxidative stress mediates the effects of Raman-active gold nanoparticles in human cells, Small 7 (1) (2011) 126–136.
[49] S.C. Pînzaru, L.M. Andronie, I. Domsa, O. Cozar, S. Astilean, Bridging biomolecules with nanoparticles: surface-enhanced Raman scattering from colon carcinoma and normal tissue, J. Raman Spectrosc. 39 (3) (2008) 331–334.
[50] O. Liang, P. Wang, M. Xia, C. Augello, F. Yang, G. Niu, H. Liu, Y.-H. Xie, Label-free distinction between p53+/+ and p53 -/- colon cancer cells using a graphene based SERS platform, Biosens. Bioelectron. 118 (2018) 108–114.
[51] X. Li, T. Yang, C.S. Li, Y. Song, H. Lou, D. Guan, L. Jin, Surface enhanced Raman spectroscopy (SERS) for the multiplex detection of Braf, Kras, and Pik3ca mutations in plasma of colorectal cancer patients, Theranostics 8 (6) (2018) 1678–1689.
[52] D. Lin, S. Feng, J. Pan, Y. Chen, J. Lin, G. Chen, S. Xie, H. Zeng, R. Chen, Colorectal cancer detection by gold nanoparticle based surface-enhanced Raman spectroscopy of blood serum and statistical analysis, Opt. Express 19 (14) (2011) 13565–13577.
[53] D. Lin, H. Huang, S. Qiu, S. Feng, G. Chen, R. Chen, Diagnostic potential of polarized surface enhanced Raman spectroscopy technology for colorectal cancer detection, Opt. Express 24 (3) (2016) 2222–2234.
[54] S. Feng, W. Wang, I.T. Tai, G. Chen, R. Chen, H. Zeng, Label-free surface-enhanced Raman spectroscopy for detection of colorectal cancer and precursor lesions using blood plasma, Biomed. Opt. Express 6 (9) (2015) 3494–3502.
[55] Y. Hong, Y. Li, L. Huang, W. He, S. Wang, C. Wang, G. Zhou, Y. Chen, X. Zhou, Y. Huang, W. Huang, T. Gong, Z. Zhou, Label free diagnosis for colorectal cancer through coffee ring assisted surface enhanced Raman spectroscopy on blood serum, J. Biophoton. (2020) e201960176.
[56] V. Moisoiu, A. Stefancu, D. Gulei, R. Boitor, L. Magdo, L. Raduly, S. Pasca, P. Kubelac, N. Mehterov, V. Chi, M. Simon, M. Muresan, A.I. Irimie, M. Baciut, R. Stiufiuc, I.E. Pavel, P. Achimas-Cadariu, C. Ionescu, V. Lazar, V. Sarafian, I. Notingher, N. Leopold, I. Berindan-Neagoe, SERS-based differential diagnosis between multiple solid malignancies: breast, colorectal, lung, ovarian and oral cancer, Int. J. Nanomed. 14 (2019) 6165–6178.
[57] K. Kalantar-zadeh, N. Ha, J.Z. Ou, K.J. Berean, Ingestible sensors, ACS Sens. 2 (4) (2017) 468–483.

CHAPTER

In vivo imaging with SERS nanoprobes

7

Chrysafis Andreou, PhD [1], Yiota Gregoriou, PhD [2], Akbar Ali[3], Suchetan Pal, PhD [3]

[1]*Department of Electrical and Computer Engineering, University of Cyprus, Nicosia, Cyprus;* [2]*Department of Biological Sciences, University of Cyprus, Nicosia, Cyprus;* [3]*Department of Chemistry, Indian Institute of Technology, Bhilai, Chhattisgarh, India*

Introduction

Medical imaging is an integral part of modern medicine and oncology in particular [1], where it is routinely used for screening, diagnosis, tumor staging, and treatment monitoring. Additional applications of medical imaging include image-guided biopsy, preoperative assessment, and intraoperative delineation of tumors, lymph nodes, vasculature, or nerves. Some imaging modalities like X-ray (projection radiography or computed tomography (CT)) and magnetic resonance imaging (MRI) provide an image of the patient's anatomy without necessitating extrinsic contrast agents. The field of molecular imaging—with positron emission tomography (PET) as its flagship modality—seeks to reveal the presence and distribution of specific *molecular* markers in the body. This information complements the imaging of anatomical features via other modalities, and these localized molecular cues help the physician decide on a suitable treatment [2]. With the advent of machine learning and computer vision in particular, the wealth of data available through annotated medical image databases has been leveraged as a valuable resource for training algorithms for computer-aided diagnosis, giving rise to the newly minted field of radiomics [3].

The need for new types of more meaningful medical images, richer in useful data, is evident by new, emerging imaging paradigms. One such paradigm is multimodal imaging: imaging the same site with different imaging technologies [4], which is evidenced through the prevalence of combined imaging modalities, such as PET-CT, MRI-CT, and more recently PET/MR. A different route to acquiring more data via imaging is multiplexed imaging, which aims at visualizing different targets in a single scan, via a single modality [5]. Both approaches depend on the development of new, "smarter" contrast agents, that are able to home to specific sites in the body, provide imaging contrast that is molecularly targeted and detectable with multiple methods, and can accommodate a variety of imaging scenarios. The challenge of developing this new generation of imaging agents has been accepted with enthusiasm by the nanomedicine community [6,7].

SERS for Point-of-care and Clinical Applications. https://doi.org/10.1016/B978-0-12-820548-8.00003-5

The field of nanomedicine is now just starting to mature. A handful of nanoparticle-based therapeutics and contrast agents have been translated successfully into the clinic, and an ever-increasing number are in clinical trials with high promises [8]. Nanomedicine is propelled by the unique properties of nanoparticles—small enough to circulate in the body and infiltrate diseased sites, yet complex enough to confer multiple and customizable functionalities. Importantly, some effects are exclusively attainable using nanoscopic geometries and cannot be achieved via traditional "small molecule" agents. One such effect is surface plasmon resonance: the ensemble oscillation of electrons on an electric conductor excited by light of suitable frequency—a phenomenon necessary for inducing surface-enhanced Raman scattering (SERS) [9].

Raman imaging with SERS nanoprobes

Raman imaging is based on the Raman effect—the inelastic scattering of photons by molecular bond vibrations [10]. Photons emitted from a monochromatic laser source exchange energy with molecules in their path, and are scattered with their energy shifted. By collecting sufficient numbers of these Raman scattered photons, we are able to compose a spectrum of all the energy shifts caused by the scattering substance. Since this spectrum depends on the bond vibrations of the molecule, it allows us to uniquely identify its molecular structure. By sampling a region of interest—typically in a pointwise grid—it becomes possible to generate chemical images (or maps) of the sample, whether biological or otherwise.

However, Raman scattering is a weak effect, and not many energy-shifted photons are produced. In the case of biological systems, it is possible to use nanoparticle-based contrast agents engineered to yield intense signal while maintaining the exquisite specificity of the Raman spectrum. These nanoparticles make use of SERS, so they require a metal core (typically gold) to enable the plasmonic enhancement and a reporter molecule to provide its specific spectrum. When the core and reporter molecule are both resonant with the excitation laser, then the enhancement is further improved by several orders of magnitude [11]; we call this effect surface-enhanced *resonance* Raman scattering (SERRS).

As various groups contribute to the field, the terminology found in the literature is not consistent. In an attempt to make this chapter clear, we will simplify the terminology in the following way: "SERS nanoparticles" will represent *all* nanoparticles that provide enhanced Raman scattering (resonant or not); when we want to stress the importance of resonance, the term SERRS will be used; and the term "nanoprobe" will be used to distinguish a nanoparticle applied to a biological system in order to answer a specific question (such as in the case of molecular targeting or tumor delineation).

Raman imaging is an optical imaging modality, and as such, it is often compared to fluorescence imaging in terms of resolution and tissue penetration. Nonetheless, two crucial differences exist between the two techniques. The first difference lies in how the imaging contrast is generated: fluorescence relies on fluorescent dyes, most often small molecules or genetically expressed proteins, whereas the contrast agent of choice for Raman imaging typically takes the form of SERS nanoparticles. The second difference is the type of signal provided: where fluorescence constructs an image based on the intensity of a single wavelength (or band) of emitted light, Raman imaging provides *spectral* data often containing more than a 1000 wavenumbers per pixel. This second difference is key to one of the vital advantages of Raman imaging: complex spectra can be unmixed to reveal their constituent components, based on the exquisitely specific molecular fingerprints of the Raman reporters used. In this way, an unparalleled high degree of multiplexing can be realized, with the potential to detect up to 10 or more contrast agents (hence targets) in a single scan [12].

In this chapter, we will briefly look at how Raman imaging with SERS nanoprobes compares to other in vivo imaging modalities, with a special note on imaging multiple markers. We will proceed to examine some of the first applications of SERS nanoprobes in vivo and how the challenges associated with this method are being addressed. Then, we will see how SERS nanoprobes can be used as multimodal imaging and theranostic agents, providing contrast for multiple different modalities as well as conferring therapy. We will close with looking at how SERS nanoprobes have been applied to push the boundaries of multiplexed imaging, and where the field of Raman imaging with SERS nanoprobes is heading.

In vivo imaging with SERS nanoparticles—multiplexing potential

The ability to image multiple molecular and anatomical markers has the potential to revolutionize medical imaging. This is exemplified by the recent trend of multimodal imaging, i.e., combining different imaging modalities for a single patient to obtain complementary information [4], and also by the drive for multiplex imaging of many contrast agents in a single scan with a single modality [5]. Raman imaging has been shown to have superior multiplexing capacity than other in vivo techniques. This capacity stems from the ability to generate a substantial number of identifiable contrast agents, in the form of differentially targeted nanoprobes with distinct spectral signatures (typically referred to as "flavors"). These nanoparticle-based contrast agents can be generated relatively easily, using the same chemical procedure and varying the Raman reporter molecule during synthesis, and are bestowed differential targeting via surface functionalization with targeting moieties [13].

This high multiplexing capacity, along with its high spatial resolution (on par with other optical methods), makes Raman imaging a highly versatile technique, with the potential to provide molecular images of multiple targets, at cellular resolution, in vivo. In fact, as illustrated in Fig. 7.1, Raman imaging is the *only* in vivo compatible imaging technique that was shown to image up to 10 targets in a single

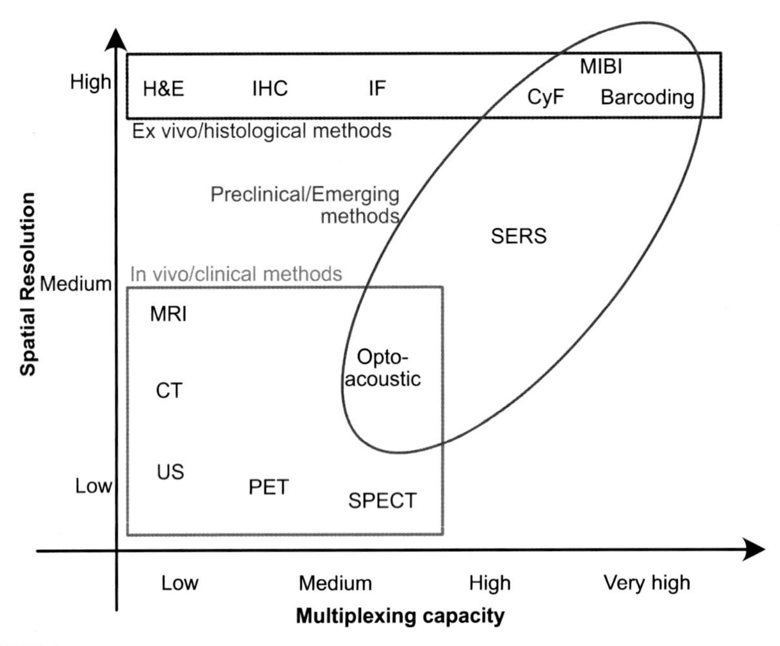

FIGURE 7.1

Clinically used and emerging imaging modalities plotted by their multiplexing capacity and resolution. Raman imaging with SERS nanoprobes has the highest multiplexing capacity among the in vivo imaging methods, comparable to techniques currently applicable ex vivo, such as mass spectrometry imaging and cyclic fluorescence. *CT*, computed tomography; *CyF*, cyclic fluorescence; *H&E*, hematoxylin and eosin; *IF*, immunofluorescence; *IHC*, immunohistochemistry; *MRI*, magnetic resonance imaging; *MIBI*, multiplex ion beam imaging; *PET*, positron emission tomography; *SERS*, surface enhanced Raman spectroscopy imaging; *SPECT*, single photon emission computed tomography; *US*, ultrasound.

scan. Although other imaging techniques (such as mass spectrometry imaging, cyclic fluorescence, and barcoding) have large-scale multiplexing capacity—well above 10—these methods are in their current form incompatible with imaging in living organisms, and are limited to imaging excised samples [5].

The multiplexing potential of Raman in vivo imaging has long been identified, since it was demonstrated in 2009. In their now-classic paper, Zavaleta et al. presented their work where 10 distinct flavors of SERS nanoparticles were imaged after subcutaneous injection in a live nude mouse [12]. As shown in Fig. 7.2, the SERS nanoparticles used in this study have identical morphology and similar composition, and only differ in the Raman reporter dye that bestows on each flavor its unique spectrum. In this proof-of-principle experiment, 10 distinct nanoparticle flavors were injected

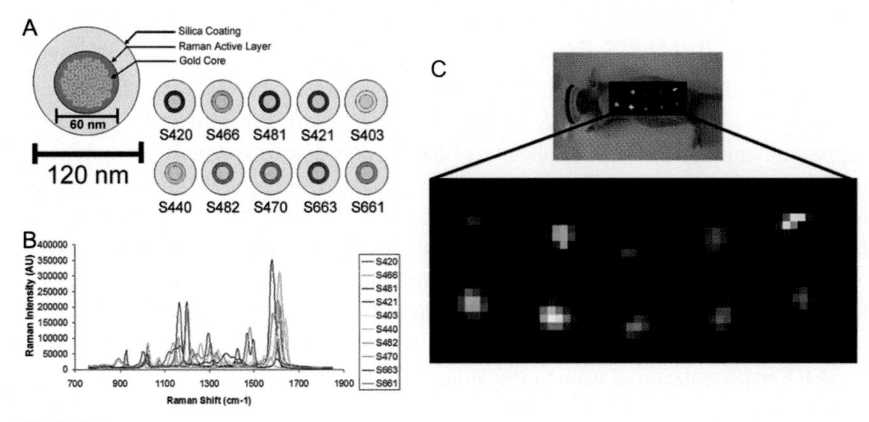

FIGURE 7.2

Early demonstration of multiplexed imaging of 10 different flavors in vivo. (A) Typical SERS nanoparticles consist of a plasmonic gold core coated with a Raman reporter molecule and encapsulated in a silica shell. (B) the different Raman reporters provide distinct spectral signatures that can be used to identify each flavor. (C) 10 flavors were detected after subcutaneous implantation of pure nanoparticle suspensions in a nude mouse under anesthesia.

Image adapted from C.L. Zavaleta, B.R. Smith, I. Walton, W. Doering, G. Davis, B. Shojaei, M.J. Natan, S.S. Gambhir, Multiplexed imaging of surface enhanced Raman scattering nanotags in living mice using noninvasive Raman spectroscopy, Proc. Natl. Acad. Sci. USA 106 (32) (2009) 13511–13516.

subcutaneously, separately (not as mixtures), in different sites along the flank of a nude mouse. The anesthetized mouse was scanned under a microspectrophotometer using a motorized stage, at a resolution of about 1 mm, with each point interrogated for 1 s with a 785 nm laser at 60 mW. An algorithm based on direct classical least squares linear fit was used for pointwise spectral unmixing. The group successfully used this method to identify the 10 different flavors in the living mouse. In a cognate experiment, four nanoparticle flavors were administered intravenously to a healthy nude mouse and were detected after accumulating in the liver.

A decade after its publication, this first example of multiplexed in vivo imaging still serves as a paradigm for Raman-based biomedical imaging. This work has outlined the main challenges that the in vivo Raman imaging community has been addressing ever since. These challenges are directing the nanoparticles to the desired location in the body, e.g., to tumors and specific molecular targets; controlling biodistribution and clearing of the SERS nanoprobes; improving tissue penetration depth and nanoparticle detection; and developing spectral unmixing techniques for quantitative and multiplexed imaging. Since then, multiple groups have addressed these challenges pushing the boundaries of the field, with many notable applications in cancer imaging. We will discuss these advances in the following sections.

Biological barriers and opportunities

To provide effective optical contrast, SERS nanoprobes must navigate the body and reach the site of interest. In a typical scenario where the aim is to delineate a cancerous tumor, the nanoprobes are administered intravenously and delivered through the bloodstream. However, these types of nanoparticles are foreign to the body, so they are faced with a series of barriers meant to defend against potential threats. Various components of the immune system present barriers to the delivery of the nanoparticles, including physical protective barriers (such as the blood–brain–barrier) as well as phagocytes of the reticuloendothelial system (RES) that sequester foreign agents from circulation. These biological barriers are not unique to SERS nanoprobes, but affect almost all nanoparticle-based agents. Consequently, considerable research has addressed the issues associated with nanoparticle delivery to tumors, especially from a drug delivery perspective. Biological barriers present a formidable challenge even to drug transport of conventional small-molecule agents, preventing effective delivery to diseased sites. Impediments to drug delivery lead to nonspecific biodistribution and inadequate accumulation of therapeutics and contrast agents to target sites causing limitations in successful diagnosis and treatment. The development of novel nanoparticle-based platforms for cancer diagnosis and therapy promises to overcome such boundaries and offers alternative means for delivery, leading to potentially superior diagnosis and treatment. The targeted delivery of nanoparticles for cancer diagnosis and treatment has recently progressed in strides, through the advancement of the field of nanomedicine [14,15]. One of the defining features of nanoparticles is their modular design [7]. This modular architecture allows the nanoparticles to be designed to circumvent most of possible biological barriers encountered upon intravenous administration. Increased circulation time is desirable, as it allows more nanoparticles to reach their target. To avoid opsonization by immune cells and extend systemic circulation time, nanoparticles are typically decorated with poly(ethylene glycol) (PEG) or even engineered with a custom protein corona [16,17]. While circulating through the vasculature, nanoparticles are known to extravasate and effectively reach the tissue surrounding growing solid tumors. This occurs partly due to their small size, which allows the nanoparticles to exit poorly formed vessels in the vicinity of the tumor by the so-called "enhanced permeability and retention" (EPR) effect, although biological variation makes this phenomenon a matter of controversy [18].

Passive tumor targeting

The EPR effect was first described by Matsumura and Maeda in 1986 [19] and further investigated and validated by Maeda and coworkers [20–24]. Their studies showed that most solid tumors have blood vessels with faulty architecture characterized by enhanced permeability due to the production of several angiogenic vascular permeability factors. Consequently, most solid tumors exhibit some degree of permeability to nanoscopic particles, which promotes increased accumulation of long-circulating macromolecules (larger than 40 kDa) by extravasation [25]. Additionally, tumors are characterized by poor lymphatic drainage, which causes increased interstitial fluid pressure. The angiogenic nature of newly formed tumor vessels in combination with the lack of lymphatics around the tumor region enhances the delivery of nanoparticles and macromolecules to the tumor well beyond that observed in normal tissues [19]. This increased accumulation in tumors can be exploited to provide superior imaging contrast, to increase the efficacy of the treatment, and to decrease side effects related to nonselective toxicity against healthy tissue. As a result, over the past few decades, this characteristic of solid tumors known as the EPR effect has been a significant incentive for researching the application of novel, long-circulating nanoparticles for therapy and diagnosis.

The EPR effect has also been identified as a major factor in the delivery of SERRS nanoprobes to tumors as even nontargeted nanoparticles (passivated with a PEG coating) were shown to reach a variety of tumor types [26]. In the work by Harmsen et al., which was the first to show such extensive application of Raman imaging with gold nanostars, passive accumulation via the EPR effect was used for in vivo imaging after intravenous injection, without any targeting moieties. The authors were able to image primary tumors in xenograft and genetically engineered mouse models, satellite tumors around the tumor bed after debulking, and even microscopic premalignant lesions, reporting the visualization of tumors as small as 100 μm. The key advancement that enabled the nanostars to delineate smaller tumors than previous studies was that they were resonant: both the star-shaped core and the reporter dye exhibit absorption maxima around the excitation laser wavelength of 785 nm. The effect of SERRS granted an exceptionally high sensitivity to the nanostars—with a limit of detection in the low femtomolar range—allowing the spectral identification even when relatively small numbers of nanoparticles were taken up by the microscopic tumors (Fig. 7.3).

Even though passively targeted nanoparticles have shown promising results for tumor delineation, they do not provide molecular information which could be useful for tumor profiling or for assessing treatment response. Additionally, relying on the EPR effect alone has proven inadequate and raised valid concerns regarding the clinical translation of nanoparticle-based agents [18]. This inadequacy is mainly due to the biological variability of the EPR effect: its intensity is not uniform within a tumor and also deviates at different stages of the tumor progression, between species, at different loci in the same patient, and between individuals. Additionally, there are limited data on the efficacy of the EPR effect in humans, hindering further clinical testing and translation [27–30].

Nevertheless, over the past few decades, EPR-based nanoparticle drug delivery and, more recently, imaging are evolving as the prevailing tactic offering many advantages in patient care over conventional approaches, such as reduced morbidity [31]. This improvement has led to the clinical approval of various nanoparticle formulations such as liposomal doxorubicin (Doxil, 1995) [32,33] and albumin-bound paclitaxel (Abraxane, 2005) [34].

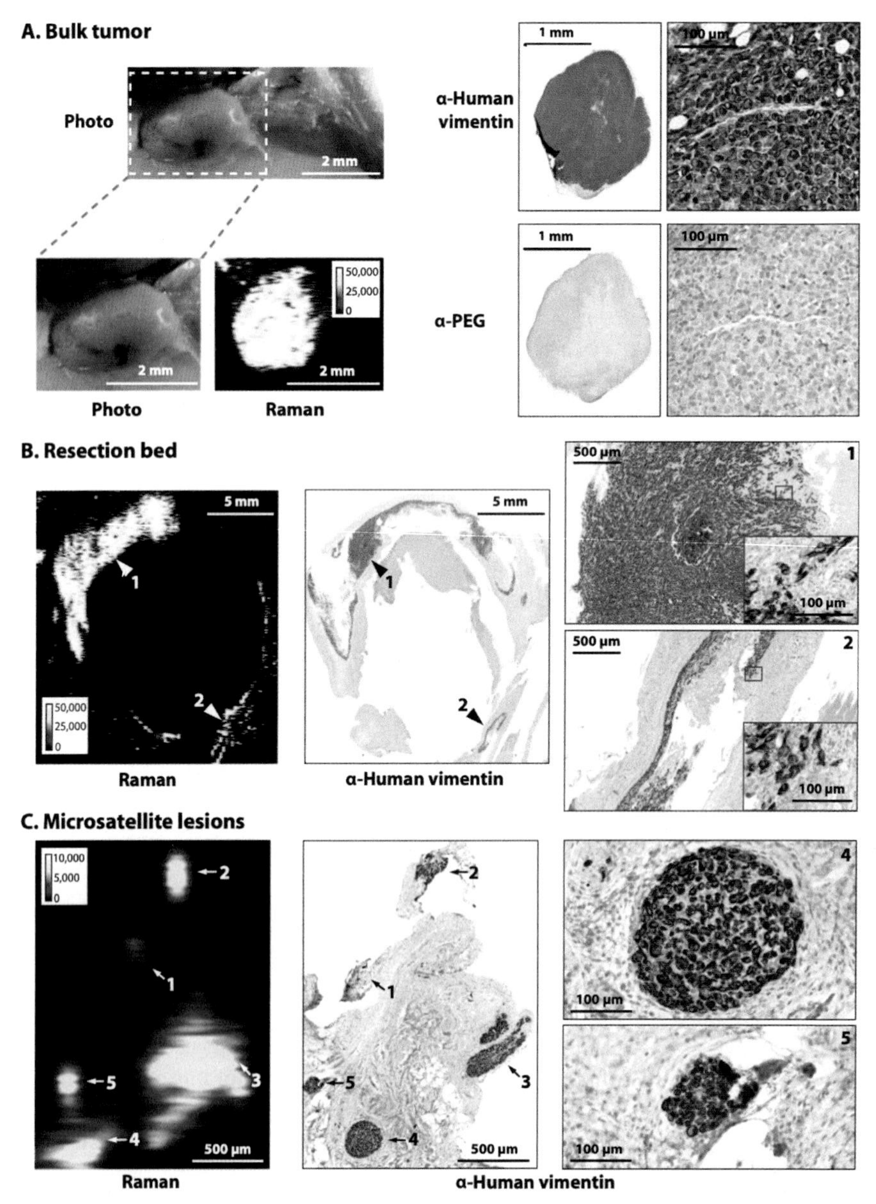

FIGURE 7.3

SERRS nanostars home to the bulk tumor and delineate microscopic metastases even in the absence of a targeting moiety—via the EPR effect—in mouse models of human cancer. The SERRS nanostars were used in an intraoperative setting to image (A) the bulk tumor, (B) residual malignancies in the tumor bed after resection, and (C) small satellite tumors in the area around the primary tumor.

Opsonization/sequestration by the mononuclear phagocyte system

Prolonged circulation times facilitate nanoparticle delivery to their target site. However, intravenously injected nanoparticles undergo nonspecific uptake by organs of the RES, which constitutes an important biological barrier in their delivery. The RES (also known as the mononuclear phagocytic system or macrophage system) is part of the immune system and consists of a collection of phagocytic cells, predominantly resident macrophages, in the spleen, lymph nodes, and chiefly in the liver. The RES is mainly responsible for the recognition and clearance of particles foreign to the body, a category that includes intravenously injected nanoparticles, causing their rapid removal from the bloodstream after injection [35]. Rapid blood clearance not only restricts nanoparticle accumulation at diseased sites but also may initiate an inflammatory response by the RES, causing toxicity in some cases [36–46].

Sequestration based on physicochemical properties

Probing the pharmacokinetic profile of intravenously administered nanoparticles allows for predictive relationships between structure and bioaccumulation patterns. It has been established that nanoparticle properties such as size, shape, surface chemistry, roughness, and surface coatings are determining factors that affect the pharmacokinetic patterns. In fact, one can tune the pharmacokinetics of nanoparticle agents simply by adjusting their size [36]. Rapid clearance from the bloodstream is generally not desirable as it limits nanoparticle bioaccumulation in the target tissue [47–49]. It is well established in the literature that nanoparticles with a diameter less than 10 nm are rapidly cleared by the kidneys and thus have a short residence span in the blood [49–51]. Additionally, nanoparticles with a diameter greater than 200 nm activate the mononuclear phagocytes of the RES and are also quickly removed from the bloodstream, accumulating in the liver and spleen [52–54]. If only considering the *size* difference alone, nanoparticles with diameters in the range around 100 nm are preferred [51]. Nanoconstructs with such sizes are advantageous due to their extended blood circulation and the relatively slow rate of uptake by the RES [55].

Nanoparticles are often removed from the circulation by opsonization. Opsonization involves the adsorption of plasma proteins (e.g., serum albumin, apolipoproteins, complement components, and immunoglobulins) onto their surface. The

Image from S. Harmsen, R. Huang, M.A. Wall, H. Karabeber, J.M. Samii, M. Spaliviero, J.R. White, S.Monette, R. O'Connor, K.L. Pitter, S.A. Sastra, M. Saborowski, E.C. Holland, S. Singer, K.P. Olive, S.W. Lowe, R.G. Blasberg, M.F. Kircher, Surface-enhanced resonance Raman scattering nanostars for high-precision cancer imaging, Sci Transl Med 7 (271) (2015) 271ra7.

adsorbed protein coating, often referred to as a "corona", has been widely known to initiate nanoparticle recognition and clearance by the RES [56] and can be engineered to extend circulation half-life [17]. The formation of the protein corona on the surface of the nanoparticle and the subsequent recognition by the host's immune system varies by nanoparticle size, surface charge, hydrophobicity, and surface chemistry [57]. These parameters are currently used to design nanoparticles that elicit different responses from the RES: to either avoid or, on the contrary, promote recognition and fast sequestration by phagocytes. Engineering the nanoparticle corona could lessen the toxicity and inflammation caused by the RES activation and improve nanoparticle delivery, enhancing imaging and drug effectiveness. For example, a common strategy is changing the capping material of the nanoparticle to prevent (or at least delay) the formation of a corona on its surface. Hydrophilic PEG has been extensively used to render nanoparticles more soluble and less prone to aggregation. Additionally, conjugation with PEG is known to provide a brush-like steric barrier that shields the nanoparticle from recognition by the RES, which results in further enhancement of circulation times in vivo [58–60]. Likewise, Dextran layers employed on commercial iron oxide MRI agent nanoparticles (known as superparamagnetic iron oxide nanoparticles (SPIONs), e.g., ferumoxytol) have shown similar benefits [61].

In some instances, it is desired to maximize the rate of nanoparticle sequestration by the RES. Using this inverse strategy and inducing rapid clearance of SPIONs by macrophages was shown to be beneficial for minimizing the background signal generated from unbound SPIONs in certain imaging applications in a mouse model [62]. Additionally, rapid sequestration of SPIONs by the resident macrophages in the liver has been exploited as a means to provide inverse contrast for liver tumors [63]. Following the same paradigm, SERS nanoparticles without a passivating PEG layer were used to image malignancies in the liver and spleen [64]. In this application, the nanoparticles provided inverse contrast, i.e., they accumulated quickly in the healthy liver parenchyma, providing strong Raman signal, while the tumor areas remained dark. The sequestration of nanoparticles from the bloodstream was reported to have a time constant of 77 s. This approach was capable of delineating large tumors (hepatocellular carcinomas) and microscopic histiosarcomas (as small as 250 μm) in genetically engineered mouse models, in the liver as well as in the spleen. The authors compared Raman imaging with SERS nanoparticles with the state-of-the-art clinically used small molecule dye indocyanine green (ICG) by injecting both agents in the same animals, shown in Fig. 7.4. It was reported that SERS nanoparticles provided higher contrast and presented superior photostability.

Similar strategies have been reported for image-guided lymph node excision. Sentinel lymph nodes are considered as a marker for metastasis of the tumor, as they are typically the first locoregional sites to be invaded. Different groups reported that by using handheld detectors it is possible to identify the infiltrated lymph nodes and remove them [65,66]. It is worth noting that as lymph nodes have resident phagocytes, they present high nanoparticle uptake in their healthy state, and inverse contrast was used in these studies to identify the compromised lymph nodes.

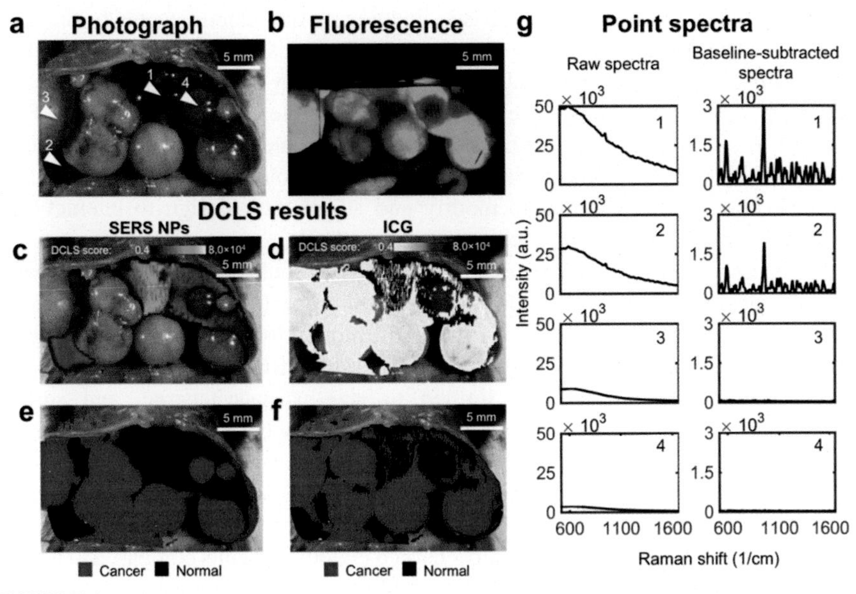

FIGURE 7.4

Comparison of SERS nanoparticles and the small molecule dye (indocyanine green (ICG)) for imaging liver tumors. Both contrast agents were administered (following the appropriate protocol for each) to the same mouse models and imaged together. (A) A mouse model of hepatocellular carcinoma was imaged in an intraoperative setting. (B) Widefield fluorescence imaging reveals the presence of ICG (green) and autofluorescence (red). Point-by-point scanning followed by spectral unmixing reveals the distribution of the Raman (C) and fluorescence (D) signals, to create masks (E and F) revealing the tumor as identified by each method. (G) Example point spectra from normal liver (1,2) and tumor (3,4) demonstrate the high contrast and specificity of the Raman signal.

Image from C. Andreou, V. Neuschmelting, D.F. Tschaharganeh, C.H. Huang, A. Oseledchyk, P. Iacono, H. Karabeber, R.R. Colen, L. Mannelli, S.W. Lowe, M.F. Kircher, Imaging of liver tumors using surface-enhanced Raman scattering nanoparticles, Acs Nano 10 (5) (2016) 5015–5026.

Blood–brain barrier

When it comes to imaging and treating malignancies in the brain and other parts of the central nervous system (CNS), one needs to consider the inadequacy of various therapeutics and contrast agents to cross the blood–brain barrier (BBB). The BBB is a specialized part of the vascular system, which is made by capillary endothelial cell tight junctions, luminal glycocalyx, basal lamina, and astrocytic foot processes. In this way, the BBB forms a highly selective barrier that protects the brain by restricting the passage of foreign substances and pathogens. However, in its role as a protective barrier, the BBB also greatly limits the efficient transport of therapeutic and diagnostic compounds, and especially nanoparticles. Overcoming the barrier's selectivity to deliver nanoparticle-based imaging or therapeutic agents to the CNS presents a major challenge to treatment and diagnosis of many brain disorders. New formulations able to penetrate the BBB are actively sought, and much research has been focused on testing the efficacy of nanoparticle delivery through the BBB and into the brain [67–69]. Although nanoparticles face the same restrictions as other exogenous species for crossing the BBB, they hold promise in such applications because of the potential to modify and adjust their design to interact with the local immune system and facilitate transport [70].

In the hopes of optimizing nanoparticle design for such a purpose, much research has been devoted to relating the physicochemical properties of various nanoparticle structures to the degree of uptake in the BBB, mostly for the purposes of drug delivery. Distinct characteristics that impact successful BBB crossing have been identified and include but are not limited to size, surface charge, and surface modifications. Nanoparticles smaller than 20 nm have a better chance of effectively crossing the BBB [71] due to the fact that at this size, nanoparticle endocytosis is facilitated through a clathrin-mediated mechanism. Similarly, positively charged nanoparticles can use the adsorptive transcytosis pathway much more efficiently due to attractive forces with negatively charged endothelial cells compared to neutral or negatively charged nanoparticles [72]. Additionally, longer nanoparticle circulation times increase bioavailability and may facilitate uptake through the BBB. Consequently, PEGylation has been used as a general strategy for overcoming the RES and, at the same time facilitating transport through the BBB. Surface modifications with various ligands have also been used as a way to increase transport and receptor-mediated transcytosis through the BBB. Ligands for targets such as GLUT1 or albumin transporters [73], Lf receptors, LRP1 (targeted by angiopep-2), or Tf receptors have been used [74].

Various other strategies have also been used in combination with optimizing nanoparticle design in the quest for improving the BBB crossing. The most successful approach at overcoming this restriction of nanoparticle transport through the BBB is to provoke a momentary permeability increase between adjacent endothelial cells. This is usually achieved by using ultrasound/microbubbles and osmotic pressure [74–76]. This enhanced BBB permeability allows for a transient increase in nanoparticle entry into the brain. However, this approach comes with many risks as it also promotes the uncontrolled entry of many other compounds into the brain, which could lead to toxic effects, including inflammation.

Promoting the binding of nanoparticles to the surface of endothelial cells via the transcytosis pathway may be a safer approach at facilitating nanoparticle entry through the BBB. Transcytosis can be either adsorptive or receptor-mediated and is facilitated by altering the design of the nanoparticle through surface modification, as previously mentioned [74,77].

When it comes to delivering SERS nanoparticles to the brain, several examples appear in the literature, mainly for imaging glioblastoma multiforme (GBM) in animal models. The reasoning behind using SERS nanoprobes as an imaging agent for brain tumors derives from the high signal-to-background ratio they provide, which in combination with their optical resolution, allow the imaging of smaller features than with other modalities and enable the reduction of surgical margins. Extensive work has been performed on SERS-based imaging of GBM by Kircher and his group.

In an initial study, SERS nanoparticles were used as a triple-modality contrast agent, detectable by MRI, photoacoustic, and Raman imaging [78]. The nanoparticles were commercially acquired, featuring the Raman reporter (trans-1,2-bis(4-pyridyl)-ethylene), and were subject to surface modification with an MRI contrast agent (gadolinium). The nanoparticles were administered intravenously to a mouse model of GBM, induced via stereotactic implantation of U87MG cells. Raman imaging was performed with a commercial microspectrophotometer with a 785 nm laser (off-resonance with the nanoparticles), and the authors reported that the nanoparticles were detected by all three modalities at the area of the tumor. The authors proceeded to perform SERS-guided tumor resection, as shown in Fig. 7.5. The nanoparticle signal delineated the primary tumor and revealed residual microscopic tumor infiltrations after debulking. The nanoparticles in this study did not employ a targeting moiety and accumulated in the tumor area via the EPR effect, made

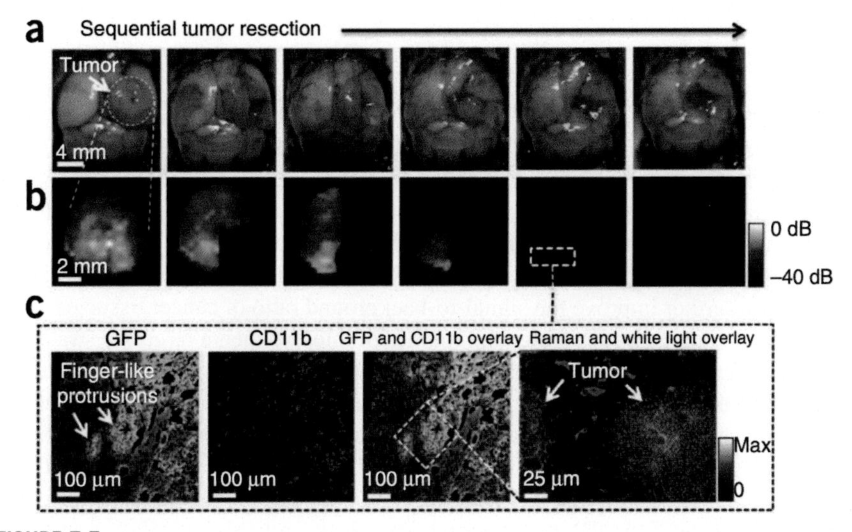

FIGURE 7.5

SERS nanoparticle-guided tumor excision of brain tumors. (A) Glioblastoma bearing mice underwent craniotomy under anesthesia, and their tumors were sequentially excised. (B) Raman imaging at each step of the excision demonstrated the remaining tumor. Once the bulk of the tumor was removed, microscopic protrusions around the tumor edges became visible. (C) Ex vivo validation confirmed that the SERS nanoparticles infiltrated the tumor, without being taken up by healthy neighboring tissue.

Reproduced from M.F. Kircher, A. de la Zerda, J.V. Jokerst, C.L. Zavaleta, P.J. Kempen, E. Mittra, K. Pitter, R. Huang, C. Campos, F. Habte, R. Sinclair, C.W. Brennan, I.K. Mellinghoff, E.C. Holland, S.S. Gambhir, A brain tumor molecular imaging strategy using a new triple-modality MRI-photoacoustic-Raman nanoparticle, Nat. Med. 18 (5) 2012 829–834.

possible as the invasive tumor compromised the BBB in its vicinity, affording excellent signal contrast.

Subsequent studies by the Kircher group expanded the use of SERS nanoprobes for delineation of brain tumors and SERS-guided GBM resection using a handheld Raman scanner [79], by introducing on-resonance molecularly targeted SERRS nanostars [80], and further developing the combined use of photoacoustic/Raman imaging [81].

Other groups developed various other strategies to enable SERS nanoparticles to cross the BBB and reveal brain tumors, using active targeting [82] or focused ultrasound [83]. In one example, smaller PEG-coated gold nanoparticles with a diameter around 10 nm were functionalized with the brain-targeting peptide angiopep-2 and were shown to successfully cross the BBB [84]. However, the small size of the individual nanoparticles was not conducive to strong plasmonic excitation, so a pH-sensitive linker was added to allow them to cluster once inside the acidic microenvironment of the tumor. In this way, the nanoparticles in tumor regions provided high contrast, whereas the ones in the physiological brain areas provided little to no signal.

Molecular targeting

It is a common strategy to functionalize SERS nanoparticles with PEG chains to evade the immune system in vivo, and in this way, achieve relatively long blood circulation time due to the well-established "stealth effect" of PEG [85]. The prolonged circulation increases bioavailability and enables extravasation and accumulation of the nanoprobes in cancer tissues, via the EPR effect as described above. However, if one aims to further improve the signal-to-background ratio of cancer tissue or detect the expression of a specific molecular marker of interest, then a targeting strategy is pursued. Nanoparticles can be rendered into actively targeted probes via surface functionalization with targeting moieties when it is desired to achieve specific engagement with antigens overexpressed on the cancer cell surface or other cells in the cancer microenvironment. Various targeting moieties have been reported to direct nanoparticles to a target, including antibodies, affibodies, small molecules, polypeptides, and DNA/RNA aptamers. Due to the reliable specificity and high selectivity of such interactions, the literature is rich with reports of molecular targeting in vitro, extensively reviewed elsewhere [86–88]. In the following section, we will only revisit molecular targeting for in vivo and freshly excised (ex vivo) human or mouse tissue samples.

Ex vivo SERS-based molecular imaging

Molecular imaging can help differentiate malignancies from healthy tissues and reduce the excision margins in the surgical theater. Surgical resection remains the mainstay treatment for most solid tumor types. Clinically, in the resection of breast cancer in particular, and other tumor types in general, there is a need for intraoperative imaging technologies to guide tissue-conserving surgeries. Currently, postoperative pathology analysis is used to verify whether the surgical margins are clear of residual tumor during excision, but this method leads to high rates of re-excision, extending patient morbidity. As an alternative, several groups are developing an ex vivo staining and imaging strategy with SERS nanoprobes, to allow the quick assessment of the presence of a molecular marker immediately after excision. This ex vivo strategy has the advantage that it is applied to tissues outside the patient without needing to address and bypass the biological barriers encountered in vivo.

The Liu group developed an efficient strategy for targeting a large number of biomarkers overexpressed in breast cancer in freshly excised tissue. The group developed an elegant method of SERS-based molecular imaging via topical application of antibody-functionalized SERS nanoparticles, and reliably and rapidly performed multiplexed Raman-encoded molecular imaging of several biomarkers concurrently using five or more SERS probes with distinct signatures and an optimized in situ staining method [89–94]. In another example, Davis et al. assessed CD47–Carbonic Anhydrase 9 antibody-based active targeting and passive targeting mechanisms by topically applying nanoparticles to ex vivo human bladder cancer tissues [95].

In vivo molecular imaging with SERS nanoprobes

Although detecting molecular targets in excised tissues is beneficial and has immediate applications, in vivo administration and imaging may help in identifying tumors more specifically, and aiding with excision via intraoperative imaging. Currently, routine intraoperative imaging does not employ traditional clinical methods such as MRI and PET, as they are too cumbersome for the operating theater and are reserved only for preoperative staging. Optical imaging methods are suitable for the intraoperative imaging scenario, as they consist of lightweight equipment, and their main weakness—their limited penetration depth—is less of a concern in the context of an invasive procedure. In fact, the most prevalent modality for intraoperative imaging is fluorescence with small-molecule infrared fluorescent dyes. This is why SERS nanoprobes are being explored as a means to provide molecularly targeted contrast in the same setting. SERS nanoprobes have been employed to provide in vivo targeted optical contrast in a multitude of preclinical studies. However, most studies employ mouse models with subcutaneous tumors and validate their targeted SERS nanoprobes via noninvasive imaging.

Qian et al. demonstrated intravenous injection of PEGylated SERS nanoparticles conjugated to tumor-targeting single-chain variable fragment (ScFv) antibodies selectively targeted tumor biomarkers such as epidermal growth factor receptors (EGFRs) in xenograft tumor models [96]. Dinish et al. employed the actively targeted SERS nanoprobes for in vitro and in vivo detection of three known cancer biomarkers—EGFR, CD44, and TGFβRII—in a subcutaneous xenograft mouse model of breast cancer [97]. Harmsen et al. conjugated EGFR antibody to resonant nanoprobes with exquisitely high SERRS signal to target A431 xenograft tumors, and reported achieving better delineation of tumor tissue compared to nontargeted SERRS nanoprobes administered to the same animals [98]. Nayak et al. functionalized similar SERRS nanoprobes with an antitissue factor (TF) monoclonal antibody (ALT-836) and visualized the overexpression of TF by pulmonary metastases in a metastatic breast cancer mouse model intraoperatively, after intravenous injection. Numerous micrometastatic lesions up to 200 μm were detected throughout the lung and corroborated by immunohistochemistry and bioluminescence imaging. To verify targeting specificity, TF was blocked in a cohort of metastatic mice with excess antibody prior to administering the nanoprobes, and successfully blocked targeting significantly reducing the signal [99]. As mentioned above, Yue et al. targeted EGFRvIII, a variant of the EGFR in a GBM mouse model. The signal to background ratio was higher for EGFRvIII + U87MG glioblastoma xenograft than the nanoprobes without targeting [82]. Li et al. developed alkyne- and nitrile-groups as anchoring points to directly fabricate novel SERS tags without the need for thiol anchoring. The SERS tags were functionalized with different antibodies for multiplexed imaging of cancer cells and human breast cancer tissues [100].

Intraoperative imaging can also follow the topical application of SERS nanoprobes. Ultrabright SERRS nanoprobes functionalized with folate receptor–targeted antibody were used to delineate micrometastases of ovarian cancer by intraperitoneal application of a cocktail of targeted and nontargeted nanoprobes. The prime advantage of this approach was that metastatic tumors overexpressing the folate receptor were made visible without a systemic injection, and therefore, uptake by RES organs was not observed, achieving in vivo imaging while completely bypassing the body's barriers [13,101]. Davis et al. used CD47 antibody-labeled SERS nanoparticles via a topical application and revealed increased binding of CD47-targeted nanoparticles on tumor relative to normal tissue in a breast tumor model [102].

Although there are many advancements in the domain of antibody targeting, potential drawbacks include the considerable immunogenicity and high cost of production of humanized monoclonal antibodies, both of which might hinder clinical translation. To alleviate these potential disadvantages, alternative targeting strategies have been investigated. One of the relatively inexpensive and convenient targeting strategy is the use of small molecules or polypeptides. For instance, Huang et al. functionalized SERS nanoprobes via an integrin targeting RGD peptide for imaging GBM. The authors reported that they were able to detect distant micrometastases (as small as five-cell clusters) efficiently with the targeted nanoprobes [80]. Feng et al. developed folate targeted SERS nanoprobes with gold nanobipyramid cores, which could selectively target folate receptor–positive cancer tissues derived from MCF7 cells in xenograft mouse models of breast cancer [103].

DNA and RNA aptamers are also being explored as a potential alternative to monoclonal antibodies as targeting agents. Advantages of aptamers over antibodies include (i) low molecular weight, (ii) low or no immunogenicity, (iii) reproducible, large-scale, and inexpensive synthesis, and (iv) custom-built chemical modifications.

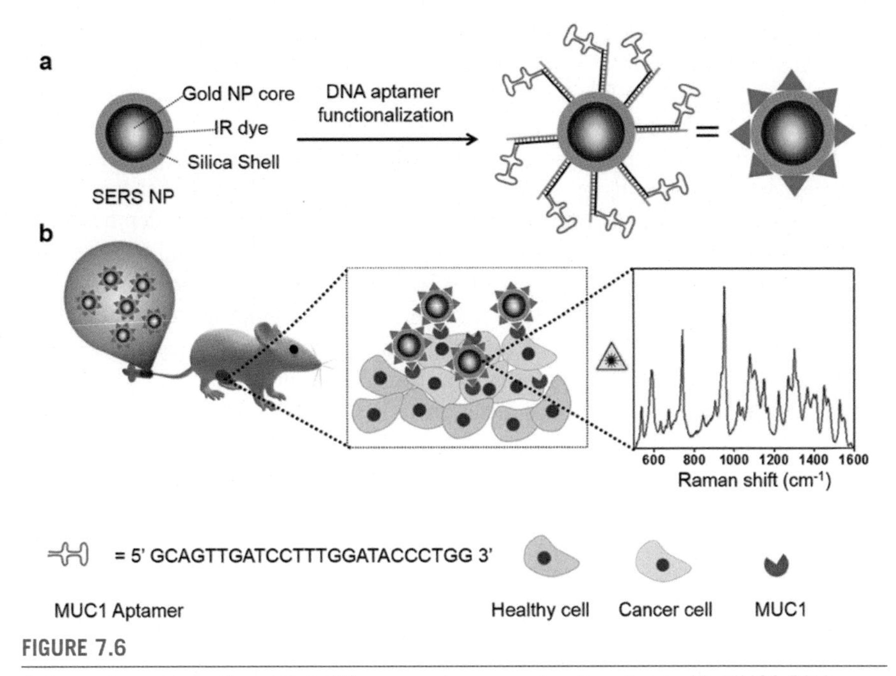

FIGURE 7.6

Aptamer-based targeting: (A) SERS nanoprobes were functionalized with MUC1 DNA aptamers and (B) administered to a mouse xenograft model of human breast cancer via intravenous injection. The MUC1-NPs selectively accumulate to tissues overexpressing MUC1, identifying the tumor.

Adapted from S. Pal, S. Harmsen, A. Oseledchyk, H.-T. Hsu, M.F. Kircher, MUC1 aptamer targeted SERS nanoprobes, Adv. Funct. Mater. 27 (32) (2017) 1606632.

Pal et al. used a well-known aptamer against Mucin1, for selective targeting of Muc1 positive triple negative breast cancer mouse models, as shown in Fig. 7.6. The Muc1-targeted SERS nanoprobes selectively targeted Muc1-positive tumor tissue but not the Muc1-negative healthy tissue and demonstrated the efficacy of the system [104]. Using a similar targeting approach, Wu et al. attached a nucleolin-binding DNA aptamer (AS1411) to SERS nanoparticles for ex vivo staining of cancer tissue and normal tissue from a mouse [105].

Multimodal imaging using SERS nanoprobes

SERS nanoprobes offer high sensitivity and specificity in cancer imaging due to their "fingerprint" like spectra, as has been demonstrated in various studies. However, currently available Raman imaging hardware is focused on research applications and employs a time-consuming point-by-point acquisition of Raman spectra, which impedes the real-time image acquisition required for clinical applications. The development of dedicated high-speed wide-field Raman scanners is an active area of research, and a small-animal imaging system was reported to improve Raman image acquisition by 10-fold [106]. Although such a system may help preclinical research based on small animal imaging, the use of Raman imaging is still very challenging for deeply embedded tumors present in clinical settings with human patients. Since Raman spectroscopy is an optical technique, it is limited by the absorbance and scattering of light by biological tissues—allowing only a few millimeters depth of penetration and restricting potential applications of SERS-based imaging to superficial imaging e.g., in intraoperative scenarios. Many preclinical studies described in literature use subcutaneously injected cancer cells in xenograft mouse models leading to very superficial tumor depth [107] to make noninvasive Raman imaging feasible.

When using current imaging paradigms for deep tissue imaging, one has to rely on surgical incision of the suspected cancerous area, followed by a lengthy imaging session. In an attempt to circumvent this limitation, contrast agents for Raman imaging are often paired with contrast for an additional modality which can provide complementary noninvasive, deep tissue imaging capabilities, as well as faster preliminary imaging for localization of cancer, followed by Raman imaging of a smaller area, or after tumor debulking. Several reports investigated multimodal nanoparticles combining SERS nanoprobes with one or more imaging modalities such as fluorescence, CT, MRI, and PET to supplement Raman imaging [108].

Additionally, the modular nature of nanoparticles allows a therapeutic modality also to be incorporated, often based on the photothermal effect afforded by the plasmonic core. Over the last decade, a plethora of studies emerged reporting the development of novel multimodal SERS nanoprobes, with a few validated in preclinical in vivo studies [109–126]. Kircher et al. developed a triple-modality nanoparticle that combines MRI, multispectral optoacoustic tomography (MSOT), and SERS contrast in the same structure. The triple modality nanoparticles accurately delineated the margins of brain tumors in GBM mouse models pre- and intraoperatively [78]. Amendola et al. developed a different design principle for the fabrication of MRI-CT-SERS multimodal nanoparticles that involved formation nanoprobes with Raman reporters sandwiched between gold nanoparticles, which were then cloaked with PEG chains. The nanoprobes were reported to selectively accumulate in tumor tissue, along with the typical sequestration by RES organs [127]. Campbell et al. reported a radiolabeling strategy of already developed SERS nanoprobes with the ^{64}Cu isotope and used it to demonstrate a time-dependent biodistribution using PET, exploiting this established and versatile noninvasive deep-tissue imaging technique [128]. Litti et al. developed an MRI and SERRS multimodal nanoparticle in which Gd(III) was loaded in high density onto the SERS nanoprobes using a DOTA chelating agent. In vivo and ex vivo MRI and SERS imaging and photothermal capabilities were further elegantly demonstrated [129]. Wall et al. developed an efficient, chelator-free strategy of radiolabeling the silica shell of SERS nanoprobes. The dual-modality nanoprobes were used for PET and SERS imaging and demonstrated efficient delineation of liver cancer [130]. Gao et al. developed a "smart" activatable type of MRI-SERS nanoprobes, which can permeate through the BBB

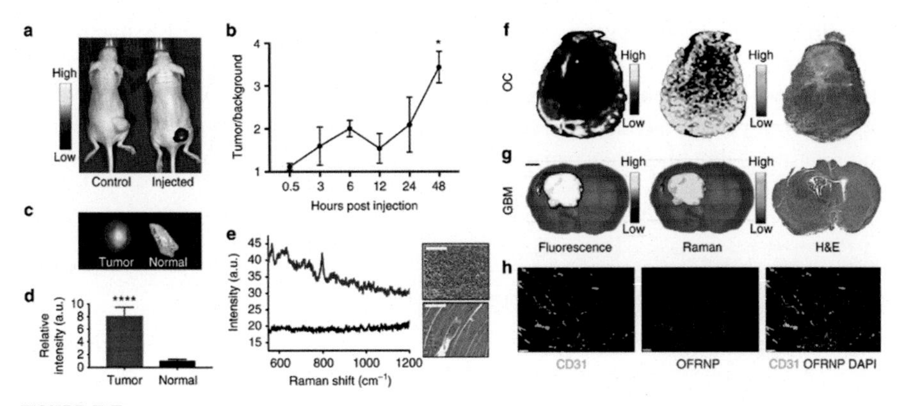

FIGURE 7.7

In vivo cancer imaging fluorescence Raman nanoprobes. (A) NIR fluorescence images of ovarian cancer xenograft mouse after 48 h i.v. administration of 200 μL of 1 μM of Dylight-780 labeled DNA, and 200 μL of 10 nM of nanoprobes (right). (B) Tumor to background ratio derived from NIR fluorescence images at different time points. The ratio attained a maximum value after 48 h postinjection ($n = 3$). (C and D) Ex vivo fluorescence images of tumor showing ~8 times higher fluorescence than muscle next to the tumor ($n = 3$). (E) Background subtracted Raman spectra of the tissues indicated the presence of the Raman signature of nanoprobes exclusively in the tumor tissue (verified by H&E imaging Scale bars are 200 μm). (F) Ex vivo fluorescence and Raman images of the excised ovarian cancer tumor with adjacent H&E histology correlation. (G) Ex vivo fluorescence and Raman images of the coronal section of a GBM mouse brain with adjacent H&E histology correlation. Scale bar is 1 mm h High magnification immunofluorescence images of ovarian cancer tumor strained with CD31 antibody (green channel) for the visualization of neovasculature. The nanoprobes (red channel) were clearly localized in and just out of neovasculature. Scale bars are 20 μm.

Adapted from S. Pal, A. Ray, C. Andreou, Y. Zhou, T. Rakshit, M. Wlodarczyk, M. Maeda, R. Toledo-Crow, N. Berisha, J.Yang, H.-T. Hsu, A. Oseledchyk, J. Mondal, S. Zou, M.F. Kircher, DNA-enabled rational design of fluorescence-Raman bimodal nanoprobes for cancer imaging and therapy, Nat Commun 10 (1) (2019) 1926.

and delineate cancer tissue in GBM mouse models [84]. Neuschmelting et al. used optoacoustic and Raman imaging nanoprobes to first noninvasively detect cancer in GBM mouse models with MSOT and then SERS to further verify the presence and microscopic extent of the malignancy. The results were in excellent agreement with the histological verification [81]. Recently, Pal et al. developed a novel nanoparticle design principle to combine two optical Raman and fluorescence light from the same source, as shown in Fig. 7.7. This combination of SERS and fluorescence imaging in the same contrast agent is otherwise challenging, as the plasmonic core required for SERS greatly diminishes fluorescence of dyes in its vicinity. To circumvent this, DNA was employed to enable a new design principle that achieves both efficient fluorescence and enhanced Raman signal. Further photothermal remediation was demonstrated using a NIR (785 nm) laser [131].

Although there are quite a few reports of such multimodal SERS nanoprobes in vivo, there is a lack of instrumentation to allow all-in-one imaging. Only recently, a system that can concurrently detect SERS and photoacoustic modalities in vivo has been developed. Conveniently, high contrast photoacoustic imaging provided microvascular imaging that can potentially be used to guide surgeons for tumor resection [132].

The future of in vivo Raman imaging

SERS nanoprobes are on course to becoming a powerful and versatile imaging platform. As an optical contrast agent for Raman imaging, these nanoparticles were shown to provide high sensitivity and unparalleled signal specificity, allowing delineation of even microscopic tumors and premalignant lesions in a variety of mouse models, providing very promising results in the preclinical arena. The technique's main strength lies in the highly specific signal of the nanoprobes, which is the key to multiplexed in vivo imaging, as probes with distinct Raman spectra can be differentially targeted against a variety of molecular targets, and detected concurrently.

However, Raman imaging with SERS nanoprobes has yet to fully live up to this promised potential in preclinical testing, and this represents a considerable impediment toward the clinical translation of the method. The major remaining limitations are threefold and they are intertwined. Firstly, the poor penetration depth of light in tissue limits the available signal from the nanoparticles. Secondly, the currently available Raman imaging hardware relies on point-by-point scanning, which requires prohibitively long scan times to acquire a reasonable signal-to-noise ratio, especially for areas of clinically relevant extent. Finally, the biodistribution profile of nanoagents administered intravenously raises concerns regarding the SERS nanoprobes' clearance and long-term toxicity effects on the patient. The issue of biocompatibility becomes ever more pressing when imaging multiple targets, as the number of nanoprobes provided will need to increase proportionately. SERS nanoprobes have so far been tested only in the preclinical arena, and their toxicity profile evaluated in mice [133]. However, other related nanoconstructs are in clinical trials. For example, silica nanoparticles with a fluorescent dye and a radiotracer showed no toxic or adverse effects in clinical trials [134], while the toxicity of gold nanoparticles is being evaluated in multiple preclinical studies [135,136]. Creative solutions are required to help this technique reach maturity in the preclinical realm and drive its clinical translation. Recent research addresses these concerns and shows that these obstacles are not prohibitive but can, in fact, be overcome.

Imaging deeper

SERS nanoprobes rely on light (in the visible and near IR bands of the spectrum) to be excited and provide their signal which is also light of the same spectral region.

However, organic tissues absorb and scatter light, limiting its penetration depth. Visible frequencies suffer the most, as they are strongly absorbed, making IR light (typically at 785 nm) better suited for in vivo imaging. Even in the IR frequencies, both the incident (excitation) and emitted (Raman) photons are scattered by tissue, limiting detection to 1–2 cm at best. This limitation has restricted the use of Raman imaging with SERS nanoprobes to superficial applications, such as intradermal/subcutaneous tumor models in mice, or invasive intraoperative scanning of internal organs.

SERS and endoscopy

Raman imaging can be used to image deep inside the body by using endoscopy. This application is most relevant to tumors of the digestive system, such as in esophageal and colorectal cancer. Currently, gastrointestinal endoscopy is used routinely in the clinic in the form of white-light imaging. In this scenario, the clinician is tasked with identifying the presence and extent of malignancies based on their color or texture alone, without molecular contrast for verification. This is typically followed by a core biopsy for molecular analysis. However, it has been reported that up to 25% of lesions may be missed in this way [137]. Raman imaging holds the promise to enable profiling of these malignancies via active/molecular or passive (EPR based) targeting with high accuracy and sensitivity, in real-time, while doing away with invasive biopsy. As Raman imaging is an optical modality, Raman endoscopes do not need to deviate in form from the standard optical-fiber endoscopes currently available, and hybrid systems combining white-light, Raman, and fluorescence can be engineered to provide multimodal optical information.

A prototype Raman endoscope was presented by Zavaleta et al. [138]. This system was designed to provide spectral (Raman) information and fit through a clinical endoscope. The system was able to identify SERS nanoparticles injected into excised human tissue samples down to a concentration of 326 fM. Another notable feature of this work is the endoscope's ability to focus at distances between 1 and 10 mm, and in this way, locate lesions that may be further due to the tissue morphology. The system's performance and multiplex detection capabilities were extensively validated using SERS nanoparticles in various combinations and concentrations in human tissues, porcine intestine, and imaging phantoms. The group even recorded Raman spectra from patients, although without application of any SERS nanoparticles, due to regulatory restrictions.

When it comes to using targeted SERS nanoprobes for endoscopic imaging and detection, a few highly promising approaches have been reported using in vivo imaging in animal models. One approach uses the topical application of SERS nanoprobes via oral gavage for imaging of esophageal cancer in rats [89]. The molecular targets pursued were EGFR and HER2, plus a nontargeted control, which was used as a standard for ratiometry. A similar approach by a different group used in-house synthesized nanoparticles with two Raman flavors and also fluorescence signal to enable multimodal (Raman/fluorescence) endoscopy [139]. The same molecular targets were investigated (EGFR and HER2), and their presence was validated using mouse xenograft models of human breast cancer. The fluorescent signal was used to locate the presence of the nanoparticles, and subsequently, their SERS signal allowed the identification of the nanoprobe flavor.

More recently, a different approach of targeted endoscopic imaging was demonstrated based on systemic (intravenous) administration of SERRS nanoprobes [140]. In this work, microscopic and even premalignant lesions were detected, signifying an exquisitely high sensitivity of detection. The nanoprobes were passivated with PEG and injected into mice models with pathologies including Barett's esophagus, esophageal adenocarcinoma, *H. pylori*–induced stomach tumors, and premalignant tumors in the intestines. Imaging was performed with a custom miniature endoscope for mice, in vivo and in ex vivo samples, and validated via histology. This paradigm reiterates that even passivated nanoparticles are taken up by tumors, and can be detected with sufficient sensitivity, without necessitating functionalization via antibodies; although systemic administration may be required.

Spatially offset optics

A different—noninvasive—paradigm that allows detection in greater depths (up to 5 cm) has been proposed by using offset optics, i.e., different light paths for the excitation and emission light, in a technique termed "spatially offset Raman spectroscopy" (SORS) [141,142]. Although SORS was initially developed for through-barrier detection of analytes based on their intrinsic Raman signals, it has now been adapted to biological applications via the use of SERS nanoparticles to give rise to "surface-enhanced spatially offset Raman spectroscopy" (SESORS) [142,143].

Recently, the first report of in vivo spatially offset targeted imaging was presented for the noninvasive imaging of glioblastoma through the skull of a genetically engineered mouse model [107]. Integrin-targeted SERRS nanoprobes were synthesized, featuring a gold nanostar core with the Raman reporter molecule IR782 (resonant with the 785 nm excitation laser), encapsulated in a silica shell, and functionalized with cyclic-RGDyK peptide. The nanoprobes were administered intravenously, and the mice were scanned under anesthesia using an in-house built system, at millimeter resolution. A single plane along the coronal direction was scanned, and signal analysis showed the presence of the nanoprobes as well as the Raman signal from bone, indicating the skull geometry, as shown in Fig. 7.8. The images produced in this way were compared to histological sections of the mouse brains, extracted post mortem, and were found to show the extent of the tumor correctly.

Additionally, the spatially offset signal was compared to the signal acquired via conventional Raman optics, where the same optical path was used for excitation and collection. The spatially offset method was able to detect 10 times more Raman photons than the conventional scan, achieving higher sensitivity and contrast. The

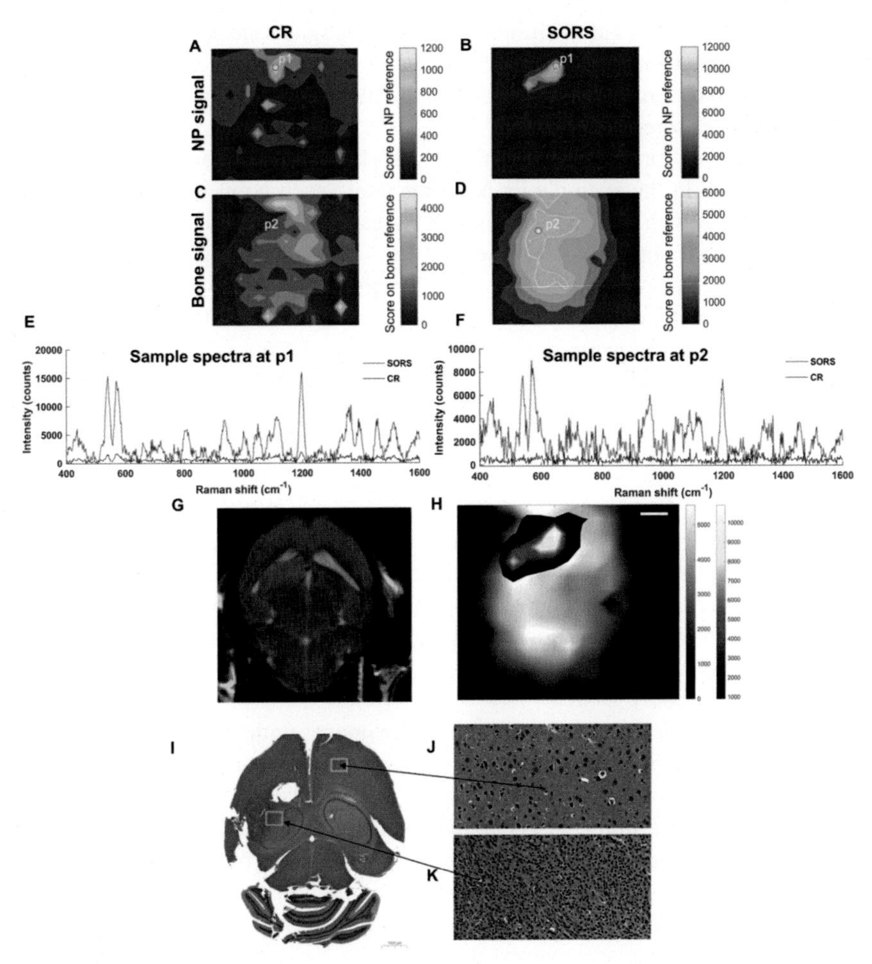

FIGURE 7.8

Imaging of GBM noninvasively in a mouse using SESORS and integrin-targeted SERRS nanoprobes. (A) Conventional Raman optics detect much lower contrast from the nanoprobes than the spatially offset configuration (B). The same trend is observed for the intrinsic Raman signal from the skull (bone, C and D), and sample spectra are shown in (E and F). The composite SORS image (H) reveals the tumor in agreement with MRI (G) and histology (I–K).

Image adapted from F. Nicolson, B. Andreiuk, C. Andreou, H.T. Hsu, S. Rudder, M.F. Kircher, Non-invasive in vivo imaging of cancer using surface-enhanced spatially offset Raman spectroscopy (SESORS), Theranostics 9 (20) (2019) 5899–5913.

success of this example lies in the combination of SORS with the exquisitely bright resonant SERRS nanoprobes, and for this reason, the technique was reported as SESORRS—surface-enhanced spatially offset *resonant* Raman spectroscopy.

As spatially offset optics allow Raman imaging through tissue up to several centimeters in-depth, this technique could be eventually translated into the clinic. The technique is especially suited for applications in tissues with low optical density, such as for breast cancer screening.

However, even for noninvasive imaging, the time required for a scan is currently unsuitably long. In the example by Nicolson et al., each point was interrogated for five acquisitions with integration times of 3 s each [107]. If this technique is to be used in a clinical scenario where several cm^2 are to be imaged, the acquisition time needs to improve dramatically. This limitation is not unique to SORS but also applies to conventional Raman imaging. For this modality to become clinically relevant, new imaging paradigms are necessary that do not rely on point-by-point scanning and provide the required spectral image in a fraction of the time.

Imaging faster

Rapid acquisition is a crucial requirement for in vivo imaging applications. In a clinically relevant scenario, areas with a size of several square decimeters need to be scanned in order to cover a whole organ on a human patient, preferably with a resolution of 1 mm or less. Additionally, the acquisition needs to be completed quickly to avoid drift artifacts due to movement, and collect enough signal to attain a robust signal to noise ratio.

In theory, Raman imaging is able to produce images with down to submicrometer resolution, as it is an optical technique with resolution theoretically limited by the diffraction of light. The amount of signal generated by SERS nanoprobes has been vastly improved in recent generations, allowing down to femtomolar sensitivities, and has become comparable to fluorescence [144]. In reality, however, Raman imaging at submillimeter resolutions remains prohibitively slow. This is not a limitation of the method itself, but rather of the design of commercially available imaging systems, which are mainly developed and marketed as instruments for general research, and not as dedicated medical imaging tools.

A possible configuration for a dedicated small-animal Raman imaging system was proposed by Bohndiek et al. [106] to provide rapid imaging for preclinical applications in mouse models. Instead of relying on a mechanical stage to translate the mouse under the microscope optics, the proposed system employed mirrors actuated by a galvanometer to scan the laser beam over a stationary sample. In this way, the laser was moved rapidly along the direction of one axis, providing excitation in a line of around 2.7 mm. The signal from the line was collected and analyzed to its constituent frequencies simultaneously to form a 2D pattern containing the Raman spectra along the excited line. This resulted in much faster acquisition compared to point-by-point scanning—by at least one order of magnitude. This fast imaging technique allowed the group to image areas as large as $6 \times 6\ cm^2$, spanning the full length of a mouse torso. To demonstrate the capabilities of their system, the group used commercially available SERS nanospheres. The system was able to perform multiplexed imaging of mixtures of 2, 3, and 4 flavors injected subcutaneously in nude mice. Additionally, noninvasive multiplexed imaging was successfully performed using SERS nanospheres accumulated in the liver after intravenous injection. One of the main strengths of this scanning configuration is that it does not compromise its spectroscopic nature; it provides rich, spectral information for every point scanned, and in this way retains its ability of spectral signal decomposition and allows for multiplexed imaging.

A different paradigm for fast imaging was presented by the Wilson group, where widefield images are acquired by using optical filters [145]. Unlike conventional point-scanning techniques, the authors opted to sacrifice the acquisition of the whole Raman spectrum in favor of rapidly acquiring widefield images at a few selected wavenumbers using a tunable optical bandpass filter. By imaging a single wavenumber band, without the full spectral information, it is not possible to know if the signal is truly Raman signal or if it has a different origin (i.e., fluorescence background). In this work, this problem is remedied by scanning not only at the wavenumbers specific to the SERS NPs used but also at the wavenumbers around them. By using the information of the immediate spectral neighborhood, it becomes possible to estimate the background and calculate the Raman signal from the wavenumber in question. The authors were able to use this technique to image mixtures of four distinct SERS NPs injected subcutaneously in a mouse, establishing that it can be used for multiplexed imaging. In a different paper, the group demonstrated that the technique could be used with topical application of targeted SERS nanoparticles to establish the presence of antigens on mouse xenograft models [120]. This mode of Raman imaging has the potential to image areas of clinically relevant size in timescales of less than 1 minute. It is, however, reliant on the SERS nanoprobes having strong specific *single* peaks, that are orthogonal to not only any other applied nanoparticles but also to intrinsic Raman signals.

Nanoprobe administration

The topical application of Raman nanoprobes offers an alternative to systemic (intravenous) injection. On the one hand, topical administration allows the nanoparticles to bypass the body's defense system, namely the RES, and reach the tumor directly. When the location of the tumor is known, the topical application allows for imaging the tumor site without exposing the patient to large amounts of nanoparticles—a highly desired milestone for clinical translation. On the other hand, systemic administration allows nanoparticles to circulate in the body, reaching any tumors that have access to the vasculature, even occult tumors, metastases, or microscopic lesions that are not readily visible.

A particularly challenging scenario is when microscopic metastases are present, that do not yet have access to the vasculature, and are also not readily visible, as is the case for ovarian cancer. Ovarian cancer is challenging to image and to treat, as upon diagnosis, most patients present with local metastases disseminated throughout the peritoneal cavity. Although the primary tumor can be excised, numerous microscopic metastases persist and are typically treated with locally infused chemotherapy, with mixed results. By making these microscopic lesions visible, it might become possible to excise them, or treat them via laser ablation, to the benefit of the patient. One method that can help in this scenario was presented by the Kircher group [101]. The authors used ratiometric detection of two SERRS nanoprobes to visualize microscopic tumors after intraperitoneal injection in mouse models of human ovarian cancer. Two flavors of SERRS nanoprobes were used, one targeted with an antibody against the folate receptor, and the other passivated with PEG. Using ratiometric imaging was necessary, as nanoparticles adhered onto the surfaces of the viscera nonspecifically and created a high background signal as shown in Fig. 7.9. By using spectral unmixing of the two probes, and taking the ratio of the targeted probe over the nontargeted, the background signal was removed, revealing the presence of the tumors. The authors were able to detect tumors as small as 370 μm with histological validation.

Conclusion

In the postgenomic era, medical imaging has been progressing in leaps and bounds. Medical imaging provides more imaging data than ever before, annotated and correlated with patient information, and increasingly with more molecular information. Artificial intelligence is used to analyze these images, extracting latent information, in ways unimaginable to radiologists of the past. Nanomedicine is also proving itself with numerous nanoparticle-based drug delivery agents already in the clinic or in clinical trials, and a small number of imaging contrast agents following suit.

In this context, Raman imaging with SERS nanoprobes is undoubtedly a highly promising modality, able to provide molecularly targeted imaging contrast, with microscopic resolution and high sensitivity, in a variety of imaging settings. This modality has been explored in preclinical studies for almost 2 decades. Although its potential for highly specific multiplexed imaging was identified early on, and proof-of-concept studies have been overall successful, technical challenges have been slow to overcome, causing doubts about the possibility of clinical translation of the method.

However, recent reports show that this imaging modality is gaining momentum, and the remaining challenges are one-by-one vanquished. Notably, the increased signal provided by engineering the combined resonances of the metallic core and the Raman reporter has been vital in revealing increasingly smaller malignancies, detecting signals from deep within tissue, and reducing acquisition times. Spatially offset optics were shown to extend the penetration depth of the method, dispelling one of its most common criticisms. The effectiveness of molecular targeting of the probes has been established, and the ability to detect multiple (more than 5) markers at once reported, although in excised tissues and not yet in living organisms. The plethora of reports on multimodal imaging and theranostic strategies shows that integration with other clinical modalities may be the preferred strategy of introducing Raman imaging and SERS to the clinical stage.

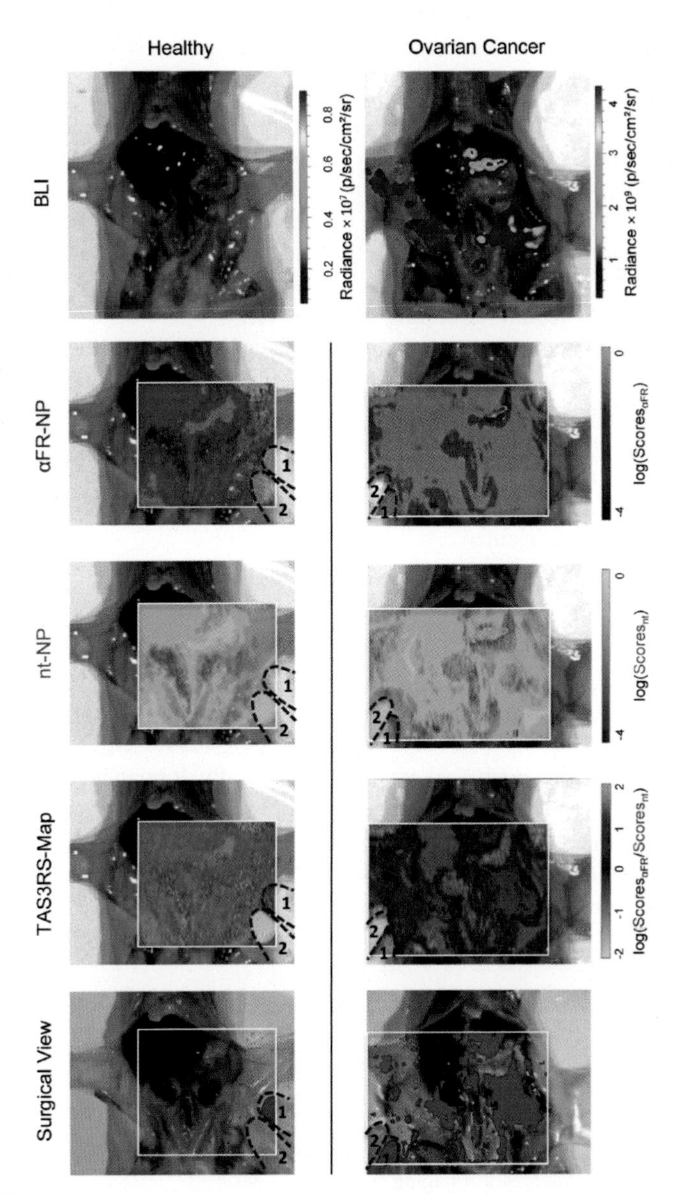

FIGURE 7.9

Topically applied surface-enhanced resonance Raman ratiometric spectroscopy for imaging of ovarian cancer in mouse models. SERS nanoprobes targeted against the folate receptor demonstrate high background signal after intraperitoneal administration (second row), as do nontargeted PEG-coated SERS nanoprobes (third row). However, the ratio of the targeted over the nontargeted probe signal reveals the tumor areas, overexpressing the folate receptor (fourth row).

Adapted from A. Oseledchyk, C. Andreou, Folate-targeted surface-enhanced resonance Raman scattering nanoprobe ratiometry for detection of microscopic ovarian cancer, ACS Nano 11 (2) (2017) 1488–1497.

Two significant obstacles remain, however, that are not directly in the hands of the scientific community to resolve:

- The first is a regulatory issue: nanoparticle agents are required to undergo extensive and rigorous testing before gaining clinical approval—and rightly so. Comparable nanoparticles, such as fluorescent silica nanodots, are already in clinical trials [146]. However, SERS nanoparticles are actively being developed, and their form, shape, and composition have not yet been standardized. This lack of uniformity complicates clinical testing, as each different nanoparticle iteration would require its testing for approval.
- The second impediment to the clinical translation of the method is a matter of economics: the lack of a rapid, widefield, clinical Raman scanner. Current Raman scanners are research-grade instruments, made for versatile analysis in a laboratory setting, but too slow for large area in vivo imaging. Although building a dedicated rapid imager for clinical use is technically possible, it currently makes no financial sense for a company to develop it—not without an approved contrast agent.

Even without resolving these two obstacles, the technique has a lot to offer, as demonstrated by the various ex vivo clinical applications with SERS nanoprobes. Additionally, the targeted nanoprobes could be employed preclinically, and as tools of basic science, revealing the molecular milieu of cancers or other pathologies, especially when it is important to image the localization of multiple targets in a living setting.

Whether clinically, in vivo, ex vivo, or preclinically, Raman imaging with SERS nanoprobes has been and will continue to be a powerful tool, integrating medical imaging and nanomedicine. By harnessing its strengths, particularly its multiplexing capabilities, it may become instrumental in acquiring real-time and dynamic molecular images, with unprecedented richness in information.

References

[1] R. Weissleder, M.J. Pittet, Imaging in the era of molecular oncology, Nature 452 (7187) (2008) 580–589.
[2] M. Ghasemi, I. Nabipour, A. Omrani, Z. Alipour, M. Assadi, Precision medicine and molecular imaging: new targeted approaches toward cancer therapeutic and diagnosis, Am. J. Nucl. Med. Mol. Imaging 6 (6) (2016) 310–327.
[3] R.J. Gillies, P.E. Kinahan, H. Hricak, Radiomics: images are more than pictures, they are data, Radiology 278 (2) (2016) 563–577.
[4] K. Heinzmann, L.M. Carter, J.S. Lewis, E.O. Aboagye, Multiplexed imaging for diagnosis and therapy, Nat. Biomed. Eng. 1 (9) (2017) 697–713.
[5] C. Andreou, R. Weissleder, M.F. Kircher, Multiplexed imaging in oncology: molecular targets in contrast, Nat. Biomed. Eng. (2021). In press.
[6] R. van der Meel, E. Sulheim, Y. Shi, F. Kiessling, W.J.M. Mulder, T. Lammers, Smart cancer nanomedicine, Nat. Nanotechnol. 14 (11) (2019) 1007–1017.

[7] C. Andreou, S. Pal, L. Rotter, J. Yang, M.F. Kircher, Molecular imaging in nanotechnology and theranostics, Mol. Imaging Biol. 19 (3) (2017) 363–372.
[8] A.C. Anselmo, S. Mitragotri, Nanoparticles in the clinic: an update, Bioeng. Transl. Med. 4 (3) (2019) e10143.
[9] J. Langer, D. Jimenez de Aberasturi, J. Aizpurua, R.A. Alvarez-Puebla, B. Auguie, J.J. Baumberg, G.C. Bazan, S.E.J. Bell, A. Boisen, A.G. Brolo, J. Choo, D. Cialla-May, V. Deckert, L. Fabris, K. Faulds, F.J. Garcia de Abajo, R. Goodacre, D. Graham, A.J. Haes, C.L. Haynes, C. Huck, T. Itoh, M. Kall, J. Kneipp, N.A. Kotov, H. Kuang, E.C. Le Ru, H.K. Lee, J.F. Li, X.Y. Ling, S.A. Maier, T. Mayerhofer, M. Moskovits, K. Murakoshi, J.M. Nam, S. Nie, Y. Ozaki, I. Pastoriza-Santos, J. Perez-Juste, J. Popp, A. Pucci, S. Reich, B. Ren, G.C. Schatz, T. Shegai, S. Schlucker, L.L. Tay, K.G. Thomas, Z.Q. Tian, R.P. Van Duyne, T. Vo-Dinh, Y. Wang, K.A. Willets, C. Xu, H. Xu, Y. Xu, Y.S. Yamamoto, B. Zhao, L.M. Liz-Marzan, Present and future of surface-enhanced Raman scattering, ACS Nano 14 (2019).
[10] C. Andreou, S.A. Kishore, M.F. Kircher, Surface-enhanced Raman spectroscopy: a new modality for cancer imaging, J. Nucl. Med. 56 (9) (2015) 1295–1299.
[11] J.R. Lombardi, R.L. Birke, A unified view of surface-enhanced Raman scattering, Acc. Chem. Res. 42 (6) (2009) 734–742.
[12] C.L. Zavaleta, B.R. Smith, I. Walton, W. Doering, G. Davis, B. Shojaei, M.J. Natan, S.S. Gambhir, Multiplexed imaging of surface enhanced Raman scattering nanotags in living mice using noninvasive Raman spectroscopy, Proc. Natl. Acad. Sci. USA 106 (32) (2009) 13511–13516.
[13] C. Andreou, A. Oseledchyk, F. Nicolson, N. Berisha, S. Pal, M.F. Kircher, Surface-enhanced resonance Raman scattering nanoprobe ratiometry for detecting microscopic ovarian cancer via folate receptor targeting, J. Vis. Exp. 145 (2019) e58389.
[14] M. Hadjidemetriou, K. Kostarelos, Nanomedicine: evolution of the nanoparticle corona, Nat. Nanotechnol. 12 (4) (2017) 288–290.
[15] J.K. Patra, G. Das, L.F. Fraceto, E.V.R. Campos, M.D.P. Rodriguez-Torres, L.S. Acosta-Torres, L.A. Diaz-Torres, R. Grillo, M.K. Swamy, S. Sharma, S. Habtemariam, H.S. Shin, Nano based drug delivery systems: recent developments and future prospects, J. Nanobiotechnol. 16 (1) (2018) 71.
[16] J.M. Harris, R.B. Chess, Effect of pegylation on pharmaceuticals, Nat. Rev. Drug Discov. 2 (3) (2003) 214–221.
[17] S. Zanganeh, R. Spitler, M. Erfanzadeh, A.M. Alkilany, M. Mahmoudi, Protein corona: opportunities and challenges, Int. J. Biochem. Cell Biol. 75 (2016) 143–147.
[18] J.W. Nichols, Y.H. Bae, EPR: evidence and fallacy, J. Contr. Release 190 (2014) 451–464.
[19] Y. Matsumura, H. Maeda, A new concept for macromolecular therapeutics in cancer chemotherapy: mechanism of tumoritropic accumulation of proteins and the antitumor agent smancs, Cancer Res. 46 (12 Pt 1) (1986) 6387–6392.
[20] H. Maeda, The enhanced permeability and retention (EPR) effect in tumor vasculature: the key role of tumor-selective macromolecular drug targeting, Adv. Enzyme Regul. 41 (2001) 189–207.
[21] H. Maeda, SMANCS and polymer-conjugated macromolecular drugs: advantages in cancer chemotherapy, Adv. Drug Deliv. Rev. 46 (1–3) (2001) 169–185.
[22] J. Fang, T. Sawa, H. Maeda, Factors and mechanism of "EPR" effect and the enhanced antitumor effects of macromolecular drugs including SMANCS, Adv. Exp. Med. Biol. 519 (2003) 29–49.

[23] H. Maeda, G.Y. Bharate, J. Daruwalla, Polymeric drugs for efficient tumor-targeted drug delivery based on EPR-effect, Eur. J. Pharm. Biopharm. 71 (3) (2009) 409–419.
[24] A.K. Iyer, G. Khaled, J. Fang, H. Maeda, Exploiting the enhanced permeability and retention effect for tumor targeting, Drug Discov. Today 11 (17–18) (2006) 812–818.
[25] H. Maeda, Vascular permeability in cancer and infection as related to macromolecular drug delivery, with emphasis on the EPR effect for tumor-selective drug targeting, Proc. Jpn. Acad. Ser. B Phys. Biol. Sci. 88 (3) (2012) 53–71.
[26] S. Harmsen, R. Huang, M.A. Wall, H. Karabeber, J.M. Samii, M. Spaliviero, J.R. White, S. Monette, R. O'Connor, K.L. Pitter, S.A. Sastra, M. Saborowski, E.C. Holland, S. Singer, K.P. Olive, S.W. Lowe, R.G. Blasberg, M.F. Kircher, Surface-enhanced resonance Raman scattering nanostars for high-precision cancer imaging, Sci. Transl. Med. 7 (271) (2015) 271ra7.
[27] D. Zardavas, A. Irrthum, C. Swanton, M. Piccart, Clinical management of breast cancer heterogeneity, Nat. Rev. Clin. Oncol. 12 (7) (2015) 381–394.
[28] A.A. Alizadeh, V. Aranda, A. Bardelli, C. Blanpain, C. Bock, C. Borowski, C. Caldas, A. Califano, M. Doherty, M. Elsner, M. Esteller, R. Fitzgerald, J.O. Korbel, P. Lichter, C.E. Mason, N. Navin, D. Pe'er, K. Polyak, C.W. Roberts, L. Siu, A. Snyder, H. Stower, C. Swanton, R.G. Verhaak, J.C. Zenklusen, J. Zuber, J. Zucman-Rossi, Toward understanding and exploiting tumor heterogeneity, Nat. Med. 21 (8) (2015) 846–853.
[29] J. Park, Y. Choi, H. Chang, W. Um, J.H. Ryu, I.C. Kwon, Alliance with EPR effect: combined strategies to improve the EPR effect in the tumor microenvironment, Theranostics 9 (26) (2019) 8073–8090.
[30] N.D. Marjanovic, R.A. Weinberg, C.L. Chaffer, Cell plasticity and heterogeneity in cancer, Clin. Chem. 59 (1) (2013) 168–179.
[31] S.K. Golombek, J.N. May, B. Theek, L. Appold, N. Drude, F. Kiessling, T. Lammers, Tumor targeting via EPR: strategies to enhance patient responses, Adv. Drug Deliv. Rev. 130 (2018) 17–38.
[32] T. Safra, F. Muggia, S. Jeffers, D.D. Tsao-Wei, S. Groshen, O. Lyass, R. Henderson, G. Berry, A. Gabizon, Pegylated liposomal doxorubicin (doxil): reduced clinical cardiotoxicity in patients reaching or exceeding cumulative doses of 500 mg/m^2, Ann. Oncol. 11 (8) (2000) 1029–1033.
[33] Y. Barenholz, Doxil(R)—the first FDA-approved nano-drug: lessons learned, J. Contr. Release 160 (2) (2012) 117–134.
[34] W.J. Gradishar, Albumin-bound paclitaxel: a next-generation taxane, Expert Opin. Pharmacother. 7 (8) (2006) 1041–1053.
[35] H.M. Patel, S.M. Moghimi, Serum-mediated recognition of liposomes by phagocytic cells of the reticuloendothelial system - the concept of tissue specificity, Adv. Drug Deliv. Rev. 32 (1–2) (1998) 45–60.
[36] K. Yaehne, A. Tekrony, A. Clancy, Y. Gregoriou, J. Walker, K. Dean, T. Nguyen, A. Doiron, K. Rinker, X.Y. Jiang, S. Childs, D. Cramb, Nanoparticle accumulation in angiogenic tissues: towards predictable pharmacokinetics, Small 9 (18) (2013) 3118–3127.
[37] J.M. Carter, K.E. Driscoll, The role of inflammation, oxidative stress, and proliferation in silica-induced lung disease: a species comparison, J. Environ. Pathol. Toxicol. Oncol. 20 (Suppl. 1) (2001) 33–43.
[38] W.S. Cho, M. Choi, B.S. Han, M. Cho, J. Oh, K. Park, S.J. Kim, S.H. Kim, J. Jeong, Inflammatory mediators induced by intratracheal instillation of ultrafine amorphous silica particles, Toxicol. Lett. 175 (1–3) (2007) 24–33.

[39] K.E. Driscoll, TNFalpha and MIP-2: role in particle-induced inflammation and regulation by oxidative stress, Toxicol. Lett. 112–113 (2000) 177–183.
[40] R.P. Nishanth, R.G. Jyotsna, J.J. Schlager, S.M. Hussain, P. Reddanna, Inflammatory responses of RAW 264.7 macrophages upon exposure to nanoparticles: role of ROS-NF kappa B signaling pathway, Nanotoxicology 5 (4) (2011) 502–516.
[41] G.A. Orr, W.B. Chrisler, K.J. Cassens, R. Tan, B.J. Tarasevich, L.M. Markillie, R.C. Zangar, B.D. Thrall, Cellular recognition and trafficking of amorphous silica nanoparticles by macrophage scavenger receptor A, Nanotoxicology 5 (3) (2011) 296–311.
[42] J. Pajarinen, V.P. Kouri, E. Jamsen, T.F. Li, J. Mandelin, Y.T. Konttinen, The response of macrophages to titanium particles is determined by macrophage polarization, Acta Biomater. 9 (11) (2013) 9229–9240.
[43] E.J. Park, K. Park, Oxidative stress and pro-inflammatory responses induced by silica nanoparticles in vivo and in vitro, Toxicol. Lett. 184 (1) (2009) 18–25.
[44] G. Sharma, D.T. Valenta, Y. Altman, S. Harvey, H. Xie, S. Mitragotri, J.W. Smith, Polymer particle shape independently influences binding and internalization by macrophages, J. Contr. Release 147 (3) (2010) 408–412.
[45] F. Velard, J. Braux, J. Amedee, P. Laquerriere, Inflammatory cell response to calcium phosphate biomaterial particles: an overview, Acta Biomater. 9 (2) (2013) 4956–4963.
[46] A.A. Clancy, Y. Gregoriou, K. Yaehne, D.T. Cramb, Measuring properties of nanoparticles in embryonic blood vessels: towards a physicochemical basis for nanotoxicity, Chem. Phys. Lett. 488 (4–6) (2010) 99–111.
[47] A. Albanese, P.S. Tang, W.C.W. Chan, The effect of nanoparticle size, shape, and surface chemistry on biological systems, Annu. Rev. Biomed. Eng. 14 (2012) 1–16.
[48] X.S. Liu, N. Huang, H. Li, Q. Jin, J. Ji, Surface and size effects on cell interaction of gold nanoparticles with both phagocytic and nonphagocytic cells, Langmuir 29 (29) (2013) 9138–9148.
[49] H.S. Choi, W. Liu, P. Misra, E. Tanaka, J.P. Zimmer, B.I. Ipe, M.G. Bawendi, J.V. Frangioni, Renal clearance of quantum dots, Nat. Biotechnol. 25 (10) (2007) 1165–1170.
[50] J.E. Zuckerman, C.H.J. Choi, H. Han, M.E. Davis, Polycation-siRNA nanoparticles can disassemble at the kidney glomerular basement membrane, Proc. Natl. Acad. Sci USA 109 (8) (2012) 3137–3142.
[51] N. Hoshyar, S. Gray, H.B. Han, G. Bao, The effect of nanoparticle size on in vivo pharmacokinetics and cellular interaction, Nanomedicine 11 (6) (2016) 673–692.
[52] S.A. Kulkarni, S.S. Feng, Effects of particle size and surface modification on cellular uptake and biodistribution of polymeric nanoparticles for drug delivery, Pharm. Res. 30 (10) (2013) 2512–2522.
[53] A.H. Faraji, P. Wipf, Nanoparticles in cellular drug delivery, Bioorgan. Med. Chem. 17 (8) (2009) 2950–2962.
[54] A.L.B. de Barros, A. Tsourkas, B. Saboury, V.N. Cardoso, A. Alavi, Emerging role of radiolabeled nanoparticles as an effective diagnostic technique, EJNMMI Res. 2 (2012).
[55] S.M. Moghimi, A.C. Hunter, T.L. Andresen, Factors controlling nanoparticle pharmacokinetics: an integrated analysis and perspective, Annu. Rev. Pharmacol. 52 (2012) 481–503.
[56] S. Tenzer, D. Docter, J. Kuharev, A. Musyanovych, V. Fetz, R. Hecht, F. Schlenk, D. Fischer, K. Kiouptsi, C. Reinhardt, K. Landfester, H. Schild, M. Maskos, S.K. Knauer, R.H. Stauber, Rapid formation of plasma protein corona critically affects nanoparticle pathophysiology, Nat. Nanotechnol. 8 (10) (2013) 772–781.

[57] A.E. Nel, L. Madler, D. Velegol, T. Xia, E.M.V. Hoek, P. Somasundaran, F. Klaessig, V. Castranova, M. Thompson, Understanding biophysicochemical interactions at the nano-bio interface, Nat. Mater. 8 (7) (2009) 543–557.

[58] M.E. Akerman, W.C. Chan, P. Laakkonen, S.N. Bhatia, E. Ruoslahti, Nanocrystal targeting in vivo, Proc. Natl. Acad. Sci. USA 99 (20) (2002) 12617–12621.

[59] L. Sercombe, T. Veerati, F. Moheimani, S.Y. Wu, A.K. Sood, S. Hua, Advances and challenges of liposome assisted drug delivery, Front. Pharmacol. 6 (2015) 286.

[60] C.D. Walkey, W.C. Chan, Understanding and controlling the interaction of nanomaterials with proteins in a physiological environment, Chem. Soc. Rev. 41 (7) (2012) 2780–2799.

[61] N.K. Banda, G. Mehta, Y. Chao, G. Wang, S. Inturi, L. Fossati-Jimack, M. Botto, L. Wu, S.M. Moghimi, D. Simberg, Mechanisms of complement activation by dextran-coated superparamagnetic iron oxide (SPIO) nanoworms in mouse versus human serum, Part. Fibre Toxicol. 11 (2014) 64.

[62] S. Tong, S. Hou, Z. Zheng, J. Zhou, G. Bao, Coating optimization of superparamagnetic iron oxide nanoparticles for high T2 relaxivity, Nano Lett. 10 (11) (2010) 4607–4613.

[63] Y. Kawamori, O. Matsui, M. Kadoya, J. Yoshikawa, H. Demachi, T. Takashima, Differentiation of hepatocellular carcinomas from hyperplastic nodules induced in rat liver with ferrite-enhanced MR imaging, Radiology 183 (1) (1992) 65–72.

[64] C. Andreou, V. Neuschmelting, D.F. Tschaharganeh, C.H. Huang, A. Oseledchyk, P. Iacono, H. Karabeber, R.R. Colen, L. Mannelli, S.W. Lowe, M.F. Kircher, Imaging of liver tumors using surface-enhanced Raman scattering nanoparticles, Acs Nano 10 (5) (2016) 5015–5026.

[65] M. Spaliviero, S. Harmsen, R.M. Huang, M.A. Wall, C. Andreou, J.A. Eastham, K.A. Touijer, P.T. Scardino, M.F. Kircher, Detection of lymph node metastases with SERRS nanoparticles, Mol. Imaging Biol. 18 (5) (2016) 677–685.

[66] P. Sun, Y. Zhang, K. Li, C. Wang, F. Zeng, J. Zhu, Y. Wu, X. Tao, Image-guided surgery of head and neck carcinoma in rabbit models by intra-operatively defining tumour-infiltrated margins and metastatic lymph nodes, EBioMedicine 50 (2019) 93–102.

[67] I. Posadas, S. Monteagudo, V. Cena, Nanoparticles for brain-specific drug and genetic material delivery, imaging and diagnosis, Nanomedicine 11 (7) (2016) 833–849.

[68] M.M. Patel, B.M. Patel, Crossing the blood-brain barrier: recent advances in drug delivery to the brain, CNS Drugs 31 (2) (2017) 109–133.

[69] X.W. Dong, Current strategies for brain drug delivery, Theranostics 8 (6) (2018) 1481–1493.

[70] S. Zanganeh, P. Georgala, C. Corbo, L. Arabi, J.Q. Ho, N. Javdani, M.R. Sepand, K. Cruickshank, L.F. Campesato, C.H. Weng, S. Hemayat, C. Andreou, R. Alvim, G. Hutter, M. Rafat, M. Mahmoudi, Immunoengineering in glioblastoma imaging and therapy, Wiley Interdiscip. Rev. Nanomed. Nanobiotechnol. 11 (6) (2019) e1575.

[71] O. Betzer, M. Shilo, R. Opochinsky, E. Barnoy, M. Motiei, E. Okun, G. Yadid, R. Popovtzer, The effect of nanoparticle size on the ability to cross the blood-brain barrier: an in vivo study, Nanomedicine 12 (13) (2017) 1533–1546.

[72] H.L. Ou, T.J. Cheng, Y.M. Zhang, J.J. Liu, Y.X. Ding, J.R. Zhen, W.Z. Shen, Y.J. Xu, W.Z. Yang, P. Niu, J.F. Liu, Y.L. An, Y. Liu, L.Q. Shi, Surface-adaptive zwitterionic nanoparticles for prolonged blood circulation time and enhanced cellular uptake in tumor cells, Acta Biomater. 65 (2018) 339–348.

[73] T.T. Lin, P.F. Zhao, Y.F. Jiang, Y.S. Tang, H.Y. Jin, Z.Z. Pan, H.N. He, V.C. Yang, Y.Z. Huang, Blood-brain-barrier-penetrating albumin nanoparticles for biomimetic drug delivery via albumin-binding protein pathways for antiglioma therapy, ACS Nano 10 (11) (2016) 9999−10012.

[74] D.Z. Liu, Y. Cheng, R.Q. Cai, W.W. Wang, H. Cui, M. Liu, B.L. Zhang, Q.B. Mei, S.Y. Zhou, The enhancement of siPLK1 penetration across BBB and its anti glioblastoma activity in vivo by magnet and transferrin co-modified nanoparticle, Nanomed. Nanotechnol. 14 (3) (2018) 991−1003.

[75] A. Burgess, K. Hynynen, Drug delivery across the blood-brain barrier using focused ultrasound, Expert Opin. Drug Deliv. 11 (5) (2014) 711−721.

[76] Q.G. He, J. Liu, J. Liang, X.P. Liu, W. Li, Z. Liu, Z.Y. Ding, D. Tuo, Towards improvements for penetrating the blood-brain barrier-recent progress from a material and pharmaceutical perspective, Cells 7 (4) (2018).

[77] Y.H. Zhang, J.B. Walker, Z. Minic, F.C. Liu, H. Goshgarian, G.Z. Mao, Transporter protein and drug-conjugated gold nanoparticles capable of bypassing the blood-brain barrier, Sci. Rep. 6 (2016).

[78] M.F. Kircher, A. de la Zerda, J.V. Jokerst, C.L. Zavaleta, P.J. Kempen, E. Mittra, K. Pitter, R. Huang, C. Campos, F. Habte, R. Sinclair, C.W. Brennan, I.K. Mellinghoff, E.C. Holland, S.S. Gambhir, A brain tumor molecular imaging strategy using a new triple-modality MRI-photoacoustic-Raman nanoparticle, Nat. Med. 18 (5) (2012) 829−834.

[79] H. Karabeber, R. Huang, P. Iacono, J.M. Samii, K. Pitter, E.C. Holland, M.F. Kircher, Guiding brain tumor resection using surface-enhanced Raman scattering nanoparticles and a hand-held Raman scanner, ACS Nano 8 (10) (2014) 9755−9766.

[80] R. Huang, S. Harmsen, J.M. Samii, H. Karabeber, K.L. Pitter, E.C. Holland, M.F. Kircher, High precision imaging of microscopic spread of glioblastoma with a targeted ultrasensitive SERRS molecular imaging probe, Theranostics 6 (8) (2016) 1075−1084.

[81] V. Neuschmelting, S. Harmsen, N. Beziere, H. Lockau, H.-T. Hsu, R. Huang, D. Razansky, V. Ntziachristos, M.F. Kircher, Dual-modality surface-enhanced resonance Raman scattering and multispectral optoacoustic tomography nanoparticle approach for brain tumor delineation, Small 14 (23) (2018) 1800740.

[82] Q. Yue, X. Gao, Y. Yu, Y. Li, W. Hua, K. Fan, R. Zhang, J. Qian, L. Chen, C. Li, Y. Mao, An EGFRvIII targeted dual-modal gold nanoprobe for imaging-guided brain tumor surgery, Nanoscale 9 (23) (2017) 7930−7940.

[83] R.J. Diaz, P.Z. McVeigh, M.A. O'Reilly, K. Burrell, M. Bebenek, C. Smith, A.B. Etame, G. Zadeh, K. Hynynen, B.C. Wilson, J.T. Rutka, Focused ultrasound delivery of Raman nanoparticles across the blood-brain barrier: potential for targeting experimental brain tumors, Nanomedicine 10 (5) (2014) 1075−1087.

[84] X. Gao, Q. Yue, Z. Liu, M. Ke, X. Zhou, S. Li, J. Zhang, R. Zhang, L. Chen, Y. Mao, C. Li, Guiding brain-tumor surgery via blood-brain-barrier-permeable gold nanoprobes with acid-triggered MRI/SERRS signals, Adv. Mater. 29 (21) (2017).

[85] N.J. Butcher, G.M. Mortimer, R.F. Minchin, Unravelling the stealth effect, Nat. Nanotechnol. 11 (4) (2016) 310−311.

[86] L.A. Lane, X. Qian, S. Nie, SERS nanoparticles in medicine: from label-free detection to spectroscopic tagging, Chem. Rev. 115 (19) (2015) 10489−10529.

[87] L. Xiao, K.A. Bailey, H. Wang, Z.D. Schultz, Probing membrane receptor-ligand specificity with surface- and tip- enhanced Raman scattering, Anal. Chem. 89 (17) (2017) 9091−9099.

[88] M. Vendrell, K.K. Maiti, K. Dhaliwal, Y.T. Chang, Surface-enhanced Raman scattering in cancer detection and imaging, Trends Biotechnol. 31 (4) (2013) 249–257.
[89] Y.W. Wang, S. Kang, A. Khan, P.Q. Bao, J.T.C. Liu, In vivo multiplexed molecular imaging of esophageal cancer via spectral endoscopy of topically applied SERS nanoparticles, Biomed. Opt. Express 6 (10) (2015) 3714–3723.
[90] S. Kang, Y. Wang, N.P. Reder, J.T. Liu, Multiplexed molecular imaging of biomarker-targeted SERS nanoparticles on fresh tissue specimens with channel-compressed spectrometry, PLoS One 11 (9) (2016) e0163473.
[91] Y.W. Wang, N.P. Reder, S. Kang, A.K. Glaser, J.T.C. Liu, Multiplexed optical imaging of tumor-directed nanoparticles: a review of imaging systems and approaches, Nanotheranostics 1 (4) (2017) 369–388.
[92] Y.W. Wang, J.D. Doerksen, S. Kang, D. Walsh, Q. Yang, D. Hong, J.T. Liu, Multiplexed molecular imaging of fresh tissue surfaces enabled by convection-enhanced topical staining with SERS-coded nanoparticles, Small 12 (40) (2016) 5612–5621.
[93] Y.W. Wang, N.P. Reder, S. Kang, A.K. Glaser, Q. Yang, M.A. Wall, S.H. Javid, S.M. Dintzis, J.T.C. Liu, Raman-encoded molecular imaging with topically applied SERS nanoparticles for intraoperative guidance of lumpectomy, Cancer Res. 77 (16) (2017) 4506–4516.
[94] Y. Wang, S. Kang, J.D. Doerksen, A.K. Glaser, J.T.C. Liu, Surgical guidance via multiplexed molecular imaging of fresh tissues labeled with SERS-coded nanoparticles, IEEE J. Sel. Top. Quantum Electron. 22 (4) (2016) 6802911.
[95] R.M. Davis, B. Kiss, D.R. Trivedi, T.J. Metzner, J.C. Liao, S.S. Gambhir, Surface-enhanced Raman scattering nanoparticles for multiplexed imaging of bladder cancer tissue permeability and molecular phenotype, ACS Nano 12 (10) (2018) 9669–9679.
[96] X. Qian, X.-H. Peng, D.O. Ansari, Q. Yin-Goen, G.Z. Chen, D.M. Shin, L. Yang, A.N. Young, M.D. Wang, S. Nie, In vivo tumor targeting and spectroscopic detection with surface-enhanced Raman nanoparticle tags, Nat. Biotechnol. 26 (1) (2008) 83–90.
[97] U.S. Dinish, G. Balasundaram, Y.-T. Chang, M. Olivo, Actively targeted in vivo multiplex detection of intrinsic cancer biomarkers using biocompatible SERS nanotags, Sci. Rep. 4 (1) (2014) 4075.
[98] S. Harmsen, M.A. Bedics, M.A. Wall, R. Huang, M.R. Detty, M.F. Kircher, Rational design of a chalcogenopyrylium-based surface-enhanced resonance Raman scattering nanoprobe with attomolar sensitivity, Nat. Commun. 6 (2015) 6570.
[99] T.R. Nayak, C. Andreou, A. Oseledchyk, W.D. Marcus, H.C. Wong, J. Massagué, M.F. Kircher, Tissue factor-specific ultra-bright SERRS nanostars for Raman detection of pulmonary micrometastases, Nanoscale 9 (3) (2017) 1110–1119.
[100] M. Li, J. Wu, M. Ma, Z. Feng, Z. Mi, P. Rong, D. Liu, Alkyne- and nitrile-anchored gold nanoparticles for multiplex SERS imaging of biomarkers in cancer cells and tissues, Nanotheranostics 3 (1) (2019) 113–119.
[101] A. Oseledchyk, C. Andreou, Folate-targeted surface-enhanced resonance Raman scattering nanoprobe ratiometry for detection of microscopic ovarian cancer, ACS Nano 11 (2) (2017) 1488–1497.
[102] R.M. Davis, J.L. Campbell, S. Burkitt, Z. Qiu, S. Kang, M. Mehraein, D. Miyasato, H. Salinas, J.T.C. Liu, A Raman imaging approach using CD47 antibody-labeled SERS nanoparticles for identifying breast cancer and its potential to guide surgical resection, Nanomaterials 8 (11) (2018).

[103] J. Feng, L. Chen, Y. Xia, J. Xing, Z. Li, Q. Qian, Y. Wang, A. Wu, L. Zeng, Y. Zhou, Bioconjugation of gold nanobipyramids for SERS detection and targeted photothermal therapy in breast cancer, ACS Biomater. Sci. Eng. 3 (4) (2017) 608−618.

[104] S. Pal, S. Harmsen, A. Oseledchyk, H.-T. Hsu, M.F. Kircher, MUC1 aptamer targeted SERS nanoprobes, Adv. Funct. Mater. 27 (32) (2017) 1606632.

[105] J. Wu, D. Liang, Q. Jin, J. Liu, M. Zheng, X. Duan, X. Tang, Bioorthogonal SERS nanoprobes for mulitplex spectroscopic detection, tumor cell targeting, and tissue imaging, Chem. Eur. J. 21 (37) (2015) 12914−12918.

[106] S.E. Bohndiek, A. Wagadarikar, C.L. Zavaleta, D. Van de Sompel, E. Garai, J.V. Jokerst, S. Yazdanfar, S.S. Gambhir, A small animal Raman instrument for rapid, wide-area, spectroscopic imaging, Proc. Natl. Acad. Sci. 110 (30) (2013) 12408.

[107] F. Nicolson, B. Andreiuk, C. Andreou, H.T. Hsu, S. Rudder, M.F. Kircher, Non-invasive in vivo imaging of cancer using surface-enhanced spatially offset Raman spectroscopy (SESORS), Theranostics 9 (20) (2019) 5899−5913.

[108] A.-I. Henry, B. Sharma, M.F. Cardinal, D. Kurouski, R.P. Van Duyne, Surface-enhanced Raman spectroscopy biosensing: in vivo diagnostics and multimodal imaging, Anal. Chem. 88 (13) (2016) 6638−6647.

[109] Z. Wang, J. Zhang, H. Wang, J. Hai, B. Wang, Se atom-induced synthesis of concave spherical Fe_3O_4@Cu_2O nanocrystals for highly efficient MRI−SERS imaging-guided NIR photothermal therapy, Part. Part. Syst. Charact. 35 (11) (2018) 1800197.

[110] T. Köker, N. Tang, C. Tian, W. Zhang, X. Wang, R. Martel, F. Pinaud, Cellular imaging by targeted assembly of hot-spot SERS and photoacoustic nanoprobes using split-fluorescent protein scaffolds, Nat. Commun. 9 (1) (2018) 607.

[111] Q. Li, A.K. Parchur, A. Zhou, In vitro biomechanical properties, fluorescence imaging, surface-enhanced Raman spectroscopy, and photothermal therapy evaluation of luminescent functionalized CaMoO4:Eu@Au hybrid nanorods on human lung adenocarcinoma epithelial cells, Sci. Technol. Adv. Mater. 17 (1) (2016) 346−360.

[112] M. Potara, T. Nagy-Simon, A.M. Craciun, S. Suarasan, E. Licarete, F. Imre-Lucaci, S. Astilean, Carboplatin-loaded, Raman-encoded, Chitosan-coated silver nanotriangles as multimodal traceable nanotherapeutic delivery systems and pH reporters inside human ovarian cancer cells, ACS Appl. Mater. Interfaces 9 (38) (2017) 32565−32576.

[113] A.K. Parchur, Q. Li, A. Zhou, Near-infrared photothermal therapy of Prussian-blue-functionalized lanthanide-ion-doped inorganic/plasmonic multifunctional nanostructures for the selective targeting of HER2-expressing breast cancer cells, Biomater. Sci. 4 (12) (2016) 1781−1791.

[114] S. Sasidharan, D. Bahadur, R. Srivastava, Albumin stabilized gold nanostars: a biocompatible nanoplatform for SERS, CT imaging and photothermal therapy of cancer, RSC Adv. 6 (87) (2016) 84025−84034.

[115] Z.A. Nima, M. Mahmood, Y. Xu, T. Mustafa, F. Watanabe, D.A. Nedosekin, M.A. Juratli, T. Fahmi, E.I. Galanzha, J.P. Nolan, A.G. Basnakian, V.P. Zharov, A.S. Biris, Circulating tumor cell identification by functionalized silver-gold nanorods with multicolor, super-enhanced SERS and photothermal resonances, Sci. Rep. 4 (1) (2014) 4752.

[116] S. Boca-Farcau, M. Potara, T. Simon, A. Juhem, P. Baldeck, S. Astilean, Folic acid-conjugated, SERS-labeled silver nanotriangles for multimodal detection and targeted photothermal treatment on human ovarian cancer cells, Mol. Pharmaceut. 11 (2) (2014) 391−399.

[117] T. Zhou, B. Wu, D. Xing, Bio-modified Fe_3O_4 core/Au shell nanoparticles for targeting and multimodal imaging of cancer cells, J. Mater. Chem. 22 (2) (2012) 470–477.

[118] J.S. Wi, E.S. Barnard, R.J. Wilson, M. Zhang, M. Tang, M.L. Brongersma, S.X. Wang, Sombrero-shaped plasmonic nanoparticles with molecular-level sensitivity and multifunctionality, ACS Nano 5 (8) (2011) 6449–6457.

[119] Y. Liu, Z. Chang, H. Yuan, A.M. Fales, T. Vo-Dinh, Quintuple-modality (SERS-MRI-CT-TPL-PTT) plasmonic nanoprobe for theranostics, Nanoscale 5 (24) (2013) 12126–12131.

[120] R.J. Mallia, P.Z. McVeigh, C.J. Fisher, I. Veilleux, B.C. Wilson, Wide-field multiplexed imaging of EGFR-targeted cancers using topical application of NIR SERS nanoprobes, Nanomedicine 10 (1) (2015) 89–101.

[121] C. Pohling, J.L. Campbell, T.A. Larson, D. Van de Sompel, J. Levi, M.H. Bachmann, S.E. Bohndiek, J.V. Jokerst, S.S. Gambhir, Smart-dust-nanorice for enhancement of endogenous Raman signal, contrast in photoacoustic imaging, and T2-shortening in magnetic resonance imaging, Small 14 (19) (2018) e1703683.

[122] Y. Cui, X.-S. Zheng, B. Ren, R. Wang, J. Zhang, N.-S. Xia, Z.-Q. Tian, Au@organosilica multifunctional nanoparticles for the multimodal imaging, Chem. Sci. 2 (8) (2011) 1463–1469.

[123] Z. Wang, S. Zong, H. Chen, C. Wang, S. Xu, Y. Cui, SERS-fluorescence joint spectral encoded magnetic nanoprobes for multiplex cancer cell separation, Adv. Healthc. Mater. 3 (11) (2014) 1889–1897.

[124] K.-Y. Ju, S. Lee, J. Pyo, J. Choo, J.-K. Lee, Bio-inspired development of a dual-mode nanoprobe for MRI and Raman imaging, Small 11 (1) (2015) 84–89.

[125] A. Carrouée, E. Allard-Vannier, S. Même, F. Szeremeta, J.-C. Beloeil, I. Chourpa, Sensitive trimodal magnetic resonance imaging-surface-enhanced resonance Raman scattering-fluorescence detection of cancer cells with stable magneto-plasmonic nanoprobes, Anal. Chem. 87 (22) (2015) 11233–11241.

[126] A.M. Fales, H. Yuan, T. Vo-Dinh, Silica-coated gold nanostars for combined surface-enhanced Raman scattering (SERS) detection and singlet-oxygen generation: a potential nanoplatform for theranostics, Langmuir 27 (19) (2011) 12186–12190.

[127] V. Amendola, S. Scaramuzza, L. Litti, M. Meneghetti, G. Zuccolotto, A. Rosato, E. Nicolato, P. Marzola, G. Fracasso, C. Anselmi, M. Pinto, M. Colombatti, Magneto-plasmonic Au-Fe alloy nanoparticles designed for multimodal SERS-MRI-CT imaging, Small 10 (12) (2014) 2476–2486.

[128] J.L. Campbell, E.D. SoRelle, O. Ilovich, O. Liba, M.L. James, Z. Qiu, V. Perez, C.T. Chan, A. de la Zerda, C. Zavaleta, Multimodal assessment of SERS nanoparticle biodistribution post ingestion reveals new potential for clinical translation of Raman imaging, Biomaterials 135 (2017) 42–52.

[129] L. Litti, N. Rivato, G. Fracasso, P. Bontempi, E. Nicolato, P. Marzola, A. Venzo, M. Colombatti, M. Gobbo, M. Meneghetti, A SERRS/MRI multimodal contrast agent based on naked Au nanoparticles functionalized with a Gd(iii) loaded PEG polymer for tumor imaging and localized hyperthermia, Nanoscale 10 (3) (2018) 1272–1278.

[130] M.A. Wall, T.M. Shaffer, S. Harmsen, D.-F. Tschaharganeh, C.-H. Huang, S.W. Lowe, C.M. Drain, M.F. Kircher, Chelator-free radiolabeling of SERRS nanoparticles for whole-body PET and intraoperative Raman imaging, Theranostics 7 (12) (2017) 3068–3077.

[131] S. Pal, A. Ray, C. Andreou, Y. Zhou, T. Rakshit, M. Wlodarczyk, M. Maeda, R. Toledo-Crow, N. Berisha, J. Yang, H.-T. Hsu, A. Oseledchyk, J. Mondal, S. Zou,

M.F. Kircher, DNA-enabled rational design of fluorescence-Raman bimodal nanoprobes for cancer imaging and therapy, Nat. Commun. 10 (1) (2019) 1926.

[132] W. Shi, R.J. Paproski, P. Shao, A. Forbrich, J.D. Lewis, R.J. Zemp, Multimodality Raman and photoacoustic imaging of surface-enhanced-Raman-scattering-targeted tumor cells, J. Biomed. Opt. 21 (2) (2016) 20503.

[133] A.S. Thakor, R. Luong, R. Paulmurugan, F.I. Lin, P. Kempen, C. Zavaleta, P. Chu, T.F. Massoud, R. Sinclair, S.S. Gambhir, The fate and toxicity of Raman-active silica-gold nanoparticles in mice, Sci. Transl. Med. 3 (79) (2011) 79ra33.

[134] E. Phillips, O. Penate-Medina, P.B. Zanzonico, R.D. Carvajal, P. Mohan, Y. Ye, J. Humm, M. Gonen, H. Kalaigian, H. Schoder, H.W. Strauss, S.M. Larson, U. Wiesner, M.S. Bradbury, Clinical translation of an ultrasmall inorganic optical-PET imaging nanoparticle probe, Sci. Transl. Med. 6 (260) (2014) 260ra149.

[135] P. Singh, S. Pandit, V. Mokkapati, A. Garg, V. Ravikumar, I. Mijakovic, Gold nanoparticles in diagnostics and therapeutics for human cancer, Int. J. Mol. Sci. 19 (7) (2018).

[136] C. Lopez-Chaves, J. Soto-Alvaredo, M. Montes-Bayon, J. Bettmer, J. Llopis, C. Sanchez-Gonzalez, Gold nanoparticles: distribution, bioaccumulation and toxicity. In vitro and in vivo studies, Nanomedicine 14 (1) (2018) 1−12.

[137] A. Stallmach, C. Schmidt, A. Watson, R. Kiesslich, An unmet medical need: advances in endoscopic imaging of colorectal neoplasia, J. Biophoton. 4 (7−8) (2011) 482−489.

[138] C.L. Zavaleta, E. Garai, J.T. Liu, S. Sensarn, M.J. Mandella, D. Van de Sompel, S. Friedland, J. Van Dam, C.H. Contag, S.S. Gambhir, A Raman-based endoscopic strategy for multiplexed molecular imaging, Proc. Natl. Acad. Sci. USA 110 (25) (2013) E2288−E2297.

[139] S. Jeong, Y.I. Kim, H. Kang, G. Kim, M.G. Cha, H. Chang, K.O. Jung, Y.H. Kim, B.H. Jun, D.W. Hwang, Y.S. Lee, H. Youn, Y.S. Lee, K.W. Kang, D.S. Lee, D.H. Jeong, Fluorescence-Raman dual modal endoscopic system for multiplexed molecular diagnostics, Sci. Rep. 5 (2015) 9455.

[140] S. Harmsen, S. Rogalla, R.M. Huang, M. Spaliviero, V. Neuschmelting, Y. Hayakawa, Y. Lee, Y. Tailor, R. Toledo-Crow, J.W. Kang, J.M. Samii, H. Karabeber, R.M. Davis, J.R. White, M. van de Rijn, S.S. Gambhir, C.H. Contag, T.C. Wang, M.F. Kircher, Detection of premalignant gastrointestinal lesions using surface-enhanced resonance Raman scattering-nanoparticle endoscopy, ACS Nano 13 (2) (2019) 1354−1364.

[141] W.J. Olds, E. Jaatinen, P. Fredericks, B. Cletus, H. Panayiotou, E.L. Izake, Spatially offset Raman spectroscopy (SORS) for the analysis and detection of packaged pharmaceuticals and concealed drugs, Forensic Sci. Int. 212 (1−3) (2011) 69−77.

[142] N. Stone, R. Baker, K. Rogers, A.W. Parker, P. Matousek, Subsurface probing of calcifications with spatially offset Raman spectroscopy (SORS): future possibilities for the diagnosis of breast cancer, Analyst 132 (9) (2007) 899−905.

[143] N. Stone, M. Kerssens, G.R. Lloyd, K. Faulds, D. Graham, P. Matousek, Surface enhanced spatially offset Raman spectroscopic (SESORS) imaging - the next dimension, Chem. Sci. 2 (4) (2011) 776−780.

[144] A. Pallaoro, G.B. Braun, M. Moskovits, Biotags based on surface-enhanced Raman can Be as bright as fluorescence tags, Nano Lett. 15 (10) (2015) 6745−6750.

[145] P.Z. McVeigh, R.J. Mallia, I. Veilleux, B.C. Wilson, Widefield quantitative multiplex surface enhanced Raman scattering imaging in vivo, J. Biomed. Opt. 18 (4) (2013).

[146] M.S. Bradbury, E. Phillips, P.H. Montero, S.M. Cheal, H. Stambuk, J.C. Durack, C.T. Sofocleous, R.J.C. Meester, U. Wiesner, S. Patel, Clinically-translated silica nanoparticles as dual-modality cancer-targeted probes for image-guided surgery and interventions, Integr. Biol. 5 (1) (2013) 74−86.

Index

'*Note:* Page number followed by "f" indicate figures and "t" indicate tables.'

D

E

P

Q

R

S

T

U

V

W

Printed in the United States
by Baker & Taylor Publisher Services